Introduction to
Numerical Methods and
FORTRAN Programming

Introduction to Numerical Methods and FORTRAN Programming

Thomas Richard McCalla

Lieutenant Commander
United States Navy
Naval Research Laboratory
Washington, D.C.

John Wiley & Sons, Inc.
New York · London · Sydney

To
Shirly
Bonnie
Rick
and
Erin

Preface

During my postgraduate studies* in applied mathematics at the Navy Postgraduate School, I searched in vain for an introductory text that combined numerical methods and computer programming. A number of excellent numerical methods texts were available, but the majority of these were oriented toward computational methods for desk calculators. Several programming texts were also available at that time, but most of these were based on the symbolic or machine-language programming of particular computers, either real or hypothetical. Indeed, there was no suitable introductory computer-oriented numerical-methods book for the working engineer, scientist, applied mathematician, or for advanced students in these disciplines.

This book was written to provide a unified development of efficient, computer-oriented numerical methods *and* the programming of these methods in a language that is not tied to a specific computer. The material in this book has been used for the past two years to teach mixed classes of applied mathematicians, engineers, and programmers an integrated approach to the computer solution of the basic classes of problems encountered in scientific and engineering applications. Most of the material can be covered in a one-semester course consisting of four lecture hours and one programming-lab hour each week. A knowledge of calculus and

* 1959–1961.

elementary matrix theory is assumed, although material of this nature is presented at an elementary level throughout the book.

The first two chapters present an introduction to FORTRAN programming and the fundamentals of numerical methods, respectively. Each of the remaining chapters presents several representative methods for handling a basic class of problem, such as the solution of non-linear equations, solution of polynomial equations, matrix methods and solution of systems of linear algebraic equations, interpolation, estimation of parameters, numerical integration, and solution of ordinary differential equations.

Each of these chapters (3 through 9) is introduced by a discussion of theory and the formulation of the mathematical model for the particular class of problem. An outline of representative algorithms (numerical methods) is then given to provide the reader insight into the various approaches for tackling that type of problem. For example, representative algorithms of Chapter 6 include the ordinate, finite-difference, and iterated forms of interpolating polynomials. In Chapter 5, methods for solving systems of linear algebraic equations include direct methods, repeated use of direct methods, and iterative methods. Methods within each of these chapters generally progress from the simpler algorithms to the more sophisticated ones, so that the reader can grasp the basic concepts of the particular class of problem before tackling the more sophisticated algorithms.

Each section after the chapter introduction* is devoted to a representative numerical method. The material in each of these sections follows a standard organization: the derivation of the algorithm, emphasizing the recursion formulas which are the heart of the method; a summary of computational formulas and "decisions" together with computer input and output; a detailed flow chart to aid the reader in the transition to the FORTRAN program; a FORTRAN program which follows the flow chart in a one-to-one correspondence; and a numerical example which illustrates the algorithm.

A standardized notation is used throughout the book, so that the interrelations between related numerical methods can be readily grasped. For example, the interpolation formulas, which are employed in numerical integration and in the solution of differential equations, are presented in a standard form wherever they appear.

A simple, straightforward approach to error analysis (inspired by McCracken and Dorn's use of process graphs in [16]) is introduced in Section 2.4, with simple examples illustrating the analysis of roundoff error in

---

* In Chapters 3 through 9.

computational formulas. This simple technique is applied in later chapters to selected algorithms in order to aid the reader in understanding the effects of roundoff error in typical algorithms.

Three types of exercises are included: computational exercises to test the reader's understanding of algorithms presented in the text; derivation exercises to allow the reader to extend concepts presented in the text; and programming exercises to provide experience in writing FORTRAN programs which are more sophisticated than those presented.

I owe a special thanks to Mr. Thomas E. Cindric of the National Security Agency for his invaluable assistance in organizing and moulding the manuscript into a readable text, and for using and improving the text while teaching a class of engineers and mathematicians.

I am also indebted to Professor George Cramer (formerly of American University) of IBM for his guidance and encouragement in the preparation of this manuscript. Thanks are due also to Mr. Frank Polkinghorn of the Naval Research Laboratory, to Dr. James T. Day of the Mathematics Research Center, University of Wisconsin, to Mr. Rue Murray and Mr. Richard Wiggins of the National Security Agency, and to Mr. Brian Zucker of Computer Sciences Corporation for reading portions of the final manuscript.

Special appreciation is due to Mr. Walker Stone and Mr. George Novotny of John Wiley and Sons for encouraging me to write this book, and to Miss Beatrice Shube, Mrs. Harriet McDougal, Mr. Jervis Anderson, and the staff of John Wiley and Sons for their patient and efficient help in shaping and producing this book.

I am indebted to my students for a number of valuable suggestions for improving the presentation of the text.

Finally, I owe a special debt to my wife and children for three years of loving patience while this manuscript was being written and revised.

Thomas Richard McCalla

United States Naval Research Laboratory,
Washington, D.C.
July, 1966

Contents

Introduction

The purpose of this book is to present a unified development of numerical methods and FORTRAN programming required for the computer solution of the basic classes of problems encountered in scientific and engineering applications. This development of applied problem-solving techniques includes, for each class of problem, the formulation of a mathematical model, the derivation of efficient numerical methods of solution or approximation, the preparation of flow charts that efficiently organize the method for programming, and a FORTRAN program for solving the problem on a digital computer.

Since its advent less than two decades ago, the digital computer has been an invaluable tool in all scientific fields. Problems that are impossible to solve by classical methods, or are too formidable for solution by manual computation, can be resolved in a minimum of time by digital computers that perform hundreds of thousands of arithmetic operations per second.

The computer, however, is only as useful as the numerical method(s) employed. If the method is inherently inefficient or inaccurate, the computer solution is worthless, no matter how efficiently the method is organized and programmed. On the other hand, an accurate, efficient numerical method will produce poor results if inefficiently organized or poorly programmed.

The proper utilization of the digital computer to solve scientific and

1

engineering problems requires the orderly development of a mathematical model, a careful selection of efficient numerical methods, and an ability to translate mathematical instructions into a computer language.

The Mathematical Model

A particular problem, regardless of its complexity, and independent of the scientific discipline from which it stems, usually must be formulated in mathematical notation before it can be quantitatively analyzed. Such a formulation is called a mathematical model; this model may be simple or complex (or something in between). However complex the model may be, its construction requires *building blocks* of not more than a dozen distinct types, each representing a basic class of problem encountered in scientific and engineering applications. A rather complex mathematical model might include a number of the following building blocks:

(a) Algebraic/transcendental equations
(b) Systems of linear algebraic equations
(c) Systems of non-linear algebraic equations
(d) Definite integrals
(e) Ordinary differential equations
(f) Partial differential equations
(g) Elementary functions (trigonometric, hyperbolic, etc.)
(h) Characteristic equations
(i) Interpolation and curve fitting

Formulation of the mathematical model is the first step toward the solution of a problem. For each building block in the mathematical model, an efficient algorithm (numerical method) must be selected for its solution or approximation.

Closed-Form Solutions

Classical closed-form solutions for many problems in each of these generic classes often do not exist or are completely impractical from the viewpoint of efficiency of solution. To illustrate these deficiencies in closed-form methods of solution, we cite the following particular examples:

1. Algebraic equations of degree 1, 2, 3, 4 can be solved by formula by classical methods. However, the solution of algebraic equations of degree 3 and 4 by classical methods is usually cumbersome and tedious, whereas the solution of higher order algebraic equations is, in general, impossible by classical methods.

2. Systems of non-homogeneous linear algebraic equations can be solved by Cramer's rule. However, for n equations in n unknowns, this

method requires the evaluation of $n + 1$ determinants of order n; such an approach is a tedious task when n is greater than 3 or 4.

3. The definite integral $\int_a^b f(x)\, dx$ can be evaluated for many functions, either by using a table of integrals or by actually performing the integration. This often requires either integration by parts or simplification of the integrand by trigonometric substitution or other transformation. In some cases, the closed-form evaluation of the definite integral is impossible.

4. An ordinary differential equation can be solved in closed form if it is in, or can be transformed to, a standard form, such as variables-separable, homogeneous, linear first order, and so forth. Other classical techniques include the use of the operational calculus, and solution by series. Systems of first-order constant-coefficient differential equations can be solved by determining the roots of the corresponding characteristic equation. These classical techniques are often impractical or cannot be applied to a particular problem.

5. The characteristic equation can be expressed in explicit polynomial equation form. However, for a high-order equation we encounter the problem that no closed-form solution exists.

These examples point up the fact that classical closed-form methods of solving problems of the various basic classes are often impracticable or impossible. For this reason, it is obvious that other methods must be used for solving such problems.

Numerical Methods and the Digital Computer

The impossibility-of-solution can almost always be eliminated by numerical-approximation methods. For example, if a closed-form solution does not exist for evaluating the definite integral

$$I \equiv \int_a^b f(x)\, dx$$

we can theoretically, by use of the definition

$$I \equiv \int_a^b f(x)\, dx = \lim_{n \to \infty} \sum_{i=0}^{n-1} f(x_i)\, \Delta x$$

calculate the value of I by computing the infinite sum. By so doing, we have replaced an impossible situation with an impractical one, since an infinite number of computations would be required to evaluate the infinite sum. To eliminate this "impracticality" we can approximate I by replacing the infinite sum by a finite sum, obtaining

$$I \doteq \sum_{i=0}^{n-1} f(x_i)\, \Delta x$$

provided that $\sum_{i=n}^{\infty} f(x_i)\, \Delta x$ is sufficiently small in magnitude.

Similarly, impractical classical methods (such as Cramer's Rule for solving a system of non-homogeneous linear algebraic equations) can generally be replaced by more efficient numerical methods.

For each basic class of problem mentioned earlier, there exists one or more practical, efficient numerical methods of solution or approximation. Even so, there still remains the problem of performing the computations involved in the particular method. Some numerical methods are simple enough that the computations can be performed manually, either by hand or using a desk calculator. Other numerical methods are either too complex or too time-consuming to accomplish manually. For example, the accurate generation of an orbit of an earth satellite by numerically integrating the equations of motion would require lengthy computations that could not possibly be performed manually to permit real-time tracking and guidance control of the satellite.

Many such problems in science and engineering are so complex or require such lengthy, time-consuming computations that the use of manual computations are precluded, even though efficient numerical methods are employed. The digital computer, with its ability to perform hundreds of thousands of computations per second, has largely eliminated this problem.

Objectives

The objectives of this text are: (1) to develop, for each basic class of problem encountered in science and engineering† a number of accurate, efficient numerical methods that are readily adapted for use on a digital computer; (2) to provide insight into the structure of numerical methods of solution (algorithms), emphasizing the *recursive* nature of each algorithm and the concept of a computer program loop for efficiently generating the elements defined by recursion formulas; (3) to emphasize the concept of successive-approximation methods, wherein a problem solution is obtained by an *iterative* (repetitive) numerical-approximation method; (4) to illustrate how numerical methods are efficiently organized for step-by-step mechanization on a digital computer; and (5) to illustrate the use of a mathematics-like computer programming language (FORTRAN) to accomplish the arithmetic computations and logical decisions required by the particular algorithms.

† The solution of partial differential equations and the generation of elementary functions will not be covered.

1

Fundamentals
of FORTRAN Programming

1.0 Introduction

The use of the digital computer to solve scientific computational problems requires more than the ability to code mathematical formulas in the FORTRAN language. It requires the ability to analyze complex problems and to formulate detailed mathematical *procedures* for their solution. FORTRAN is a procedure-oriented computer language that is especially suited for scientific computations. Before one discusses the elements of the FORTRAN language in detail, it will be advantageous to examine a typical computational problem. This will point out the features which are desired in a computer language that is oriented for computational work.

1.1 Setting Up a Typical Computational Procedure

If a numerical method is to be employed efficiently, it must be organized in an unambiguous, ordered sequence of computational steps, with decision instructions provided to include all foreseeable possibilities. Such a step-by-step computational procedure is called an *algorithm.*

To illustrate the process of setting up an algorithm, let us consider the following special case of computing the roots of an algebraic equation:

Problem. Compute the roots of the quadratic equation

$$(1.1.1) \qquad\qquad ax^2 + bx + c = 0.$$

This equation is the statement of a problem of a particular generic class, i.e., quadratic equations which are a subclass of algebraic equations.

The solution of any quadratic equation of the form (1.1.1) can be accomplished by the quadratic formula (provided $a \neq 0$)

$$(1.1.2) \qquad x = (-b \pm \sqrt{b^2 - 4ac})/(2a).$$

Depending on the value of the discriminant $d = b^2 - 4ac$, there are three distinct forms of the root pair of equation (1.1.1):

1. If $d < 0$, there are 2 complex conjugate roots

$$x_1 = \frac{-b + i\sqrt{-d}}{2a}, \qquad x_2 = \frac{-b - i\sqrt{-d}}{2a}$$

2. If $d = 0$, there are 2 real and equal roots

$$x_1 = x_2 = \frac{-b}{2a}$$

3. If $d > 0$, there are 2 real, unequal roots

$$x_1 = \frac{-b + \sqrt{d}}{2a}, \qquad x_2 = \frac{-b - \sqrt{d}}{2a}.$$

Obviously, the solution of a single quadratic equation can be found quite simply by computing the root pair by (1), (2), or (3), as determined by the value of d. However, the solving of a great many of these quadratic equations could be a tedious, time-consuming job if done by manual means. The digital computer is an untiring robot that can be used to perform a multitude of such computations quickly and efficiently. Digital computers are often attributed human characteristics, and are frequently referred to as "giant brains." However, though the digital computer is but a machine—it cannot think or act on its own—it can rapidly and efficiently perform computations and make "decisions" via a detailed set of instructions.

Preliminary Analysis of Instructions

To employ a digital computer to solve a set of quadratic equations†, certain preliminary information must be prepared for computer use: *how many* quadratic equations are to be solved; *what* the values of the coefficients a_k, b_k, c_k are for each equation. The *parameter*, N, can be used to represent the number of quadratic equations, and the coefficients a_k, b_k, c_k

† Such as the set of N quadratic equations

$$(1.1.3) \qquad a_k x^2 + b_k x + c_k = 0 \qquad (k = 1, N)$$

($k = 1, N$), constitute the *data* of the problem. Note that the roots of the kth equation will be determined by the values of the kth set of coefficients. The parameter and data are stored in the memory of the computer for ready access when the computer calculates the roots of the quadratic equations.

A counter instruction is needed to keep track of which equation is being solved. The value of this counter will be the same as the index k of equation (1.1.3); i.e., for the first equation ($k = 1$), the counter is set to 1; for the second equation ($k = 2$), the counter is set to 2, etc. In this way, the counter is *incremented* (in this case, increased by 1) each time the computer completes one equation and is ready to start solving another.

For each quadratic equation, there will be two roots, each of which may have both real and imaginary parts. The computer must be instructed to store the individual parts of both roots for each equation. The following nomenclature can be used:

Nomenclature	Meaning
$XR1_k$	Real part of first root of equation k
$XI1_k$	Imaginary part of first root of equation k
$XR2_k$	Real part of second root of equation k
$XI2_k$	Imaginary part of second root of equation k

Next to be prepared is a detailed set of unambiguous instructions to illustrate *how* the computer is to solve the quadratic equations. It should be noted that the set of instructions for solving each equation will be the same. Only the data (coefficients) change from equation to equation. The computer must therefore be instructed to *repeat* the same set of instructions N times to solve a set of N equations.

Detailed Set of Instructions for the Digital Computer

Step 0. Instruct computer to save memory space for the given coefficients and the roots to be computed.

Input values of N and $a_k, b_k, c_k, (k = 1, N)$ into the computer's memory. The coefficients are given in the following table, together with the labels to be applied to the root parts computed in the Execution phase. The value of the counter (index k) is given in the leftmost column:

Counter	Coefficients	Labels for Roots to be Computed
$k = 1$	a_1, b_1, c_1	$XR1_1, XI1_1, XR2_1, XI2_1$
$k = 2$	a_2, b_2, c_2	$XR1_2, XI1_2, XR2_2, XI2_2$
$\cdots$	$\cdots$	$\cdots$
$k = N$	a_N, b_N, c_N	$XR1_N, XI1_N, XR2_N, XI2_N$

Initialize equation counter k (i.e., set $k = 1$).

Step 1. Compute the discriminant for the kth equation

$$d_k = b_k{}^2 - 4a_k c_k.$$

Step 2. Test the discriminant d_k (compare algebraically with 0).

If $d_k < 0$, go to Step 2a.
If $d_k = 0$, go to Step 2b.
If $d_k > 0$, go to Step 2c.

2a. Compute and store:

$$\text{XR1}_k = \frac{-b_k}{2a_k}, \qquad \text{XI1}_k = \frac{\sqrt{-d_k}}{2a_k}$$

$$\text{XR2}_k = \frac{-b_k}{2a_k}, \qquad \text{XI2}_k = -\frac{\sqrt{-d_k}}{2a_k}, \qquad \text{go to Step 3.}$$

2b. Compute and store:

$$\text{XR1}_k = \frac{-b_k}{2a_k}, \qquad \text{XI1}_k = 0.0$$

$$\text{XR2}_k = \frac{-b_k}{2a_k}, \qquad \text{XI2}_k = 0.0, \qquad \text{go to Step 3.}$$

2c. Compute and store:

$$\text{XR1}_k = \frac{-b_k + \sqrt{d_k}}{2a_k}, \qquad \text{XI1}_k = 0.0$$

$$\text{XR2}_k = \frac{-b_k - \sqrt{d_k}}{2a_k}, \qquad \text{XI2}_k = 0.0, \qquad \text{go to Step 3.}$$

Step 3. Have all N equations been solved? This question is answered by algebraically comparing counter k with N.

If $k < N$, increase k by 1, and return to Step 1.
If $k \geq N$, go to Step 4.

Step 4. Output the list of computed roots (together with corresponding coefficients).

Flow Chart Symbols

The adage "a picture is worth a thousand words" is quite applicable to our situation of describing the step-by-step instructions required to unambiguously define the procedure for solving N quadratic equations. This procedure can be illustrated by a diagram that shows in detail the order of the steps to be accomplished. Such a diagram is called a flow chart, and can be constructed using a few basic symbols. The most commonly used flow-chart symbols and their respective meanings are described in Figure 1-1.

A representative flow chart for the computation of the solutions of N quadratic equations can be constructed as shown in Figure 1-2.

1.2   Basic Elements of FORTRAN

The *communication of instructions* to the computer is facilitated if the programmer is provided a language that is closely associated with the language in which his problems are formulated. The language of FORTRAN is quite similar to algebraic notation, and is therefore ideally suited for scientific computations. In this section, a functional description of FORTRAN is presented, starting with the definition of the basic elements and operations of the language, and followed by the development of

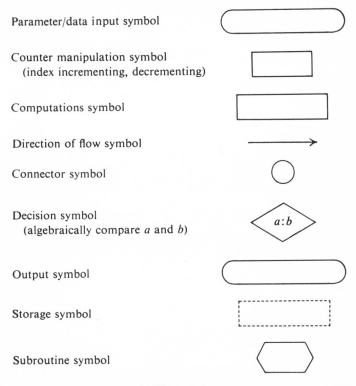

Parameter/data input symbol

Counter manipulation symbol
   (index incrementing, decrementing)

Computations symbol

Direction of flow symbol

Connector symbol

Decision symbol
   (algebraically compare a and b)

Output symbol

Storage symbol

Subroutine symbol

Figure 1-1.

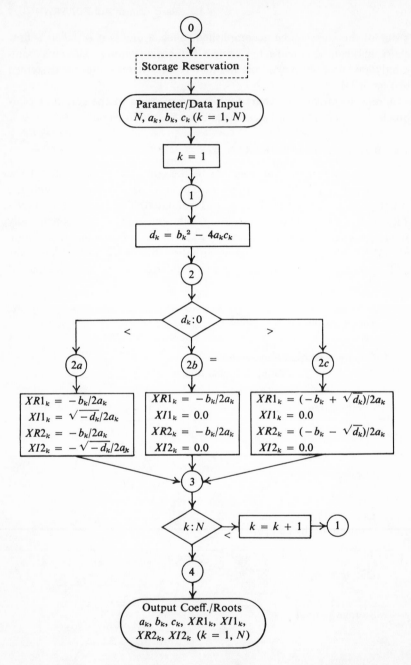

Figure 1-2. The connector numbers in this flow chart correspond to the step numbers in the "Detailed Set of Instructions for the Digital Computer."

instructions required for the input, processing, and output of information by the computer.

A digital computer can accept and store instructions and information, perform basic arithmetic operations, make logical "decisions," and output information. It is analogous to a mechanical robot, which, on signal from its operator, blindly executes a sequence of instructions. Such a sequence of instructions for the digital computer is called a *program*.

If a program is to be communicated to a digital computer, it must be in a language suitable to that computer, i.e., a language that the computer is equipped to "understand." The basic machine language of a digital computer is a highly specialized nomenclature that differs greatly from the language of mathematics. To bridge the gap between the mathematical language of the scientist and the machine language of the computer, higher order *procedure-oriented* languages have been developed; such languages are called *source* languages. One widely used source language is FORTRAN (FORmula TRANslation) originally developed by IBM. A FORTRAN compiler program translates FORTRAN *source* programs into *object* programs in the basic machine language of the computer. The programs in this text are written in FORTRAN, version IV. The programs have been written so that they can be converted to FORTRAN version II programs with a minimum of modification.

An input device must be used to convey the FORTRAN program to the digital computer. This device may be a card reader that accepts punch cards and transfers the information thereon to the computer. If the FORTRAN program is to be used repeatedly, it is convenient to "copy" the program from punch cards onto a magnetic tape. The magnetic tape then transfers the instructions to the computer.

The punch cards on which computer instructions and data are punched, consisting of 80 columns, are illustrated in Figure 1-3. Card formats for parameters/data and for FORTRAN instructions will be discussed in detail in the following sections of this chapter.

Outline

The FORTRAN language, like the language of mathematics, requires familiarization with its vocabulary as well as with certain rules and procedures. Use of the FORTRAN language (and other source languages) provides the digital computer with the capability to:

1. *Combine* a set of *basic elements* (constants, variables, subscripted variables, and functions) *into arithmetic expressions using* a set of *basic arithmetic operations* (addition, subtraction, multiplication, division, and exponentiation).

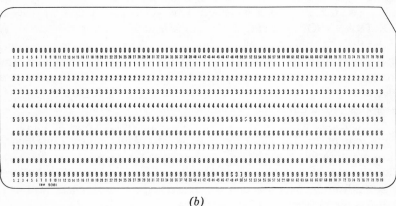

(a)

(b)

Figure 1-3. (a) FORTRAN punchcard for FORTRAN statements. (b) FORTRAN punchcard for data parameters.

2. Carry out instructions of two basic types: (a) Passive (non-executable) instructions that accomplish *house-keeping* tasks associated with the reservation of storage for arrays, and with the format specification for information to be input to the computer and output from the computer. (b) Active (executable) instructions, which, on the command of the user, execute: *Input Operations*—to transfer parameters, data, and instructions to the computer; *Processing Operations*—as directed by computational instructions (involving arithmetic expressions of §1), control instructions (transfer instructions, decision instructions, etc.); *Output Operations*—to transfer information (results) from the computer.

The FORTRAN source program consists of an ordered sequence of instructions for the computer, each of these instructions being a FOR-TRAN *statement* of a particular type, depending on whether it performs a passive or active operation, or some subcategory thereof.

The essentials of FORTRAN are developed in the following sections, in the order of the functional description outlined above.

To be able to program a mathematical problem on the digital computer, one must have some basic materials with which to work. In problem-solving courses in mathematics, physics, and engineering, the basic materials consist of *constants, variables, subscripted variables, and functions*. The counterparts of these quantities constitute the basic elements of the FORTRAN language; these elements are:

Constants. A constant is a number written in explicit numerical form. An integer constant is a whole number consisting of up to 4(14)† decimal digits. It is written *without* a decimal point. If the integer constant is used as a subscript or index, its range of values is more limited.

EXAMPLES   7
           81
           3148
           500

A floating-point‡ constant is a decimal number that can be written in either of two basic forms:

1. *Decimal Form.* A decimal number of up to 8(13) digits, consisting of an integer part and a decimal-fraction part, separated by a decimal point.

EXAMPLES   17.
           17.0
           0.975
           3.14159
           1.

2. *Exponent Form.* A decimal number of up to 8(13) digits, followed by the letter E (indicating exponent of number to base 10) and a signed or unsigned number of 2(3) digits (indicating the actual value of the exponent).

EXAMPLES   0.5E − 2 $(0.5 \times 10^{-2} = 0.005)$
           82.1E + 2 $(82.1 \times 10^{2} = 8210.0)$
           −0.314159E1 $(-0.314159 \times 10^{1} = -3.14159)$
           0.42E0 $(0.42 \times 10^{0} = 0.42)$

† The number of digits varies from one computer type to another. The maximum number of digits for typical small computers is indicated before the parenthesis; the number for typical large computers is indicated within the parentheses.

‡ In FORTRAN IV, floating-point numbers are called *real* numbers. We prefer the nomenclature of floating-point numbers.

It should be noted that when a floating-point constant is input into the computer, it is converted to a single form, regardless of its external form (decimal or exponent). This internal form is dependent on the characteristics of the particular computer.

Variables. A variable is a symbol or name which represents a particular quantity which may take on different numerical values at different times, just as the variable x in mathematics may represent any real number. The variable can be represented by one or more letters of the alphabet. It is often convenient to use integer numbers in the variable name; however, they may *never* be used as the first character of that variable name.

An integer variable consists of up to 5(7) alphanumeric† characters, the first being one of the letters I, J, K, L, M, N.

EXAMPLES   INDEX
           J
           K23
           MAX5
           N

A floating-point variable name consists of up to 5(7) alphanumeric characters, the first being a letter *other* than I, J, K, L, M, N.

EXAMPLES   A
           DIST
           CMAX5
           R12
           TIME

Subscripted Variables. A subscripted variable is one that may have up to 2(3) subscripts (as defined below), depending on the number of dimensions of the array bearing the same name.

Subscripts. Subscripts are used to indicate individual members of arrays of single-, double-, or triple-subscripted (indexed) variables. Allowable forms of subscripts are: $I, \#, I \pm \#, \# * I, \# * I \pm \#'$, where I is an unsigned, non-subscripted integer variable, and $\#$ and $\#'$ are unsigned integer constants. (In FORTRAN, the symbols $+$ and $-$ denote addition and subtraction, respectively, while the symbol $*$ indicates multiplication (see p. 16).)

An integer-subscripted variable is an integer variable name followed by 1, 2, or 3 subscripts in parentheses.

---

† An alphanumeric character is either a letter (capital) of the English language or one of the decimal digits 0, 1, 2, ..., 9. Special symbols are not included.

EXAMPLES   MAX(I)
         MIN(I, J)
         NBR(I, J, K)
         L2(J1 + 3)

A floating-point subscripted variable is a floating-point variable name followed by 1, 2, or 3 subscripts in parentheses.

EXAMPLES   X(I)
         A(I, J)
         RMIN(I, J, K)
         ARRAY(M1, M2)

Functions. Common elementary mathematical functions such as sine, cosine, tangent, exponential, logarithm, square root, and absolute value are essentially treated as basic elements in the FORTRAN language. The functions available depend on the particular version of FORTRAN and on the particular computer being used. Every FORTRAN function has a preassigned name that is followed by an argument in parentheses. The following examples illustrate the more commonly used FORTRAN library functions.†

EXAMPLES   SQRT(X)   —meaning the square root of the floating-point number x.
         COS(ALPHA)—meaning the cosine of an angle alpha (in radians).
         SIN(BETA)   —meaning the sine of an angle beta (in radians).
         ABS(Y)   —meaning the absolute value of the floating-point number y.
         ATAN(V)   —meaning the arctangent of the floating-point number v (answer in radians).
         TANH(U)   —meaning the hyperbolic tangent of the floating-point number u.

The Constants, Variables, Subscripted Variables, and Functions just described are the basic elements of the FORTRAN language. To combine these elements into *expressions* required for the formulation of mathematically represented problems, the FORTRAN language is provided with a set of arithmetic operations as described on the next page.

Basic Arithmetic Operations in FORTRAN

The FORTRAN language is provided with counterparts of the basic arithmetic operations. The FORTRAN *arithmetic operations*, together

† These library functions are programmed as FUNCTION subprograms (see Sec. 1.5).

with the symbols that represent them, are given in the following table:

Operation	Symbol
Addition	+
Subtraction	−
Multiplication	*
Division	/
Exponentiation	**

In FORTRAN, arithmetic symbols are used to combine basic elements into *arithmetic expressions.* Parentheses can be used to eliminate any ambiguity of meaning. Additionally, the FORTRAN arithmetic operations are performed in a preassigned order of precedence, unless this order is altered by the use of parentheses. The order of precedence (from first to last) is as follows:

First	**
Second	*, /
Third	+, −

Operations of the same precedence are performed from the left to right, unless the order is altered by the use of parentheses. For example, the operation order for the formation of the expression

$$A * X ** 2 + B * X + C$$
$$2 \quad 1 \quad\quad 4 \quad 3 \quad 5$$

is indicated by the numbers under the individual operation symbols.

There is no ambiguity concerning the symbol **, because FORTRAN operation symbols (+, −, *, /, **) cannot be juxtaposed. That is, the expression A + −B is invalid because the addition and subtraction symbols are juxtaposed. The symbol ** is considered as a single symbol, and not as two juxtaposed symbols.

Arithmetic Expressions in FORTRAN

In FORTRAN the arithmetic operations +, −, *, /, and ** are used to combine the constants, variables, subscripted variables, and functions into ARITHMETIC EXPRESSIONS. The following are examples of mathematical expressions and their FORTRAN counterparts:

Mathematical Expression	FORTRAN Expression
$ax^2 + bx + c$	A * X ** 2 + B * X + C
$x^2 + y^2 + z^2$	X ** 2 + Y ** 2 + Z ** 2
$1 - \sin^2 x$	1.0 − SIN(X) ** 2

The following two rules are important in the construction of FORTRAN arithmetic expressions:

Rule 1. Integer variables and constants and floating-point variables and constants cannot be combined by the operation symbols $+$, $-$, $*$, $/$ in the same FORTRAN arithmetic expression. That is, floating-point variables and constants must be added, subtracted, multiplied, and divided by other floating-point variables and constants. Integer variables and constants must be added, subtracted, multiplied, and divided by other integer variables and constants.

EXAMPLES   A + B + C   valid
              A − I   not valid
              J * E   not valid
              R / T * A   valid
              S + 2.3   valid
              F * 2   not valid

Rule 2. A floating-point variable or constant may have either an integer or a floating-point exponent. However, an integer variable or constant can have only an integer exponent.

EXAMPLES   A ** 2.0   valid
              A ** 2   valid
              I ** 2   valid
              I ** 2.0   not valid
              (C ** 2) ** 2   valid
              C ** 2 ** 2   not valid (only single exponents are per-
                               mitted except when set off by parentheses)

1.3   Non-executable Instructions for Housekeeping Tasks

A programmer conventionally uses a FORTRAN coding form on which to prepare FORTRAN instructions prior to key-punching them on punch cards. The coding form has 80 columns corresponding to the 80 columns on the punch card. The formatting of instructions and information on a punch card is a simple, yet important, aspect of the proper use of the FORTRAN language. The reader is advised to pay particular attention to the format procedures illustrated in the various examples throughout this chapter.

Before instructions are formulated to input information, to process, and to output information, it is necessary to attend to the following *housekeeping* tasks:

1. Storage in the computer memory must be reserved for any arrays of subscripted variables that will be used;

2. The input format of each parameter or data item must be specified;

3. A *list* of the parameters and data to be input must be prepared in the format specified in 2 (above).

Storage Reservation

The memory of a digital computer is like a large hotel. Each location in memory has a "room number" called an address. Arrays of subscripted variables must be allocated blocks of storage so that a particular member of the array can be located by specifying its array name and subscripts (like identifying numbers for visiting conventioneers). On the other hand, each constant or unsubscripted variable takes only one location and is automatically stored (provided there is room).

Storage reservation for arrays of subscripted variables is accomplished by DIMENSION statements which indicate the variable name, the number of subscripts, and the maximum size of each subscript.

Arrays of subscripted variables are always stored in the memory of the computer in column order.

EXAMPLE.   Given the array A containing the nine elements A_1, A_2, $A_3, \ldots, A_9$, storage can be reserved by the DIMENSION statement:

$$\text{DIMENSION A(9)}$$

The elements of this array will be stored in the computer's memory in column order $A_1, A_2, A_3, \ldots, A_9$.

EXAMPLE.   Given the array B containing the elements B_{ij}, where i is the row index and j is the column index:

$$
\begin{array}{cc}
B_{11} & B_{12} \\
B_{21} & B_{22} \\
B_{31} & B_{32}
\end{array}
$$

Storage for this two-dimensional array can be reserved by the statement

$$\text{DIMENSION B(3, 2)}$$

The elements of this array will be stored in memory in column order: $B_{11}, B_{21}, B_{31}, B_{12}, B_{22}, B_{32}$.

Note that the DIMENSION statement in the above examples defined the *maximum size* of each subscript of the array.

The general form of the DIMENSION statement is

$$\text{DIMENSION } S_1(U_1), S_2(U_2), \ldots, S_n(U_n)$$

where the S_i are subscripted variable names and the U_i consist of 1, 2, or 3 unsigned integer constants (depending on the number of dimensions of each array S_i).

If more than one array is used in a computer program, storage can be reserved for all the arrays in a single DIMENSION statement. Further, the DIMENSION statement *must* appear before the first occurrence of the named variable(s) in the program.

EXAMPLE. Given the arrays $A(I)$, $I = 1, 20$; $B(I, J)$, $I = 1, 15$, $J = 1, 30$; $C(I, J, K)$, $I = 1, 5$, $J = 1, 5$, $K = 1, 5$, we can reserve storage by the single statement

DIMENSION A(20), B(15, 30), C(5, 5, 5)

Triple-subscripted arrays are stored such that the first subscript varies most rapidly, the second subscript next, and the third subscript least rapidly. To illustrate this, the three-dimensional array can be thought of as a set of wafers sandwiched one behind the other. Elements of the front wafer are stored in column order, i.e., by columns of the first wafer; then the elements of the second wafer are stored in column order, and so on until all the elements of the last wafer are stored.

EXAMPLE. The array $C(3, 3, 2)$ is stored in the following order: First wafer:

$$C_{111}, C_{211}, C_{311}, C_{121}, C_{221}, C_{321}, C_{131}, C_{231}, C_{331},$$
$$\xrightarrow{\hspace{8cm}}$$

Second wafer:

$$C_{112}, C_{212}, C_{312}, C_{122}, C_{222}, C_{322}, C_{132}, C_{232}, C_{332}.$$
$$\xrightarrow{\hspace{8cm}}$$

Two or more programs can share storage in the computer by use of the COMMON statement that has the general form†

COMMON $S_1, S_2, \ldots, S_n$

where the S_i are subscripted or unsubscripted variable names. In FORTRAN IV, subscripted variables can be dimensioned in the COMMON statement. The COMMON statement permits the programmer to establish a correspondence between arrays in different programs. Suppose, for example, that arrays A and B in program 1 are to correspond to arrays C and D in program 2, where A has the same dimensions as C and B has the same dimensions as D. This correspondence is established by the COMMON statements

COMMON A, B   (in program 1)
COMMON C, D   (in program 2).

Specifying Input Form

The rules concerning field specifications and format statements for the output of information are identical to those governing input of parameter and data items. In this section, however, these rules will be discussed from the viewpoint of *input* specifications since they have a direct effect on the preparation of data cards.

† For a more general form that employs blocks of COMMON the reader is referred to the manufacturers' FORTRAN manuals.

The reader will recall that there are but two basic forms of constants, variables, and subscripted variables. These forms are *integer* and *floating-point*, with the floating-point category further subdivided into decimal and exponent forms.

To input information in one of the above forms, an appropriate field specification must be designated and communicated to the computer via a *format statement*.

Field specifications can be of the following types:

(Iw)	for integer form
(F$w.d$)	for decimal-form floating-point
(E$w.d$)	for exponent-form floating-point

where w is the field width in card columns, and d is the number of decimal places to the right of the decimal point.

Iw Field Specification. The field width, w, determines the number of card columns the computer will scan for the input information. Examples:

	Punched		
Col.	1 2 3 4 5 ...	Field	Value Read
.....	3 2 5	I3	3 2 5
	3 2 5	I4	3 2 5 0
	3 2 5	I2	3 2
	3 2 5	I5	0 0 3 2 5

If the number to be input is right-justified (as in the last example), the field width specified may be larger than needed.

Fw.d Field Specification. The field width, w, designates the number of card columns scanned by the computer; it must include spaces for the decimal point† and for the sign‡ of the floating-point number. Examples:

	Punched		
Col.	1 2 3 4 5 6 ...	Field	Value Read
.....	+ 3 . 1 4 5	F6.3	+ 3 . 1 4 5
	− 7 2 .	F6.0	− 0 0 7 2 .
	+ 4 9 . 2	F6.2	+ 4 9 . 2 0
	− 3 2 .	F4.0	− 3 2 .
	+ 7 2 1 4 .	F6.0	+ 7 2 1 4 .

† The decimal point can be omitted, in which case it is positioned by the FORMAT statement. If the decimal point is punched on the card, its position over-rides that specified by the FORMAT statement.

‡ The sign of a number (or exponent) can be omitted if it is positive.

Because of the requirement that the field width, w, contain a space for the sign and a space for the decimal point, w must be greater than or equal to $d + 3$.

Ew.d Field Specification. The field width, w, must be greater than or equal to $d + 7$, since spaces must be included for:

1. the sign of the number
2. the decimal point
3. the letter E
4. the sign of the exponent
5. the exponent (2 spaces)

Some examples of these are:

Punched		
1 1 1 2 3 4 5 6 7 8 9 0 1 ...	Field	Value Read
+ 0 . 2 2 E + 0 2	E9.2	+ 2 2 . 0
+ 0 . 1 2 3 4 E + 0 2	E11.4	+ 1 2 . 3 4
− 0 . 1 2 3 4 E − 0 1	E11.4	− 0 . 0 1 2 3 4
− 5 2 7 5 . E − 0 3	E10.0	− 5 . 2 7 5
− 5 2 7 . 5 E − 0 2	E10.1	− 5 . 2 7 5
+ 5 2 7 . 5 E − 0 4	E10.1	+ 0 . 0 5 2 7 5

Format Statement. The field specification for input is communicated to the computer via format statements. The general form of the Format statement is:

$$\# \text{ FORMAT } (S_1, S_2, \ldots, S_n)$$

where each S_i is the field specification for the input item i referred to in an input statement (p. 23), and *# is the statement number* used as a cross-reference between a particular format statement and the corresponding input statement.

Examples of format statements:

	1 1 1 1 1 1 1 1 1 1 2 2 2 2 2	
Col.	1 2 3 4 5 6 7 8 9 0 1 2 3 4 5 6 7 8 9 0 1 2 3 4 ...	
. . . . .	1 0 0	F O R M A T (I 3)
	1 0 1	F O R M A T (F 1 0 . 5 , I 7)
	1 0 2	F O R M A T (2 E 1 2 . 6)

The last example illustrates that two items are to be input which have the same field specification (E12.6).

Format statements need not immediately follow their corresponding input statement. Since they perform a house-keeping task, there is no requirement that they appear in a particular position in the program. All sample programs in this text follow the convention of placing format statements at the end of the source program.

The word "format," as in the above examples, is always punched beginning in column 7. The statement number can be placed anywhere in columns 2 through 5. It is normal convention, however, to right-justify the statement number.

Preparation of Data and Parameters

The list of data and parameters to be read into the computer must conform to the field specifications of the format statement (which is cross-referenced by # in the input statement). The user must prepare this list of parameters and data in the *format* and *order* specified in the format statement.

EXAMPLE. If we wish to read in the parameters PI $= +3.14159$ and N $= 7$ and the data array A, where $A_1 = +6.25$, $A_2 = -7.31$, $A_3 = +1.47$, $A_4 = +9.05$, and $A_5 = -2.04$, we can accomplish this by the format statements

$$\#_1 \text{ FORMAT (F8.5, 2X, I1)}$$
$$\#_2 \text{ FORMAT (F5.2)}$$

The format and order of the parameter card and data cards are:

	Col.	1 2 3 4 5 6 7 8 9	1 1 1 0 1 2
Parameter card	.....	+ 3 . 1 4 1 5 9	7
Data card 1		+ 6 . 2 5	
Data card 2		− 7 . 3 1	
Data card 3		+ 1 . 4 7	
Data card 4		+ 9 . 0 5	
Data card 5		− 2 . 0 4	

Note that parameter and data cards are punched beginning in column 1.

1.4 Executable Instructions in FORTRAN

Now that the housekeeping tasks of storage reservation, format specification, and parameter/data list preparation have been accomplished, FORTRAN instructions can be formulated which will *execute* the three principal phases of the program: (1) *Input*, (2) *Computation and Control*, and (3) *Output*. Each of these three phases will be described in detail in the following pages, along with the development of the FORTRAN statements that pertain to each phase.

Input

The actual transfer of the parameters and data to the computer memory is accomplished by *input statements* that cause the input device to read each item of input into the proper location in memory. The type of input statement used in the program depends on the type of device(s) employed with the particular digital computer. Input statements are of the general form indicated in the following table:

Type Device	Input Statement Form
Card reader	Read F #, LIST$_i$
Magnetic tape	READ (T #, F #) LIST$_i$

where F# is the format statement number; T# is the magnetic tape number; and LIST$_i$ is the list of parameters/data read by that statement.

EXAMPLES

    READ 101, N
101 FORMAT (I3)
    READ 102, (A(K), K = 1, 7)
102 FORMAT (7F10.4)
    READ 103, M, N, (B(I), I = 1, 5), PI
103 FORMAT (2I4, 5F8.4, F7.5)
    READ (5, 105) A(K), B(K), C(K)
105 FORMAT (F10.5, F10.6, F10.7)

In the first three examples (card reader input), note the use of the comma to separate the format statement number from the list.

We come now to the description of the instructions that are the heart of the program, i.e., those instructions that carry out the required computations and make the decisions necessary to consider all foreseeable possibilities.

These instructions can be grouped into either of two major categories: (a) computational, (b) control.

Computational Instructions

The basic computational instruction is the *substitution statement*, which is of the following general form:

$$\boxed{\text{VARIABLE} = \text{ARITHMETIC EXPRESSION}}$$

This statement is interpreted to mean: "The value of the arithmetic expression on the right replaces (is substituted for) the value of the variable on the left." The "=" sign in FORTRAN is *not* a mathematical equal symbol, as illustrated by the substitution statement

$$X = X + A$$

which states that the value of the arithmetic expression on the right replaces the value of the variable on the left. For example, if X has value 10.5 and A has value 2.0 before the execution of the substitution statement above, then the value of the expression X + A (which is 10.5 + 2.0) replaces the value of X. Hence, after the execution of the statement above, X has value 12.5. The value of A remains unchanged. Only the value of the variable on the left of a substitution statement is changed by its execution.

EXAMPLES   QUAD = A * X ** 2 + B * X + C
R2 = X ** 2 + Y ** 2 + Z ** 2
N = X + Y + 4.0
A = N + 1

In the third example above, the arithmetic expression on the right is in floating-point form, whereas the variable on the left is in integer form. This instruction is *valid*. If X = 1, and Y = 1.2, then the arithmetic expression has the value 6.2. The variable N assumes the value 6, i.e., the decimal fraction 0.2 is truncated.

In the fourth example above, the arithmetic expression on the right is in integer form, and the variable on the left is in floating-point form. If N = 6, then N + 1 = 7; when the statement A = N + 1 is executed, the value of N + 1 is converted to floating-point form and replaces the previous value of A. In this case, A assumes the value 7.0 when the instruction is executed.

The "=" symbol used in the flow charts throughout this book in statements such as

$$K = K + 1$$

or

$$SUMSQ(X, Y) = X * X + Y * Y$$

is the FORTRAN substitution symbol as defined in the substitution statement.

Control Instructions

The statements in a FORTRAN program are written sequentially, i.e., one after another in a single string, and they are normally executed in the order in which they appear. However, the order of execution can be modified, as required, by FORTRAN *control* statements, such as IF, GO TO, DO. To illustrate the various control statements, let us examine a few cases in which we need to alter the order of execution of a sequence of statements.

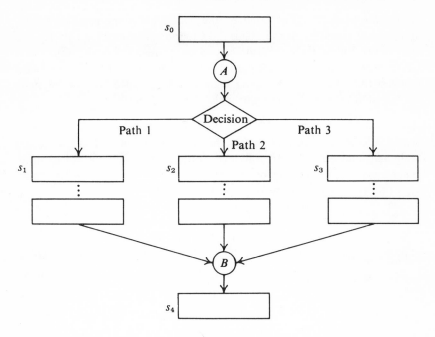

Figure 1-4.

Case I. If the formulation of an algorithm is such that its flow chart requires more than one path between any two connecting points *A* and *B*, then we must make some provision to alter the single string of statements so that the statements in any path can be executed when required. To illustrate this point, consider the flow chart (Figure 1-4) which has three paths between connectors *A* and *B*.

Since the program statements appear in a single string, it is obvious that the statement before connector *A* can be immediately followed by only one statement, and the statement after *B* can be *immediately* preceded by only one statement. We are therefore confronted with the following two problems:

1. How to go from *A* to the first statement in any one of the paths 1, 2, or 3.

2. How to go to *B* from the last instruction in any one of the paths 1, 2, or 3.

Problem (1) can be resolved by a control statement that makes the decision as to which path of instructions must be executed at a particular time.

This decision is normally based on the result(s) of some prior computation(s). For example, in the quadratic equation flow chart on p. 10, the decision to follow the complex-roots path, the real-and-equal roots path, or the real-and-unequal roots path was based on whether the discriminant was less than, equal to, or greater than zero. Such a decision can be made by the FORTRAN IF statement which has the form

$$\text{IF (EXPR) } s_1, s_2, s_3$$

where EXPR is a FORTRAN arithmetic expression and s_1, s_2, s_3 are statement numbers. The IF statement transfers program control to statement s_1, s_2, s_3, depending on whether the value of EXPR is less than, equal to, or greater than zero, respectively.

EXAMPLE   IF $(K - 99)$ 1, 1, 2

If K is less than 99 or equal to 99, program control is transferred to statement 1; if K is greater than 99, control is transferred to statement 2.

Problem (2) can be resolved by a GO TO statement that has the form

$$\text{GO TO } s$$

where s is a statement number. Execution of this statement transfers program control to the statement numbered s. Hence, after the last computational statement in each of the parallel paths, a GO TO s_4 statement can be inserted to transfer the program control from the end of that path to the statement at connector B.

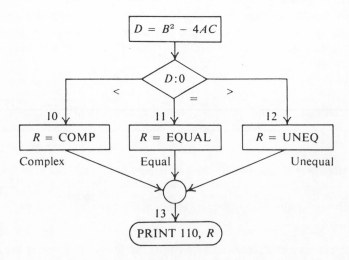

Figure 1-5.

EXAMPLE   . . .

. . .

```
     D = B ** 2 − 4. * A * C
     IF (D) 10, 11, 12
10   R = COMP
     GO TO 13
11   R = EQUAL
     GO TO 13
12   R = UNEQ
13   PRINT 110, R
```

. . .

See Figure 1–5.

In the foregoing example, if D = 0.0, program control is transferred to statement 11 (skipping statement 10 and the first GO TO 13 statement). Statement 11 is executed, i.e., the value of EQUAL replaces the value of R. The next sequential instruction, GO TO 13, is executed, transferring program control to statement 13 (skipping statement 12). Statement 13 is then executed, and so forth.

Another commonly used control instruction is the *computed* GO TO statement, which has the form

$$\text{GO TO } (s_1, s_2, \ldots, s_n), \text{ I}$$

where $s_1, s_2, \ldots, s_n$ are statement numbers, and I is an unsubscripted integer variable. This statement transfers program control to statement s_1 if I = 1, to statement s_2 if I = 2, and so on.

EXAMPLE.   GO TO (10, 11, 12), I

Program control is transferred to statement 10 if I = 1, to statement 11 if I = 2, and to statement 12 if I = 3.

Case II. The ability to *repeatedly execute an instruction* (or a sequence of instructions) is one of the most powerful features of a stored-program digital computer. Suppose for example that we desire to linearly interpolate between two points (x_1, y_1) and (x_2, y_2) at the 9 evenly spaced abscissas $\bar{x}_k$ between x_1 and x_2, i.e., where

$$\bar{x}_k = x_1 + k \frac{(x_2 - x_1)}{10}$$

This linear interpolation can be accomplished by evaluating the recursion formula

$$\bar{y}_k = \frac{y_2 \lceil \bar{x}_k - x_1 \rceil - y_1 [\bar{x}_k - x_2]}{x_2 - x_1}$$

for each of the values $k = 1, 2, \ldots, 9$.

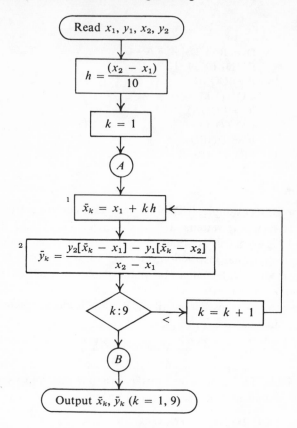

Figure 1-6.

A typical flow chart of this process is given in Figure 1-6.

Examining the flow chart, we see that we can compute the 9 values of $\bar{y}_k$ ($k = 1, 9$) by simply evaluating the formulas in blocks 1 and 2 for each of the 9 values of index k between 1 and 9.

The process of repeatedly evaluating the same formulas for successive values of an index is called looping. The instructions between connectors A and B are said to comprise a loop.

Looping and the DO statement. The index initialization (setting $k = 1$), testing (comparing k with 9), and incrementing (replacing k by $k + 1$) in the foregoing example could be accomplished by FORTRAN statement types which have already been defined. A partial program illustrating the statements required for this looping process could be written as follows:

```
      . . .
      K = 1
1   XBAR(K) = · · ·
2   YBAR(K) = · · ·
      IF (K − 9) 4, 5, 5
4   K = K + 1
      GO TO 1
5   . . .
```

However, a single FORTRAN *DO statement* can accomplish all three aspects of indexing required for the looping process, i.e., initializing, testing, and incrementing. The DO statement has the general form

$$DO\ S\ K = m_1, m_2, m_3$$

where S is a statement number, k is a non-subscripted integer variable, and m_1, m_2, m_3 are either unsigned integer constants or non-subscripted integer variables. If m_3 does not appear in the statement, then it is assumed to be equal to 1.

The DO statement causes the execution of all statements following it, up to and including statement S, to be repeated for these values of index k: $m_1, m_1 + m_3, m_1 + 2m_3, m_1 + 3m_3, \ldots, m_1 + jm_3, \ldots$, up to and including the highest value of this sequence which does not exceed m_2.

Using the DO statement for the previous example, we can write the following partial program, which will accomplish looping thru the instructions as k ranges from 1 to 9 with increments of 1:

```
      . . .
      DO 2 K = 1, 9
      XBAR(K) = · · ·
2   YBAR(K) = · · ·
5   . . .
```

Nested DO loops. DO statements can be "nested" to accomplish the looping for multi-dimensional arrays. For example,† to add the corresponding members of two arrays (A_{ij}) and (B_{ij}) $(i = 1, n; j = 1, m)$, *two* DO statements can be nested as follows:

```
      . . .
      DO 1   I = 1, N, 1
      DO 1   J = 1, M, 1
1   C(I, J) = A(I, J) + B(I, J)
      . . .
```

The nesting process can be best illustrated by a flow chart (Figure 1-7)

† See Matrix Addition, Sec. 5.2.

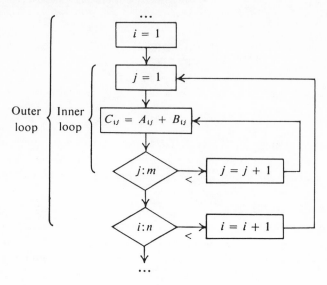

Figure 1-7.

which outlines the procedure for adding corresponding members of 2 two-dimensional arrays.

The elements of array (C_{ij}) are computed by rows. Initially, row index i is set to 1, and the elements $C_{11}, C_{12}, C_{13}, \ldots, C_{1m}$ of row 1 are computed by cycling through the inner loop for $j = 1, 2, 3, \ldots, m$. Next, the elements of row 2 are computed by setting row index i to 2 and again cycling through the inner loop for $j = 1, 2, 3, \ldots, m$. Hence, in the ith pass of the outer loop, elements of row i are computed by cycling through the inner loop for column index $j = 1, 2, 3, \ldots, m$.

When using nested DO statements, all instructions of the inner DO loop must be contained within the instruction range of the outer DO loop. In this example, this requirement is fulfilled, since statement 1 was the end statement of both DO loops. Further requirements are that GO TO statements and IF statements may not be used as the terminating statement of any DO loop (see use of CONTINUE statement), nor may statements that change the value of the index of a DO loop be used within that loop.

The CONTINUE statement. In this text, the CONTINUE statement is used as the terminating statement of a DO loop. It is advantageous to use this statement as a transfer point for continuation of a DO loop which would otherwise have a GO TO or an IF statement as the last statement in the DO loop.

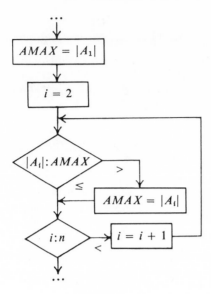

Figure 1-8.

EXAMPLE. Find the element of maximum magnitude in the array (A_i) $(i = 1, n)$. This can be accomplished by first assuming that the absolute value of A_1 is the greatest magnitude (which we denote by $AMAX$), and then comparing the absolute value of each succeeding element. When an element of greater magnitude is encountered, its value replaces the value of $AMAX$. This procedure is continued until every element of the array has been compared to the current value of $AMAX$. The use of the CONTINUE statement for this problem is as follows:

```
      AMAX = ABS(A(1))
      DO 10 I = 2, N
      IF (ABS(A(I)) − AMAX) 10, 10, 5
    5 AMAX = ABS(A(I))
   10 CONTINUE
```
      . . .

See Figure 1–8.

The STOP and END statements. The STOP statement is a control statement that signals the computer to halt the execution of the program and return control to the computer's operating system which then calls in and executes other programs.

The END statement is the *last* statement in the FORTRAN source program, and is used to terminate the compilation of the program. The

END statement is followed only by machine control cards and data and parameter cards, as required for the particular program.

The PAUSE statement. The PAUSE statement is used to temporarily halt the execution of a program, so that the operator may manually select options for continuation of the program, etc. The program can then be continued by pressing the computer start key.

Output Instructions

The actual transfer of information from the computer to the various types of output devices is accomplished by OUTPUT statements. These statements cause the information to be printed on a page printer, punched on cards or paper tape, or written on magnetic tape. The type of OUTPUT statements used in the program depends on the type of devices employed with the computer. Output statements for the printer and for the magnetic tape unit are of this general form:

Type Device	Output Statement Form
Printer	PRINT F#, LIST
Magnetic tape	WRITE (T#, F#) LIST

where F# is the format statement number; T# is the magnetic tape number; LIST is the output list (see LIST description under input).

EXAMPLE 1   PRINT 101, C, D
        101   FORMAT (1H0, 2F9.4)

EXAMPLE 2   PRINT 102, A, B
        102   FORMAT (1X, 2F10.5)

EXAMPLE 3   WRITE (5, 110) X, Y, Z
        110   FORMAT (1H1, F15.5, 2F15.6)

The rules governing the *field specification* for output information are identical with those previously discussed for data and parameter input. However, the *output format statement* must reserve the first space of each line of output for printer carriage control; this was not required for input format statements.

When using PRINT statements, output information is printed "on-line" during the execution of the program. The printer can single space or double space between lines of output, or it can print information on a separate page. The first character of each output line is reserved for such printer carriage controls. To specify the desired control, the output format statement must include the control form, 1Hc, where c can be:

Blank	for single spacing of output
0 (zero)	for double spacing of output
1	for page eject.

The 1H designates that the *first* space is for carriage control; the *c* designates *which* carriage control is to be used. The "1H*c*" carriage control command must always appear before the field specifications in the output format statement as seen in Example 1.

The SKIP field, 1X (see p. 34), can be used interchangeably with 1H_ (1Hblank), as shown in Example 2.

The reader is also reminded that data can be read into the computer in one format and printed out in a different format. For example, a programmer may find it time-saving to prepare his input data in floating-point *decimal form*; however, he may also desire to have his results printed out in floating-point *exponent form* for quick analysis.

Hollerith fields. In many cases, the solution to problems in numerical analysis leads not to a single solution, but to a number of solutions. For example, the simple problem of solving three simultaneous equations in x, y, z leads to an x-answer, a y-answer, and a z-answer. With the format and field specification rules that have been discussed so far, valid output formats could be written and utilized in a program. However, upon receipt of the results of that program, one might be confused as to which numerical answer corresponds to x, to y, or to z.

In such cases, Hollerith fields can be used in the output format statement to give additional information about the results. The general form of the Hollerith field is

$$sH$$

where s is the number of spaces in which the message will be positioned (blanks are included); H is the Hollerith symbol. Since the first space of every output line is reserved for carriage control, the Hollerith field (if used as the first item in the format statement) takes on the general form

$$sHc$$

where c designates the desired carriage control.

EXAMPLE   PRINT 110, X, Y, Z
      110 FORMAT (3H1X=, F5.2, 4H__Y=, F5.2,
            4H__Z=, F5.2)

The following would be printed:

Control		1 1 1 1 1 1 1 1 1 1 2 2 2 2 2 2 2
Col.	1 2 3 4 5 6 7 8 9	0 1 2 3 4 5 6 7 8 9 0 1 2 3 4 5 6

X = + 0 . 7 8   Y = + 0 . 9 2   Z = + 0 . 8 0

In this example, "3H" designates the message field of 3 spaces containing "1X=." Since this Hollerith field is the first item of the format statement, the "1" designates the carriage control command to page eject. The next Hollerith field is "4H." It designates the message "bbY =," (where *b* indicates a blank). Since this Hollerith field is not the first item in the format statement, the blank following the "4H" does not designate carriage control. *Only* the *first* character of the Hollerith field used as the *first item* of the format statement constitutes a carriage-control command.

It should be noted that the comma can be omitted after a Hollerith field specification. This is the only exception to the rule that two successive field specifications be separated by a comma.

Skip fields. The general form of the SKIP field is

$$cX$$

where "*c*" represents the number of columns to be skipped in the printing of output information. The SKIP field is often used *between* other field specifications in an output format statement for the convenient spacing of results.

EXAMPLE. The example used in the discussion of Hollerith fields could have been written as follows:

```
         PRINT 110, X, Y, Z
110      FORMAT (3H1X=, F5.2, 2X, 2HY=, F5.2,2X,2HZ=, F5.2)
                         1 1 1 1 1 1 1 1 1 1 2 2 2 2 2 2 2
Col.     1 2 3 4 5 6 7 8 9 0 1 2 3 4 5 6 7 8 9 0 1 2 3 4 5 6
```
―――
         X = + 0 . 7 8   Y = + 0 . 9 2   Z = + 0 . 8 0

Use of the slash. In programs in this text, the slash (/) is used only in *output* format statements. It specifies the *line spacing* of output information. A single slash designates single spacing; two slashes, double spacing; etc.

1.5 Functions and Subroutines

The solution of a typical scientific or engineering problem may require the computation of square roots or logarithms; and the evaluation of various trigonometric, hyperbolic, or exponential functions. It may also require algorithms for solving equations of various types, for interpolating, for curve fitting, or for evaluating definite integrals. Each of these entities can be considered as a *building block*. The mathematical model for a problem of any complexity can then be "constructed" by putting together the required building blocks in the proper form and order.

One of the most powerful features of FORTRAN (and other source languages as well) is the capability to either provide such building blocks or permit the programmer to "construct" his own. In FORTRAN, these take the form of functions or subroutines.

Functions produce a *single value* and are referenced by their appearance in an arithmetic expression within the source program; they can be categorized as either (a) built-in functions; (b) arithmetic statement functions (A.S.F.); or (c) FUNCTION subprograms.

SUBROUTINE subprograms can return *more than one value* and are referenced by CALL statements in the program that refers to them.

Built-in functions are predefined functions that are provided by the FORTRAN processor. Typical built-in functions include the following: absolute value of an argument (ABS); conversion from integer to floating-point (FLOAT); conversion from floating-point to integer with truncation (FIX); and a number of other specialized functions.

An *arithmetic statement function* (A.S.F.) is *defined* by a single SUBSTITUTION statement that must precede the first executable statement of the source program containing it; the A.S.F. has the general form

ARITH $(A_1, A_2, \ldots, A_n)$ = Any allowable arithmetic expression
containing $A_1, A_2, \ldots, A_n$, constants,
unsubscripted variables, and other
previously defined A.S.F's.

where ARITH is the A.S.F. name;† $A_1, A_2, \ldots, A_n$ are its arguments, or dummy variables, which must be unsubscripted variables. An arithmetic statement function is *referenced* by its appearance in any executable FORTRAN statement containing it and is executed during the execution of the source program.

The following simple program, which locates the points (X, Y) within or on a unit circle, illustrates the definition and use of an arithmetic statement function *within* the FORTRAN program that references it.

```
      SUMSQ(X, Y) = X * X + Y * Y
      DO 16 K = 1, 50
    1 READ (5, 101) X, Y
      IF (SUMSQ(X, Y) − 1.0) 5, 5, 16
    5 WRITE (6, 110) X, Y, SUMSQ(X, Y)
   16 CONTINUE
      STOP
  101 FORMAT (2F10.6)
  110 FORMAT (1X, 3E10.6)
      END
```

A FUNCTION subprogram is defined using the FORTRAN FUNCTION statement according to the general form

† The arithmetic statement function name consists of from one to six alphanumeric characters, the first of which must be a letter. In FORTRAN II, the last character of the name must be the letter F.

FUNCTION NAME (A$_1$, A$_2$, ..., A$_n$)
(Sequence of FORTRAN statements that specify computations
required to evaluate the FUNCTION)
RETURN
END

where   NAME is the name of any single-valued function,

A$_1$, A$_2$, ..., A$_n$ are the arguments, or dummy variables, which are
either unsubscripted variables or the name of another FUNC-
TION or SUBROUTINE program.

RETURN statement returns control to the program that calls the
FUNCTION subprogram.

END statement terminates the compilation of the FUNCTION
subprogram.

The following simple program, which computes the value of the intersection of the line $Y = AX + B$ with the x-axis, illustrates the definition
and use of a FUNCTION subprogram.

```
        FUNCTION POINT (A, B)
        EPS = 0.00001
        IF (ABS(A) − EPS) 10, 10, 5
   5    POINT = −B/A
   7    RETURN
  10    PRINT 113
        GOT TO 7
 113    FORMAT (19H LINE IS HORIZONTAL)
        END
```

A FUNCTION subprogram is referenced by its appearance in an arithmetic expression in the main program that refers to it. For example, if the
main program requires the computation of the line-axis intersection
described above, the name POINT (A, B) would be included in an arithmetic expression of the program at the required point, e.g.,

```
        . . .
        XNEW = XOLD + POINT (C, D)
        . . .
```

Note that the arguments A, B in the definition statement are merely
dummy variables that are replaced by the actual argument values at the
time the FUNCTION subprogram is executed by the main program. In
this example, A and B are replaced by C and D, respectively; i.e., the
intersection point of the line $Y = CX + D$ is calculated and added to
XOLD to get XNEW.

It should also be noted that the FUNCTION subprogram can be
compiled independent of the main program that references it.

FORTRAN processors are provided with a set of predefined "library" subprograms for evaluating standard mathematical functions such as square root, sine, cosine, arc tangent, logarithms, and exponential functions. These library routines are programmed as FUNCTION subprograms, as defined earlier.

A SUBROUTINE subprogram is *defined* using the FORTRAN SUBROUTINE statement according to the general form

> SUBROUTINE NAME (A_1, A_2, ..., A_n)
> DIMENSION statement(s) and/or COMMON statement(s)
> (Sequence of FORTRAN statements that accomplish
> algorithm to be performed by the subroutine)
> RETURN
> END

where   NAME is the SUBROUTINE subprogram name consisting of from one to six alphanumeric characters;

A_1, A_2, ..., A_n are its arguments, or dummy variables, which are either unsubscripted variables or the name of a FUNCTION or SUBROUTINE subprogram;

The DIMENSION statements establish the arrays required by the subroutine; COMMON statements establish arrays in the subroutine which are common to the subroutine and the calling program.

RETURN causes control to be returned to the calling program, END causes termination of compilation of the SUBROUTINE subprogram.

For example, the following SUBROUTINE subprogram could be defined to accomplish the linear interpolation described on page 27.

> SUBROUTINE LININT (X1, X2, Y1, Y2)
> COMMON   XBAR(9), YBAR(9)
> XH = (X2 − X1)/10.0
> DO 15 K = 1, 9
> XBAR(K) = X1 + FLOAT(K) * XH
> 15   YBAR(K) = (Y2 * (XBAR(K) − X1) − Y1 * (XBAR(K) − X2))/
> 1   (X2 − X1)
> RETURN
> END

A SUBROUTINE subprogram is referenced by a CALL statement in the program which calls it. The dummy variables in the SUBROUTINE statement are replaced by the actual values of the arguments in the CALL statement at the time the SUBROUTINE subprogram is executed. In this example, values of the arguments X1, X2, Y1, Y2 are communicated to the SUBROUTINE by the CALL statement, and output results XBAR(K),

YBAR(K), K = 1, 9, are returned to the calling program through COMMON. Note that the calling program must also include the arrays XBAR(9), YBAR(9) in the same relative position as in the SUBROUTINE subprogram.

A number of additional features are provided in most FORTRAN processors. They include

> Double-precision arithmetic;
> Complex arithmetic;
> Logical operations and expressions;
> Equivalence statements;
> Variable-type specifications; and
> Data statements.

For a description of these and other specialized and sophisticated features, the reader is referred to the manufacturers' FORTRAN manuals.

The COMMENT statement is very useful for inserting comments and explanations wherever they are required. Any statement that has the letter "C" in column 1 is considered a COMMENT statement by the compiler, and is ignored during the compilation of the source program. These comment statements are, however, printed out in the program listing as an aid to the programmer.

The following is an example of a typical FORTRAN program. See page 10 for the corresponding flow chart.

```
      PROGRAM QUAD
C     PROGRAM TO COMPUTE ROOTS OF QUADRATIC EQUATIONS
C     MAXIMUM NUMBER OF EQUATIONS = 100
      DIMENSION A(100),B(100),C(100),XR1(100),XI1(100),XR2(100),XI2(100)
      READ (5,100) N
      READ (5,101) (A(K), B(K), C(K), K=1,N)
      DO 20 K = 1,N
      D = B(K)**2 - 4. * A(K) * C(K)
      IF (D) 5, 10, 15
C     ROOTS ARE COMPLEX CONJUGATE
    5 XR1(K) = - B(K)/(2.0 * A(K))
      XI1(K) = SQRT(-D)/(2. * A(K))
      XR2(K) = XR1(K)
      XI2(K) = -XI1(K)
      GO TO 20
C     ROOTS ARE REAL AND EQUAL
   10 XR1(K) = -B(K)/(2.0 * A(K))
      XI1(K) = 0.0
      XR2(K) = XR1(K)
      XI2(K) = XI1(K)
      GO TO 20
C     ROOTS ARE REAL AND UNEQUAL
   15 XR1(K) = (-B(K) + SQRT(D))/(2. * A(K))
      XI1(K) = 0.0
      XR2(K) = (-B(K) - SQRT(D))/(2. * A(K))
      XI2(K) = 0.0
   20 CONTINUE
      WRITE (6,110) (A(K),B(K),C(K),XR1(K),XI1(K),XR2(K),XI2(K),K=1,N)
      STOP
  100 FORMAT (I3)
  101 FORMAT (3F15.7)
  110 FORMAT (1X, 7E15.7)
      END
```

Comments. Note the use of the GO TO 20 statements to transfer from the end of each path to common point 20; the last path does not require a GO TO 20 statement because statement 20 immediately follows the last statement in that path.

Note the use of the CONTINUE statement as the last statement of a DO Loop which would otherwise end with a GO TO statement (for 2 of the 3 paths in the DO Loop).

2

Fundamentals

of Numerical Methods

2.0 Introduction

Two concepts fundamental to an understanding of numerical methods are *recursion relations* and *successive approximations*. It is the recursive nature of numerical methods (algorithms) that facilitates problem solution by computer programs in concise form. The concept of successive approximations, wherein a problem solution is obtained by an iterative (repetitive) approximation method, emphasizes the value of the digital computer in solving problems once considered impracticable or impossible.

The steps of the problem-to-solution process introduced in Sec. 2.3 are fundamental to the efficient organization of algorithms for solution using a digital computer. The example included therein illustrates the basics of FORTRAN programming and numerical methods, and serves as a model for the development and organization of the algorithms discussed in this text.

Section 2.4 presents an elementary introduction to error analysis, and includes a simple mnemonic device for determining the propagation of errors in arithmetic expressions. This mnemonic device is used at points throughout the text to determine the propagation of errors in specific algorithms.

2.1 Basic Recursion Formulas

One of the most important concepts in numerical methods is that of the *recursion relation* or *recursion formula*. A recursion formula relates successive terms of a particular sequence of numbers, functions, polynomials, etc., and provides a means of computing successive quantities in terms of previously computed quantities. This concept can be best illustrated by a simple definition and examples.

Single-Recursion Formula

Let the elements (terms) of a sequence be denoted by $t_1, t_2, t_3, \ldots, t_n$. If the first element is known, and if any succeeding element t_k $(k = 2, n)$ of the sequence can be expressed in terms of t_{k-1} and other known quantities, then that expression for t_k is called a *single-recursion formula*. It should be noted that the range of the sequence index k can be other than from 1 to n. That is, it could as easily be from 0 to N, depending on the range of the sequence to be computed.

EXAMPLE. Find a recursion formula for computing the sum S of a set of n real numbers $a_1, a_2, a_3, \ldots, a_n$.

This sum can be computed by calculating the sequence of partial sums $S_1, S_2, S_3, \ldots, S_n$, where the kth partial sum is defined by the relation

$$S_k = a_1 + a_2 + a_3 + \cdots + a_k \qquad (k = 1, n).$$

The successive partial sums then can be written as

$$
\begin{aligned}
S_1 &= a_1 \\
S_2 &= a_1 + a_2 & &= S_1 + a_2 \\
S_3 &= a_1 + a_2 + a_3 & &= S_2 + a_3 \\
&\;\;\vdots & &\;\;\vdots \\
S_k &= a_1 + a_2 + a_3 + \cdots + a_{k-1} + a_k &&= S_{k-1} + a_k \\
&\;\;\vdots & &\;\;\vdots \\
S_n &= a_1 + a_2 + a_3 + \cdots + a_{n-1} + a_n &&= S_{n-1} + a_n.
\end{aligned}
$$

The first partial sum is calculated by the formula

$$(2.1.1) \qquad\qquad S_1 = a_1.$$

The formula for computing each of the remaining partial sums can be expressed in the form

$$(2.1.2) \qquad\qquad S_k = S_{k-1} + a_k \qquad (k = 2, n).$$

Formula (2.1.1) is called the *starter formula*, while (2.1.2) is called the *recursion formula* for generating the sequence $S_1, S_2, \ldots, S_n$ of partial sums, where $S_n = S$.

Flow Chart

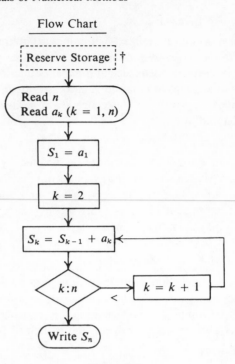

Figure 2-1.

A simple flow chart and FORTRAN program for computing S can be prepared using the starter and recursion formulas, as in Figure 2-1.

Multi-Recursion Formula

If starter elements of a sequence are known, and if each succeeding element t_k can be expressed in terms of preceding elements and other known quantities, then that expression for t_k (in terms of $t_{k-1}, t_{k-2}, \ldots$) is called a *multi-recursion* formula.

EXAMPLE. The well-known Fibonacci sequence $1, 1, 2, 3, 5, 8, \ldots$ can be generated by the multi-recursion formula

$$t_{k+2} = t_{k+1} + t_k \qquad (k = 1, 2, \ldots)$$

using starter terms $t_1 = 1$, and $t_2 = 1$.

In this text, single- or multi-recursion formulas will normally be referred to simply as recursion formulas.

† It is assumed that a maximum of 100 numbers $a_1, a_2, \ldots$ will be used for the purpose of this illustration.

Recursion Formulas for Simple Sequences

Arithmetic Sequences. Every student of mathematics is familiar with the sequence of odd positive integers $1, 3, 5, 7, 9, \ldots$. This sequence is an example of an arithmetic sequence of real numbers. A general arithmetic sequence can be expressed in the form

$$a, a + h, a + 2h, a + 3h, \ldots, a + kh, \ldots, a + nh, \ldots.$$

For example, if the values $a = 1$, $h = 2$ are substituted into the general sequence, we obtain the sequence of odd positive integers. Likewise, if we substitute the values $a = 2$, $h = 2$ into the general sequence, we obtain the sequence of even positive integers: $2, 4, 6, 8, \ldots$.

Suppose that we want to write a simple program that will generate various arithmetic sequences: This can be done by determining the *recursion formula* for terms of a general arithmetic sequence; elements of a specific sequence can then be generated by specifying values of a and h and the range of the index (e.g., $k = 0, N$). To find the recursion formula, we first denote by t_k the term $a + kh$ in the general sequence, so that

(2.1.3) $t_0 = a, t_1 = a + h, t_2 = a + 2h, t_3 = a + 3h, \ldots,$
$$t_k = a + kh, \ldots.$$

The single formula

(2.1.4) $\qquad t_k = a + kh \qquad (k = 0, 1, 2, \ldots)$

could be used to generate the terms of the general arithmetic sequence. Suppose, however, that we desire to generate t_k in terms of t_{k-1} using a simple recursion formula: To do this, we can apply the associative law of addition to the terms of (2.1.3) and rewrite these formulas in the form

$$
\begin{aligned}
t_0 &= a \\
t_1 &= a + h &&= t_0 + h \\
t_2 &= (a + h) + h &&= t_1 + h \\
t_3 &= (a + 2h) + h &&= t_2 + h \\
&\;\;\vdots &&\;\;\vdots \\
t_k &= (a + [k - 1]h) + h &&= t_{k-1} + h \\
&\;\;\vdots &&\;\;\vdots \\
t_N &= (a + [N - 1]h) + h &&= t_{N-1} + h.
\end{aligned}
$$

(2.1.5)

The individual formulas for generating $t_1, t_2, t_3, \ldots, t_k, \ldots, t_N$ can be *grouped into a single recursion formula*

(2.1.6) $\qquad t_k = t_{k-1} + h \qquad (k = 1, 2, 3, \ldots)$

which uses formula $t_0 = a$ as a starter.

A *flow chart* (Figure 2-2) for generating the terms of a general arithmetic sequence is obtained simply by "boxing-in" the starter and recursion

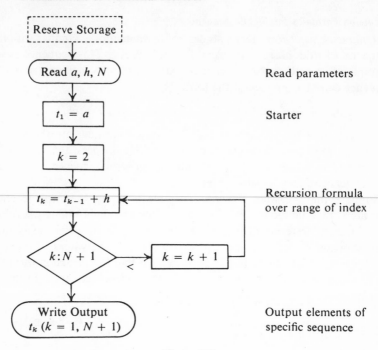

Figure 2-2.

formulas (with appropriate range of index k specified for the recursion formula), and adding storage, input, and output blocks. Since most FORTRAN processors do not permit a zero value for a subscript, it is necessary to shift the index range from $(0, N)$ to $(1, N + 1)$.†

EXAMPLE. To generate the first 100 terms of the arithmetic sequence of odd positive integers, we need only input $a = 1$, $h = 2$, $N = 99$.

```
      PROGRAM ARITHSEQ
C     FORTRAN PROGRAM FOR GENERATING ARITHMETIC SEQUENCES
      DIMENSION T(100)
      READ (5,100) A, H, N
      NP1 = N + 1
      T(1) = A
      DO 10 K = 2,NP1
   10 T(K) = T(K-1) + H
      WRITE (6,110) (T(K), K=1,NP1)
      STOP
  100 FORMAT (2F10.5, I2)
  110 FORMAT (1X, E15.7)
      END
```

† The terms of the sequence are then $t_1 = a$, $t_2 = a + h$, $t_3 = a + 2h, \ldots,$ $t_k = a + (k - 1)h, \ldots, t_{N+1} = a + Nh$.

Geometric sequences. The general geometric sequence consists of the terms a, ar, ar^2, ar^3, ..., ar^k, ..., ar^n, ... Denoting these terms, respectively, by t_0, t_1, t_2, t_3, ..., t_k, ..., t_n, ..., and applying the associative law of multiplication, we obtain

(2.1.7)

$$
\begin{aligned}
t_0 &= a \\
t_1 &= (a)r & &= t_0 r \\
t_2 &= (ar)r & &= t_1 r \\
t_3 &= (ar^2)r & &= t_2 r \\
&\;\;\vdots & &\;\;\vdots \\
t_k &= (ar^{k-1})r &&= t_{k-1} r \\
&\;\;\vdots & &\;\;\vdots \\
t_n &= (ar^{n-1})r &&= t_{n-1} r.
\end{aligned}
$$

The recursion formula for generating terms t_k $(k = 1, N)$ of the general geometric sequence is

(2.1.8) $$t_k = t_{k-1} r \qquad (k = 1, N)$$

with formula $t_0 = a$ as a starter.

Suppose that we are required to generate the partial sums S_0, S_1, S_2, S_3, ..., of the *geometric series* $S = \sum_{i=0}^{\infty} ar^i$, where the kth partial sum is

(2.1.9) $$S_k = \sum_{i=0}^{k} ar^i = \sum_{i=0}^{k} t_i.$$

That is,

(2.1.10)

$$
\begin{aligned}
S_0 &= t_0 \\
S_1 &= (t_0) + t_1 & &= S_0 + t_1 \\
S_2 &= (t_0 + t_1) + t_2 & &= S_1 + t_2 \\
S_3 &= (t_0 + t_1 + t_2) + t_3 & &= S_2 + t_3 \\
&\;\;\vdots & &\;\;\vdots \\
S_k &= (t_0 + t_1 + \cdots + t_{k-1}) + t_k &&= S_{k-1} + t_k \\
&\;\;\vdots & &\;\;\vdots \\
S_N &= (t_0 + t_1 + \cdots + t_{N-1}) + t_N &&= S_{N-1} + t_N.
\end{aligned}
$$

The recursion formula for the partial sums S_k $(k = 1, N)$ is simply

(2.1.11) $$S_k = S_{k-1} + t_k \qquad (k = 1, N)$$

with formula $S_0 = t_0$ as a starter.†

It is a simple matter to generate concurrently the terms of a geometric sequence *and* the partial sums of the corresponding geometric series. A

† Since

$$\sum_{i=0}^{k} t_i = \sum_{i=0}^{k-1} t_i + t_k$$

it follows directly from definition (2.1.9) that $S_k = S_{k-1} + t_k$.

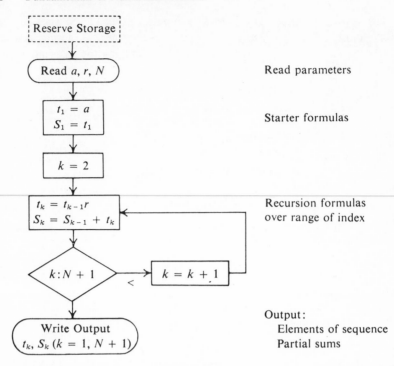

Figure 2-3.

representative flow chart is given in Figure 2-3. The range of the index has been shifted from $(0, N)$ to $(1, N + 1)$.†

The terms of a specific geometric sequence and the partial sums of the corresponding geometric series can then be generated by reading in the required values of a, r, and index limit N. For example, the first ten terms of the geometric sequence $1, 1/2, 1/4, 1/8, \ldots$ and corresponding series

```
      PROGRAM GEOSEQ
   C  FORTRAN PROGRAM FOR GENERATING GEOMETRIC SEQUENCES
      DIMENSION T(100), S(100)
      READ (5,100) A, R, N
      NP1 = N + 1
      T(1) = A
      S(1) = T(1)
      DO 10 K = 2,NP1
      T(K) = T(K-1) * R
   10 S(K) = S(K-1) + T(K)
      WRITE (6,110) (T(K), S(K), K = 1,NP1)
      STOP
  100 FORMAT (2F10.5, I2)
  110 FORMAT (1X,2E15,7)
      END
```

† I.e., $a \leftrightarrow t_1$, $ar \leftrightarrow t_2$, $ar^2 \leftrightarrow t_3, \ldots$.

1, 3/2, 7/4, 15/8, ... are generated by reading in $a = 1$, $r = 1/2$, $N = 9$.

The foregoing examples using well-known sequence types should give a clear understanding of simple recursion formulas. In Sec. 2.2 we will review the convergent properties of infinite sequences and series to illustrate the concept of iterative methods of solving the equation $f(x) = 0$ by generating a convergent sequence of successive approximations to a true solution.

2.2 Successive Approximations

Perhaps the single most important concept in numerical analysis is that of *successive approximations*. This concept should be familiar to every student of mathematics. Indeed, it is a discrete-variable analog of the continuous-variable limit process that forms the basis for differential and integral calculus.

In calculus the derivative of a function $y = f(x)$ is defined as

$$\frac{dy}{dx} = \lim_{\Delta x \to 0} \frac{f(x + \Delta x) - f(x)}{\Delta x}$$

while the definite integral is defined as

$$\int_a^b f(x)\,dx = \lim_{n \to \infty} \sum_{i=0}^{n-1} f(x_i)\,\Delta x.$$

In either case, the increment Δx is allowed to *approach zero continuously*. These definitions based on the continuous-variable limit process provide closed-form solutions to many problems in differentiation and integration, allowing us to compute derivatives and definite integrals by formulas.

However, for many real-life problems, closed-form solutions either do not exist or are extremely cumbersome to handle. Many such problems can be solved by numerical methods of successive approximations. All practical numerical computations are limited to a finite number of significant digits, and even on the largest digital computer the continuous variable is an intangible that cannot be quantitatively handled. We must therefore content ourselves with discrete values of the continuous variable, the separation between adjacent values being determined by the word size of the particular digital computer.

The concept of successive approximations can be nicely illustrated by examining the discrete-variable analogs of the derivative and definite integral definitions. To compute the derivative of a function without recourse to the definition, it is possible to take a number of discrete positive increments Δx_k ($k = 1, 2, 3, \ldots$) and successively form the difference quotients

$$\frac{\Delta y_k}{\Delta x_k} \equiv \frac{f(x + \Delta x_k) - f(x)}{\Delta x_k}.$$

If the Δx_k form a monotonic decreasing sequence, i.e., if $\Delta x_{k+1} < \Delta x_k$, then the sequence of $\Delta y_k / \Delta x_k$ are successive approximations of the derivative dy/dx. If $f(x)$ is continuous in the interval around x, and if k is allowed to become sufficiently large (but still finite), then the sequence of successive approximations $\Delta y_k / \Delta x_k$ should approach the value of the derivative dy/dx (see Figure 2-4).

Similarly, the area under the graph of the function $y = f(x)$ between limits $x = a$ and $x = b$ can be approximated by calculating successively the sums of the rectangular areas of width Δx_k ($k = 0, 1, 2, \ldots$) defined by the relation

$$I_k = \sum_{n=0}^{2^k - 1} f(a + n\,\Delta x_k)\,\Delta x_k.$$

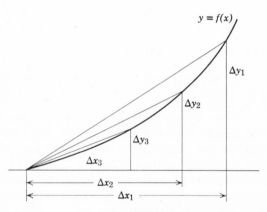

Figure 2-4.

Again, if the sequence of positive Δx_k is monotonic decreasing, then the sequence of successive approximations I_k ($k = 0, 1, 2, \ldots$) should approach the value of the definite integral given by the continuous-variable definition (see Figure 2-5).

The derivative of a function cannot be computed accurately using simple difference quotients, although this technique must sometimes be resorted to in practice. In general, numerical differentiation is inherently unstable, and should be avoided if at all possible. The reader is referred to Milne [18, pp. 96–100], wherein a number of numerical differentiation formulas are presented.

The foregoing examples have been included solely to illustrate the

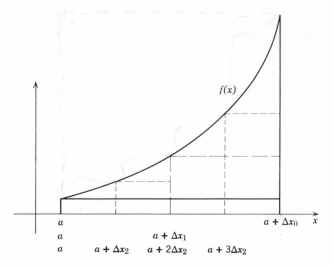

Figure 2-5.

nature of successive approximations. Methods of successive approximations used in practice are developed throughout this book to solve particular classes of problems such as:

Solution of algebraic and transcendental equations;
Solution of systems of linear algebraic equations;
Interpolation;
Numerical integration;
Solution of ordinary and partial differential equations;
Generation of elementary functions;
Solution of matrix eigenproblems (characteristic equations).

Methods of successive approximation that generate *convergent* sequences
are of especial importance in solving problems of the types listed above.
For this reason, it will be advantageous to review a few of the more
important properties relating to the convergence of sequences and series.

Sequences

Definition. An ordered collection of terms $t_0, t_1, t_2, \ldots, t_k, \ldots$ is called
a sequence, and is denoted by $\{t_k\}$. If the range of the sequence index k is
finite, the sequence is finite. If the range of k is infinite, the sequence is
infinite.

Definition. A sequence is *recursive* if its terms satisfy some recursion
relation. For example, the geometric sequence $\{t_k = ar^k\}$ is a recursive

sequence whose terms satisfy the recursion relation $t_k = t_{k-1}r$, with starter formula $t_0 = a$.

Definition. An infinite sequence $\{t_k\}$ is convergent if there exists a finite T such that† $\lim_{k \to \infty} t_k = T$.

The sequence is said to converge to the limit T. A sequence $\{t_k\}$ is divergent if $\lim_{k \to \infty} t_k$ is infinite or if $\lim_{k \to \infty} t_k$ does not exist.

EXAMPLES. Sequence $\{t_k = k/(k+1)\}$ is a convergent sequence with $T = 1$.

Sequence $\{t_k = k\}$ is divergent because $\lim_{k \to \infty} t_k = \infty$.

Sequence $\{t_k = (-1)^k\}$ is divergent because $\lim_{k \to \infty} t_k$ does not exist.

Cauchy test for convergence of an infinite sequence. If for every $\varepsilon > 0$ there exists an integer K such that $n > K$ and $m > K$ implies that $|t_n - t_m| < \varepsilon$, then the infinite sequence $\{t_k\}$ converges to a finite limit T.

EXAMPLE. For the sequence $\{t_k = 1/2^k \ (k = 1, 2, \ldots)\}$ and $\varepsilon = 0.001$, determine the value of K such that for $n > K$ and $m > K$ the value of $|t_n - t_m| < \varepsilon$.

Solution. Terms of the sequence $\{t_k = 1/2^k\}$ are as follows: 1/2, 1/4, 1/8, 1/16, 1/32, 1/64, 1/128, 1/256, 1/512, 1/1024, 1/2048, 1/4096, ...

The required value is $K = 9$. This value gives $|t_n - t_m| < 0.001$ for all $n > 9$ and $m > 9$.

Infinite Series

Definition. The sum $S = \sum_{k=0}^{\infty} t_k$ is the *infinite series* corresponding to the infinite sequence $\{t_k\}$.

Definition. The sum $S_k = \sum_{i=0}^{k} t_i$ is the kth *partial sum* of the infinite series S.

Definition. The *sequence* $\{S_k\}$ *of partial sums* of infinite series S is the sequence $S_0, S_1, S_2, \ldots$.

Definition. Convergent Series—The infinite series $S = \sum_{k=0}^{\infty} t_k$ *converges* if the sequence $\{S_k\}$ of partial sums converges. The infinite series S *diverges* if the sequence $\{S_k\}$ diverges.

EXAMPLE. The geometric series $S = \sum_{k=0}^{\infty} ar^k$ converges to limit $a/(1-r)$ if $|r| < 1$.

Proof. The kth partial sum of the geometric series is

$$S_k = a + ar + ar^2 + \cdots + ar^{k-1} + ar^k.$$

† The notation $\lim_{k \to \infty} t_k = T$ means the following: If for every $\varepsilon > 0$ there exists an integer K such that when $k > K$, then $|t_k - T| < \varepsilon$.

Multiplying S_k by r, we get

$$rS_k = ar + ar^2 + ar^3 + \cdots + ar^k + ar^{k+1}.$$

Subtracting the second sum from the first, we get

$$S_k - rS^k = a - ar^{k+1}.$$

This can be rewritten in the form

$$S_k = \frac{a - ar^{k+1}}{1 - r}$$

Taking the limit of partial sum S_k written in this form, we get

$$S = \lim_{k \to \infty} S_k = \lim_{k \to \infty} \frac{a - ar^{k+1}}{1 - r} = \frac{a}{1 - r} \qquad \text{if } |r| < 1.$$

Definition. The sum S of a convergent infinite series is equal to the limiting value of its kth partial sum, as k increases indefinitely, i.e.,

$$S = \lim_{k \to \infty} S_k.$$

NUMERICAL EXAMPLE. The geometric series with $a = 1/2$ and $r = 1/2$ is $S = (1/2) + (1/4) + (1/8) + (1/16) + (1/32) + (1/64) + \cdots$ and the sequence S_k of partial sums of this series is

$$0.5, \ 0.75, \ 0.875, \ 0.9375, \ 0.96875, \ 0.984375, \ldots.$$

This sequence of partial sums converges to the limit

$$T = 1 = \frac{a}{1 - r}.$$

Taylor Series

The well-known Taylor series representation of a function $f(x)$ can be easily derived by the *method of undetermined coefficients*, as follows: Suppose that we are given a function $f(x)$ which has derivatives of all orders up to and including n at some point $x = x_0$. We want to obtain a representation of $f(x)$ from which we can evaluate the function $f(x)$ for any x in the neighborhood of x_0.

Assume that $f(x)$ can be expressed as a *convergent infinite series* of the form

$$(2.2.1) \qquad f(x) = a_0 + a_1(x - x_0) + a_2(x - x_0)^2 + a_3(x - x_0)^3 \\ + \cdots + a_n(x - x_0)^n + \cdots$$

Calculating successively the derivatives of series (2.2.1), we get

$$f'(x) = a_1 + 2a_2(x - x_0) + 3a_3(x - x_0)^2 + \cdots + na_n(x - x_0)^{n-1} + \cdots$$
$$f''(x) = \qquad 2a_2 \qquad + \qquad 3!a_3(x - x_0) + \cdots + n(n-1)a_n(x - x_0)^{n-2} + \cdots$$
$$f'''(x) = \qquad\qquad\qquad 3!a_3 + \cdots + n(n-1)(n-2)a_n(x - x_0)^{n-3} + \cdots$$
$$\cdots$$
$$f^{(n)}(x) = \qquad\qquad\qquad\qquad\qquad\qquad n!a_n + \cdots.$$
(2.2.2)

The unknown coefficients a_k can then be determined by evaluating series (2.2.1) and its derivatives (2.2.2) at $x = x_0$, as follows:

(2.2.3)
$$f(x_0) = a_0$$
$$f'(x_0) = a_1$$
$$f''(x_0) = 2a_2$$
$$f'''(x_0) = 3!a_3$$
$$\vdots$$
$$f^{(n)}(x_0) = n!a_n.$$

Substituting into (2.2.1) the values of a_k determined by (2.2.3) we obtain the following series representation for function $f(x)$:

$$(2.2.4) \quad f(x) = f(x_0) + f'(x_0)(x - x_0) + \frac{f''(x_0)}{2!}(x - x_0)^2$$

$$+ \frac{f'''(x_0)}{3!}(x - x_0)^3 + \cdots + \frac{f^{(n)}(x_0)}{n!}(x - x_0)^n + \cdots.$$

We see then that if the function $f(x)$ can be represented by a convergent infinite series of the form (2.2.1), then that series necessarily has the form of (2.2.4). Equation (2.2.4) is the classical form of a Taylor series expansion of $f(x)$ in the neighborhood of a point $x = x_0$.

Partial Sums of Taylor Series

Given an infinite series (Taylor) in the form

$$(2.2.5) \qquad\qquad f(x) = \sum_{k=0}^{\infty} \frac{f^{(k)}(x_0)}{k!}(x - x_0)^k$$

we can approximate $f(x)$ by the nth partial sum

$$(2.2.6) \qquad\qquad S_n(x) = \sum_{k=0}^{n} \frac{f^{(k)}(x_0)}{k!}(x - x_0)^k.$$

Expression (2.2.6), the nth partial sum of the Taylor series, is referred to as a *truncated Taylor series*.

Truncation Error

If the infinite series (2.2.5) representing $f(x)$ is replaced by the nth partial sum, then we say that the infinite series has been truncated after the term of degree n. The error T caused by this truncation is

(2.2.7) $$T = \frac{f^{(n+1)}(x_1)}{(n+1)!}(x - x_0)^{n+1}, \qquad \text{where } x_1 \in (x_0, x)$$

and

(2.2.8) $$f(x) = S_n(x) + T.$$

2.3 Steps of the Problem-to-Solution Process

In Chapters 3 through 9 representative numerical methods will be presented for solving each of the following classes of problems:

Calculating the roots of algebraic/transcendental equations;
Calculating the roots of polynomial equations;
Matrix methods and solution of systems of linear equations;
Interpolation;
Estimation of parameters;
Numerical integration;
Solution of ordinary differential equations.

Each topic will be organized into a logical step-by-step development of the problem-to-solution process. This will ensure uniformity of approach in each of the various subjects covered, and aid the reader in making the transitions between succeeding steps of the development. The principal steps in the problem-to-solution process include the following:

(1) Statement of problem and formulation of math model;
(2) Methods of solution—derivation of computational formulas;
(3) Computational summary;
(4) Flow chart;
(5) Computer program;
(6) Evaluation of results—comparison of methods.

In addition, numerous examples and exercises will be included which illustrate the particular method. Programming exercises are included to allow the reader the opportunity to test his understanding of a particular method by modifying the programs to conserve storage, improve efficiency, accelerate convergence, and so forth.

These principal steps of the problem-to-solution process can be best illustrated by considering a simple computational problem and developing a logical order of required steps for the successful solution of the particular problem. *An excellent example of this process is the problem of evaluating an nth degree polynomial for a given value of x.*

Steps of the Problem-to-Solution Process (Description and Example)

1. *Statement of the Problem.* The problem to be solved must be clearly and unambiguously stated, together with any constraints or conditions that are to be imposed to ensure that the results are physically realizable. The type(s) and amount of data and control parameters should be specified, as well as the nature and form of the desired results.

EXAMPLE. Evaluate the nth degree polynomial with real coefficients $a_0, a_1, a_2, \ldots, a_n$, where a_k is the coefficient of x^{n-k}, for a given value of x.

MATH MODEL. If a problem of any complexity is to be quantitatively analyzed, it should first be formulated in mathematical notation. Such a formulation is called a math model. One such representation for the nth degree polynomial is given in the following example.

EXAMPLE $P_n(x) = a_0 x^n + a_1 x^{n-1} + a_2 x^{n-2} + \cdots + a_{n-1} x + a_n$.

2. *Method of solution—deriving computational formulas.* Most computational problems can be solved in more ways than one. To determine the best method of solution, we must establish some criteria for deciding what is best. If the problem is to be solved by using a digital computer, common criteria employed include the following: accuracy, computational efficiency, speed, storage requirements, and simplicity of recursion formulas for programming.

Computational formulas should be derived that take advantage of the characteristics of the digital computer.

EXAMPLE. The polynomial $P_n(x)$ *could* be evaluated by forming each of the products $t_k = a_k x x \ldots x$ (with $n - k\ x$ factors), for $k = 0, 1, 2, \ldots$ and adding these $n + 1$ products to obtain

$$P_n(x) = t_0 + t_1 + t_2 + \cdots + t_n.$$

This method of evaluating the polynomial requires $n(n + 1)/2$ multiplications and n additions. Although this is probably the most obvious way of evaluating $P_n(x)$, it is not the best.

A better method of solving this problem employs recursion formulas such as described earlier, and requires only n multiplications and n additions. This method is described in the following subsection.

Evaluating a Polynomial by Recursion (Horner's Nesting Procedure)

An efficient method of evaluating a polynomial

(2.3.1) $P_n(x) = a_0 x^n + a_1 x^{n-1} + a_2 x^{n-2} + \cdots + a_{n-1} x + a_n$

is to rewrite the polynomial in nested form as

(2.3.2) $P_n(x) = (\cdots(\{[(a_0)x + a_1]x + a_2\}x + a_3)x + \cdots + a_{n-1})x + a_n$.

The successive nested terms, which we denote by p_k $(k = 0, 1, \ldots, n)$, from innermost to outermost, constitute the sequence $p_0, p_1, p_2, \ldots, p_n$ of polynomials from degree 0 to degree n, respectively. Note that the sequence index ranges from 0 to n.

The elements p_k of this sequence of nested terms can be computed as follows:

$$
\begin{aligned}
p_0 &= a_0 \\
p_1 &= p_0 x + a_1 &&= (a_0)x + a_1 \\
p_2 &= p_1 x + a_2 &&= [(a_0)x + a_1]x + a_2 \\
&\;\vdots &&\;\vdots \\
p_k &= p_{k-1}x + a_k &&= (\cdots([(a_0)x + a_1]x + a_2)x + \cdots + a_{k-1})x + a_k \\
&\;\vdots &&\;\vdots \\
p_n &= p_{n-1}x + a_n &&= (\cdots([(a_0)x + a_1]x + a_2)x + \cdots + a_{n-1})x + a_n.
\end{aligned}
$$
(2.3.3)

From (2.3.2) and (2.3.3) we see that

(2.3.4) $$P_n(x) = p_n.$$

We can group the formulas in (2.3.3) for computing $p_1, p_2, \ldots, p_n$ into the single formula $p_k = p_{k-1}x + a_k$, as k ranges from 1 to n. Using starter formula $p_0 = a_0$, we can calculate the terms p_k of the nested polynomial by the recursion formula

(2.3.5)
$$
\begin{aligned}
p_0 &= a_0 && \text{starter formula} \\
p_k &= p_{k-1}x + a_k \quad (k = 1, n) && \text{recursion formula.}
\end{aligned}
$$

This problem of evaluating a polynomial $P_n(x)$ for a given x provides an excellent illustration of *the utility of the concept of recursion formulas in numerical methods*. The recursion formula is the heart of almost every numerical method, and once it is obtained for a particular numerical method, the problem of flow charting and programming that method is essentially solved.

3. *Computational summary.* Once the computational formulas have been derived, it is necessary to outline or summarize the steps involved in the selected method. These steps must then be organized into an unambiguous, ordered sequence of computations, with logical decisions provided to include all foreseeable possibilities which may arise.

EXAMPLE. Computation Summary for Evaluating a Polynomial.

Step 0. Input. Read n = degree of polynomial, x = value of abscissa at which $P_n(x)$ is to be evaluated, a_k $(k = 1, 2, \ldots, n + 1)$† = coefficients of polynomial.

Step 1. Compute successive nested terms. The recursion formulas resulting from the index shift are

$$p_1 = a_1$$
$$p_k = p_{k-1}x + a_k \qquad (k = 2, n + 1).$$

Step 2. Output results (value of $P_n(x)$). Write out the value of the polynomial

$$P_n(x) = p_{n+1}.$$

4. *Flow chart of method of solution.* The flow chart is a diagram that shows the logical sequence of basic computational steps and the decisions required to accomplish the solution. The flow chart should be a straightforward logical parallel of the computational summary, and should be in sufficient detail that a programmer can write the FORTRAN program directly from it, even if the programmer is unfamiliar with the specific method employed.

A representative flow chart for evaluating the polynomial $P_n(x)$ at a given x is shown in Figure 2-6.

† To eliminate the zero subscript that is not permitted by most FORTRAN processors, the index range is shifted from $(0, n)$ to $(1, n + 1)$. The polynomial then takes the form

$$P_n(x) = a_1x^n + a_2x^{n-1} + a_3x^{n-2} + \cdots + a_nx + a_{n+1}.$$

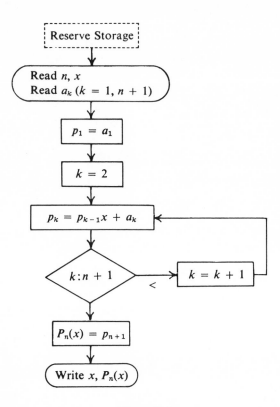

Figure 2-6.

5. *Computer program.* The computer program can be written directly from a detailed flow chart that includes the specific description of input data and parameters, the precise formulation of the computational formulas and logical decisions, and the description of the desired output. The reader can test his understanding of FORTRAN (or any other compiler language) by writing his own programs directly from the flow charts included in this book.

EXAMPLE

```
      PROGRAM POLYNOM
   C  FORTRAN PROGRAM FOR EVALUATING POLYNOMIAL
      DIMENSION A(21), P(21)
      READ (5,100) N,X
      NP1 = N + 1
      READ (5,101) (A(K), K=1,NP1)
      P(1) = A(1)
      DO 10 K = 2,NP1
   10 P(K) = P(K-1) * X + A(K)
      POLY = P(NP1)
      WRITE (6,110) X, POLY
      STOP
  100 FORMAT (I2, F15,7)
  101 FORMAT (F15,7)
  110 FORMAT (2X, 2E15,7)
      END
```

2.4   Errors in Computation

The accuracy of numerical computations is the single most important criterion for selecting a particular method for solving a given problem. For what good are the results if a method quickly and efficiently produces inaccurate answers? Before we launch into the specific numerical methods covered in Chapters 3 through 9, it will be advantageous to discuss briefly the errors in computations and how they affect the accuracy of our results.

Digital computers are highly accurate devices that are capable of handling numbers ranging from the extremely small to the extremely large. For example, a typical large-scale digital computer can handle numbers whose magnitudes range from 10^{-38} to 10^{+38}. And these numbers can be represented accurately to perhaps 10, 12, or more significant figures. Why then are we concerned with errors (as opposed to mistakes) in computation?

There are a number of reasons. First, a given set of data may contain errors† before any computation takes place. There may be errors that are due to the mathematical model used; e.g., a differential equation may only approximate the physical system it represents. Also, approximate methods of solution are employed extensively to handle problems that do not lend themselves to solution by analytic methods. Finally, errors usually crop up during the computations; these include roundoff errors (owing to limitations in the number of significant figures which the computer can handle) and truncation errors (caused, for example, by truncating an infinite series after a given number of terms). We must also include the *propagation* of *all of these types* of errors during the computations.

Floating-Point Numbers

The reader will readily recall the rules for adding, subtracting, multiplying, and dividing numbers of the form

$$N_i = a_i b^{e_i} \qquad (i = 1, 2, \dots)$$

where the a_i are called coefficients, b the base of the number system, and e_i the exponents:

$$N_1 \cdot N_2 = (a_1 b^{e_1}) \cdot (a_2 b^{e_2}) = (a_1 a_2) b^{e_1 + e_2}$$

$$\frac{N_1}{N_2} = \frac{a_1 b^{e_1}}{a_2 b^{e_2}} = \frac{a_1}{a_2} b^{e_1 - e_2}$$

$$N_1 \pm N_2 = (a_1 b^{e_1}) \pm (a_2 b^{e_2}) = (a_1 \pm a_2 b^{e_2 - e_1}) b^{e_1}.$$

† These errors may be due to such causes as errors in observation, measurement, recording, transmission, conversion, and in preprocessing.

Numbers expressed in the form $a\, b^e$ are called *floating-point numbers*. For example, the decimal number 245.3 can be expressed in floating-point form as $0.2453(10^3)$, where 0.2453 is the coefficient, 10 is the base of the decimal-number system, and 3 is the exponent. It is also true that

$$245.3 = 24.53(10^1) = 2.453(10^2) = 0.02453(10^4) = 0.002453(10^5)$$
$$= 2453.0(10^{-1}) = 24530.0(10^{-2}) = \cdots.$$

If the coefficient a is a proper fraction F in the base system such that $1/b \le |F| < 1$, then the number $F\, b^e$ is called a *normalized* floating-point number. For example, in the decimal system (base 10), the number $0.2453(10^3)$ is a normalized floating-point number, while other floating-point representations of the number 245.3 are not normalized. The fraction F is also referred to as the *mantissa*.

Computations in FORTRAN programs are carried out in normalized floating-point arithmetic because the computer automatically handles the positioning of the decimal point (assuming we are working in a decimal system).

Arithmetic Operations with Normalized Floating-Point Numbers

Let the numbers X and Y have the normalized floating-point representations $X = F_x \cdot 10^{e_x}$ and $Y = F_y \cdot 10^{e_y}$, so that $0.1 \le |F_x| < 1$, and $0.1 \le |F_y| < 1$. Then the numbers X and Y may be combined by the arithmetic operations $+$, $-$, $\times$, $/$ according to the foregoing rules for combining numbers of the form $N_i = a_i b^{e_i}$.

If $Z = X \oplus Y$, where $\oplus$ is any one of the operations $+$, $-$, $\times$, $/$, then the *normalized* floating-point representation $Z = F_z \cdot 10^{e_z}$, with $0.1 \le |F_z| < 1$, can be summarized as follows:

1. *Multiplication:*

$$Z = XY = (F_x \cdot 10^{e_x})(F_y \cdot 10^{e_y}) = F_x F_y \cdot 10^{e_x + e_y}$$

(a) If $|F_x F_y| < 0.1$,

$$F_z = 10 F_x F_y, \qquad e_z = e_x + e_y - 1$$

(b) If $0.1 \le |F_x F_y| < 1$,

$$F_z = F_x F_y, \qquad e_z = e_x + e_y$$

2. *Division:*

$$Z = \frac{X}{Y} = \frac{F_x \cdot 10^{e_x}}{F_y \cdot 10^{e_y}} = \frac{F_x}{F_y} \cdot 10^{e_x - e_y}$$

(a) If $0.1 \le \left|\dfrac{F_x}{F_y}\right| < 1$,

$$F_z = \frac{F_x}{F_y}, \qquad e_z = e_x - e_y$$

(b) If $\left|\dfrac{F_x}{F_y}\right| > 1$,

$$F_z = \frac{F_x/F_y}{10}, \qquad e_z = e_x - e_y + 1$$

3. *Addition, Subtraction:*†

$$Z = X \pm Y = (F_x \cdot 10^{e_x}) \pm (F_y \cdot 10^{e_y})$$
$$= (F_x \pm F_y \cdot 10^{e_y - e_x}) \cdot 10^{e_x}$$

Let $\hat{F}_y = F_y \cdot 10^{e_y - e_x}$. (a) If $|F_x \pm \hat{F}_y| < 0.1$,

$$F_z = 10^m (F_x \pm \hat{F}_y), \qquad e_z = e_x - m$$

where m is such that $0.1 \le |F_z| < 1$

(b) If $0.1 \le |F_x \pm \hat{F}_y| < 1$,

$$F_z = F_x \pm \hat{F}_y, \qquad e_z = e_x$$

(c) If $|F_x \pm \hat{F}_y| > 1$,

$$F_z = \frac{F_x \pm \hat{F}_y}{10}, \qquad e_z = e_x + 1.$$

Examples of Arithmetic Operations Using Normalized Floating-Point Numbers

The following examples illustrate the principles of combining normalized floating-point numbers using addition, subtraction, multiplication, and division. In these examples, the number $F_x \cdot 10^{e_x}$ will be represented in the form $F_x(e_x)$.

1. *Multiplication:* $Z = X \cdot Y$

(a) $X =$ 0.3184(2)   (b) $X =$ 0.4836(3)
   $Y =$ 0.2512(2) $Y =$ 0.5123(2)

   $Z = 0.07998208(4)$ $Z = 0.24774828(5)$
   $\quad = 0.7998208(3)$

2. *Division:* $Z = \dfrac{X}{Y}$

(a) $X =$ 0.3333(2)   (b) $X =$ 0.4251(3)
   $Y =$ 0.4251(1) $Y =$ 0.2465(2)

   $Z = 0.78405081(1)$ $Z = \;\;1.72454361(1)$
   $\qquad\qquad\qquad\qquad\qquad = 0.172454361(2)$

† Let $e_x \ge e_y$. If not, the roles of X and Y can be interchanged.

3. *Addition:* $Z = X + Y$

(a) $X = 0.4322(3)$ (b) $X = 0.5723(2)$ (c) $X = 0.6123(4)$
$Y = -0.3611(3)$ $Y = 0.2344(2)$ $Y = 0.5243(4)$

$Z = 0.0711(3)$ $Z = 0.8067(2)$ $Z = 1.1366(4)$
$ = 0.711(2)$ $ = 0.11366(5)$

If $e_x > e_y$, then the mantissa F_y must be shifted to the right with the Y exponent increased by 1 for each position shifted until the two exponents become equal. For example, to add $X = 0.2844(3)$ and $Y = 0.4162(1)$, F_y is shifted so that $\hat{F}_y = F_y \cdot 10^{e_y - e_x} = 0.004162$

$$X = 0.2844 \quad (3)$$
$$Y = 0.004162(3)$$

$$Z = 0.288562(3)$$
$$= 0.2885 \quad (3) + 0.6200(-1)$$

4. *Subtraction:* $Z = X - Y$

(a) $X = 0.2846(4)$ (b) $X = 0.7533(2)$ (c) $X = 0.7428(3)$
$Y = 0.1922(4)$ $Y = 0.2411(2)$ $Y = -0.4111(3)$

$Z = 0.0924(4)$ $Z = 0.5122(2)$ $Z = 1.1539(3)$
$ = 0.9240(3)$ $ = 0.11539(4)$

The rules for subtracting numbers with different exponents are analogous to the rules for adding numbers with different exponents.

Roundoff Error in Floating-Point Arithmetic Operations

The mantissa F_z of the result $Z = X \oplus Y$, where $\oplus = +, -, \times, /$, and where F_x and F_y are proper fractions containing d digits, can have up to $2d$ digits if it is formed in a double-length accumulator. However, if we can only store numbers having mantissas with a maximum of d digits, then it is necessary to round off the mantissa F_z before storing Z. The leftmost d digits in the double-length accumulator are the most significant, and are denoted F, while the rightmost d digits are the least significant, and are denoted by f. Then Z can be expressed in the form

$$Z = F \cdot 10^{e_z} + f \cdot 10^{e_z - d}.$$

For example, assuming $d = 4$, we obtain the product

$$X = 0.4836(10^3)$$
$$Y = 0.5123(10^2)$$

$$Z = 0.24774828(10^5) = [0.2477 + 0.00004828]10^5$$
$$= 0.2477(10^5) + 0.4828(10^{5-4})$$
$$= F \cdot 10^{e_z} + f \cdot 10^{e_z - d}.$$

The general rule for rounding the result $Z = X \oplus Y$, where F_x and F_y contain d digits, is as follows: If $|f| < 0.5$,

$$|Z| = |F| \cdot 10^{e_z}$$

if $|f| \geq 0.5$,

$$|Z| = |F| \cdot 10^{e_z} + 10^{e_z - d}.$$

In the example above, $|f| = 0.4828 < 0.5$, so that the result Z is rounded such that $|Z| = |F| \cdot 10^{e_z}$, with the result $Z = 0.2477(10^5)$. The maximum error in rounding can be easily determined as follows. Let ε_z denote the error in the rounded value of Z. It is obvious then that if $|f| < 0.5$,

$$|\varepsilon_z| = |f| \cdot 10^{e_z - d}$$

if $|f| \geq 0.5$,

$$|\varepsilon_z| = |1 - f| \cdot 10^{e_z - d}.$$

In either case, the maximum error in rounding a number Z (with a mantissa of d digits) by the foregoing rule is

$$|\varepsilon_z| \leq 0.5 \cdot 10^{e_z - d}.$$

The relative error in rounding the floating-point number Z is the ratio of the error to the number; i.e., the relative error, denoted $r_z = \varepsilon_z / Z$. The absolute value of the relative error in rounding Z is largest when the magnitude of the mantissa F of number Z is smallest (when $|Z| = 0.1 \cdot 10^{e_z}$) and the rounding-error magnitude largest ($0.5 \cdot 10^{e_z - d}$), i.e.,

$$|r_z| = \left| \frac{\varepsilon_z}{Z} \right| \leq \left| \frac{0.5 \cdot 10^{e_z - d}}{0.1 \cdot 10^{e_z}} \right| = 5 \cdot 10^{-d}.$$

Propagation of Errors in Arithmetic Expressions

There are two basic types of errors to contend with in computational work: (1) *inherent errors* and (2) *acquired errors*. The inherent errors exist in the data before we begin computations; the acquired errors are either truncation errors (e.g., those errors produced by replacing an infinite series by a finite sum) or roundoff errors produced by preceding computations. The effects of the propagation of errors, either inherent or acquired, can be quite devastating in computations.

Before we can quantitatively analyze the propagation of errors in an arithmetic expression, we must define what we mean by error. The error ε_x in a quantity x is defined as the difference between its true value x^t and its measured or computed value x; i.e.,

(2.4.1) $$\varepsilon_x = x^t - x.$$

The relative error r_x in a quantity x is defined as the ratio of its error ε_x to its true value x^t. More often than not, however, we do not know the true value, but only the measured or computed value x. Hence, we define the relative error as

$$r_x = \frac{\varepsilon_x}{x}.$$

It is much more meaningful to speak of the relative error r_x than the error ε_x. To see this, consider the following two cases in which each measured value differs by one unit from the corresponding true value:

$$x^t = 9, \qquad x = 10$$
$$y^t = 999, \qquad y = 1000.$$

Notice that although $\varepsilon_x = \varepsilon_y = -1, r_x = -1/10$ and $r_y = -1/1000$. That is, x is in error by one part in 10, while y is in error by one part in 1000, even though their errors ε_x and ε_y are equal.

Let the symbol $\oplus$ denote any of the arithmetic operations of addition $(+)$, subtraction $(-)$, multiplication $(\cdot)$, or division $(/)$. The error in a quantity $x \oplus y$ will be denoted by $\varepsilon_{x \oplus y}$, where x and y are known quantities that contain errors ε_x and ε_y, respectively, which may be either inherent or acquired errors. By definition (2.4.1), the error in the quantity $x \oplus y$ is

$$\varepsilon_{x \oplus y} = (x^t \oplus y^t) - (x \oplus y).$$

Let us assume for the moment that we can compute the quantity $x \oplus y$ without roundoff error. The error in this quantity for each of the arithmetic operations $+$, $-$, $\cdot$, and $/$ can be easily determined as follows:

1. *Addition:*
$$\begin{aligned}
\varepsilon_{x+y} &= (x^t + y^t) - (x + y) \\
&= (x^t - x) + (y^t - y) \\
&= \varepsilon_x + \varepsilon_y
\end{aligned}$$

2. *Subtraction:*
$$\begin{aligned}
\varepsilon_{x-y} &= (x^t - y^t) - (x - y) \\
&= (x^t - x) - (y^t - y) \\
&= \varepsilon_x - \varepsilon_y
\end{aligned}$$

3. *Multiplication:*
$$\begin{aligned}
\varepsilon_{x \cdot y} &= (x^t \cdot y^t) - (x \cdot y) \\
&= (x + \varepsilon_x)(y + \varepsilon_y) - (xy) \\
&= xy + \varepsilon_x y + \varepsilon_y x + \varepsilon_x \varepsilon_y - xy \\
&= \varepsilon_x y + \varepsilon_y x, \qquad \text{ignoring } \varepsilon_x \varepsilon_y
\end{aligned}$$

4. *Division:*

$$\varepsilon_{x/y} = \frac{x^t}{y^t} - \frac{x}{y}$$

$$= \frac{x^t y - y^t x}{y^t y}$$

$$= \frac{(x + \varepsilon_x)y - (y + \varepsilon_y)x}{(y + \varepsilon_y)y}$$

$$= \frac{xy + \varepsilon_x y - xy - x\varepsilon_y}{y^2(1 + \varepsilon_y/y)}$$

$$\doteq \frac{\varepsilon_x y - \varepsilon_y x}{y^2}, \qquad \text{if } \left| \frac{\varepsilon_y}{y} \right| \ll 1.$$

In practice it is necessary (usually) to round off the result $x \oplus y$. For example, if x and y are each ten-digit numbers, then the result $x \oplus y$ can contain more than ten digits. If this result is to be used in a succeeding computation, assuming we are using a ten-digit machine, then it must be rounded off to ten digits before it can be entered into the next computation. For this reason, the foregoing error formulas must be modified to include a roundoff error term.

Let $\bar{\alpha}$, $\bar{\sigma}$, $\bar{\mu}$, $\bar{\delta}$ denote the errors in rounding the quantities $x + y$, $x - y$, $x \cdot y$, x/y, respectively. The error formulas for the error in the computed quantity $x \oplus y$, with the applicable roundoff term included, then become

1. *Addition:*

$$\varepsilon_{x+y} = \varepsilon_x + \varepsilon_y + \bar{\alpha}$$

2. *Subtraction:*

$$\varepsilon_{x-y} = \varepsilon_x - \varepsilon_y + \bar{\sigma}$$

3. *Multiplication:*

$$\varepsilon_{x \cdot y} = \varepsilon_x y + \varepsilon_y x + \bar{\mu}$$

4. *Division:*

$$\varepsilon_{x/y} = \frac{1}{y^2} [\varepsilon_x y - \varepsilon_y x] + \bar{\delta}.$$

Since the relative errors in computed quantities are more meaningful than the errors, we will present the relative-error formulas before attempting to analyze the errors in more complicated arithmetic expressions. As it turns out, this analysis is simpler to perform using the relative-error formulas.

It should be carefully noted that the magnitude of the roundoff error does not depend on the arithmetic operation used to compute the quantity $z = x \oplus y$. The quantity z is just a number that is rounded off because of the machine limitation to carry a specified number of digits in the computations. This rounding is independent of the manner in which z was obtained or computed. The symbols $\bar{\alpha}$, $\bar{\sigma}$, $\bar{\mu}$, $\bar{\delta}$ are employed solely to aid in the analysis of the error in an arithmetic expression.

Relative-Error Formulas

The absolute-error equations can be transformed to relative-error equations by using the relation $r_{x \oplus y} = (\varepsilon_{x \oplus y})/(x \oplus y)$. The resulting equations can be written in the form

(1)   (2)   (3)   (4)   (5)†   (6)

$$r_x \cdot \frac{x}{x + y} + r_y \cdot \frac{y}{x + y} + \alpha = r_{x+y} \quad \text{(Addition)}$$

$$r_x \cdot \frac{x}{x - y} + r_y \cdot \frac{-y}{x - y} + \sigma = r_{x-y} \quad \text{(Subtraction)}$$

$$r_x \cdot 1 + r_y \cdot 1 + \mu = r_{x \cdot y} \quad \text{(Multiplication)}$$

$$r_x \cdot 1 + r_y \cdot (-1) + \delta = r_{x/y} \quad \text{(Division)}$$

$$N \cdot O + N \cdot O + R = N$$

Note that columns (1) through (5) have been labeled N, O, N, O, R, respectively, to indicate *number, operation, number, operation, roundoff.* That is, corresponding terms in the error equations have been written in the same column so that the contributing causes can be identified. The components of the error in $x \oplus y$ are identified as follows:

Column (1) contains r_x, the relative error in the *number* x, regardless of the arithmetic operation, so that the column (1) element is due to the *number* x. Hence, we label that column N for number.

Column (2) elements are associated with the particular arithmetic operation being used to combine x and y. Hence, we label that column O for *operation.*

Column (3) contains r_y, the relative error in the *number* y, regardless of the arithmetic operation. Column (3) is labeled N to indicate that the elements of that column are contributed by the *number* y.

Column (4) elements are associated with (i.e., peculiar to) the particular arithmetic operation used to combine x and y. Hence the label O for operation.

After the sum, difference, product, or quotient of the numbers x and y is formed, the result must be rounded if we are limited to a given number

† The relative roundoff errors for addition, subtraction, multiplication, and division are denoted by a, σ, μ, and δ, respectively. As noted earlier, the relative roundoff error magnitude does not exceed 5.10^{-d} where d is the number of digits retained after rounding.

of digits. Consequently, a relative-error term must be added for this *roundoff*. Hence the label R for column (5).

Column (6) contains the relative error in the expression $x \oplus y$. If $x \oplus y$ is part of a more complicated expression, $r_{x \oplus y}$ becomes the error term associated with the *number* $x \oplus y$ and is labeled N accordingly.

First, let us illustrate *the use of the mnemonic NO NOR* in a few simple cases, so that the reader can acquire some facility in *analyzing error propagation in arithmetic expressions*. Later, the same device will be employed to analyze error propagation in the particular algorithm under consideration.

EXAMPLE. Form the sum of the real numbers a_0, a_1, a_2, a_3 which have relative errors r_0, r_1, r_2, r_3, respectively. Let α_k denote the relative roundoff error in the kth addition. Then the sum

$$S = \{[(a_0 + a_1) + a_2] + a_3\}$$

is formed by first computing the sum in parentheses, then the sum in square brackets, and finally the sum in braces.

The corresponding relative-error expression with terms set off by corresponding parentheses, brackets, and braces, is

$$r_s = \left\{ \left[\left(r_0 \frac{a_0}{a_0+a_1} + r_1 \frac{a_1}{a_0+a_1} + \alpha_1 \right) \frac{a_0+a_1}{a_0+a_1+a_2} + r_2 \frac{a_2}{a_0+a_1+a_2} + \alpha_2 \right] \frac{a_0+a_1+a_2}{a_0+a_1+a_2+a_3} + r_3 \frac{a_3}{a_0+a_1+a_2+a_3} + \alpha_3 \right\}$$

Now, this expression could have been written more simply by using the partial sums $S_1 = a_0 + a_1$, $S_2 = a_0 + a_1 + a_2$, $S_3 = a_0 + a_1 + a_2 + a_3$. Using this notation, we write

$$r_s = \left\{ \left[\left(\frac{r_0 a_0 + r_1 a_1 + \alpha_1 S_1}{S_1} \right) \frac{S_1}{S_2} + \frac{r_2 a_2 + \alpha_2 S_2}{S_2} \right] \frac{S_2}{S_3} + \frac{r_3 a_3 + \alpha_3 S_3}{S_3} \right\}.$$

Using the definition $\varepsilon_s = S r_s$ and the fact that $S_3 = S$, we obtain

$$\varepsilon_s = r_0 a_0 + r_1 a_1 + \alpha_1 S_1 + r_2 a_2 + \alpha_2 S_2 + r_3 a_3 + \alpha_3 S_3$$

$$= \sum_{k=0}^{3} r_k a_k + \sum_{k=1}^{3} \alpha_k S_k.$$

If the $|r_k|$ are bounded by $K \cdot 10^{-d}$, where $K \geq 5$, and the $|\alpha_k|$ by $5 \cdot 10^{-d}$, then it follows that

$$|\varepsilon_s| \leq \left[K \sum_{k=0}^{3} |a_k| + 5 \sum_{k=1}^{3} |S_k| \right] \cdot 10^{-d}$$

In general, the error incurred in forming the sum $S = \sum_{k=0}^{n} a_k$ with relative error r_k in term a_k, and with relative roundoff α_k in the kth addition, can be expressed in the form

$$\varepsilon_s = \sum_{k=0}^{n} r_k a_k + \sum_{k=1}^{n} \alpha_k S_k$$

where

$$S_k = a_0 + a_1 + a_2 + \cdots + a_k.$$

If the a_k are all of the same sign, the partial sums S_k for the ordering $|a_0| \leq |a_1| \leq |a_2| \leq \cdots \leq |a_n|$ are smaller in magnitude than the corresponding partial sums for the ordering $|a_0| \geq |a_1| \geq |a_2| \geq \cdots \geq |a_n|$. Hence, the error bound

$$|\varepsilon_s| \leq \sum_{k=0}^{n} |r_k a_k| + \sum_{k=1}^{n} |\alpha_k S_k|$$

is smaller for the first ordering than for the second. For this reason, when adding a sequence of numbers of varying magnitude (but of the same sign), it is wise to add the numbers in order of magnitude from the smallest to the largest.

EXAMPLE. Consider now the error incurred in forming the product of the real numbers a_0, a_1, a_2, a_3 with respective relative errors r_0, r_1, r_2, r_3; let μ_k denote the relative roundoff error in the kth multiplication.

$$P = a_0 \cdot a_1 \cdot a_2 \cdot a_3 = \{[(a_0 \cdot a_1) \cdot a_2] \cdot a_3\}.$$

The corresponding relative-error expression is

$$r_p = \{[(r_0 \cdot 1 + r_1 \cdot 1 + \mu_1) \cdot 1 + r_2 \cdot 1 + \mu_2] \cdot 1 + r_3 \cdot 1 + \mu_3\}$$

$$\underbrace{\quad}_{N\ 0\quad N\ 0\quad R}$$

$$\underbrace{\qquad\qquad}_{N}\quad 0\quad N\ 0\quad R$$

$$\underbrace{\qquad\qquad\qquad}_{N}\quad 0\quad N\ 0\quad R$$

$$\underbrace{\qquad\qquad\qquad}_{N}$$

where the parentheses, brackets, and braces in the relative-error expression correspond to those in the original expression. We see then that

$$r_p = r_0 + r_1 + \mu_1 + r_2 + \mu_2 + r_3 + \mu_3 = \sum_{k=0}^{3} r_k + \sum_{k=1}^{3} \mu_k$$

and it follows that

$$\varepsilon_p = P\left[\sum_{k=0}^{3} r_k + \sum_{k=1}^{3} \mu_k\right].$$

In general, the error incurred in forming the product of $n + 1$ real numbers

$$P = \prod_{k=0}^{n} a_k$$

can be expressed in the form

$$\varepsilon_p = P\left[\sum_{k=0}^{n} r_k + \sum_{k=1}^{n} \mu_k\right]$$

where r_k denotes the relative error in factor a_k and μ_k denotes the relative error in the kth multiplication. Further, if $r = \max_k |r_k|$, and $\mu = \max_k |\mu_k|$, then the error bound for the product is

$$|\varepsilon_p| \le |P|[(n + 1)r + n\mu].$$

EXAMPLE. If the coefficients a_0, a_1, a_2 in the polynomial

$$P_2(x) = a_0 x^2 + a_1 x + a_2$$

have relative errors r_0, r_1, r_2, respectively, determine the error for the value of $P_2(x)$, where x has relative error r_x.

It was shown in the preceding section that Horner's nesting procedure is an efficient method of evaluating a polynomial. The quadratic polynomial $P_2(x)$ can be expressed in nested form as

$$P_2(X) = \{([(a_0 \cdot x) + a_1] \cdot x) + a_2\}$$

and the corresponding relative-error equation is†

$$r_p = \left\{\left(\left[(r_0 \cdot 1 + r_x \cdot 1 + \mu_1)\frac{a_0 x}{a_0 x + a_1} + r_1\frac{a_1}{a_0 x + a_1} + \alpha_1\right] \cdot 1 + r_x \cdot 1 + \mu_2\right)\frac{(a_0 x + a_1)x}{P_2(x)} + r_2\frac{a_2}{P_2(x)} + \alpha_2\right\}$$

$$\underbrace{N \quad O \quad N \quad O \quad R}$$
$$\underbrace{N \qquad O \qquad N \quad O \qquad R}$$
$$\underbrace{N \qquad\qquad\qquad O \quad N \quad O \quad R}$$
$$\underbrace{N \qquad\qquad\qquad O \qquad N \quad O \qquad R}.$$

Collecting terms, we obtain

$$r_p = \left[\frac{(r_0 + r_x + \mu_1)a_0 x + r_1 a_1 + (\alpha_1 + r_x + \mu_2)(a_0 x + a_1)}{a_0 x + a_1}\right]$$

$$\cdot \frac{(a_0 x + a_1)x}{P_2(x)} + \frac{r_2 a_2 + \alpha_2 P_2(x)}{P_2(x)}.$$

† The α_k and μ_k denote the relative roundoff in the kth addition and multiplication, respectively.

Multiplying through by $P_2(x)$, and collecting terms, we find that

$$\varepsilon_p = a_0x^2(r_0 + r_x + \mu_1 + \alpha_1 + r_x + \mu_2 + \alpha_2)$$
$$+ a_1x(r_1 + \alpha_1 + r_x + \mu_2 + \alpha_2) + a_2(r_2 + \alpha_2).$$

If the $|r_k|$, $|\alpha_k|$, and $|\mu_k|$ are bounded by $5 \cdot 10^{-d}$, it follows that

$$|\varepsilon_p| \le [7|a_0x^2| + 5|a_1x| + 2|a_2|]5 \cdot 10^{-d}.$$

It should be noted that the smaller $|x|$ is, the smaller the error bound for $|\varepsilon_p|$. This fact will be used later in the Birge-Vieta method to show why roots of smallest magnitude should be extracted first when successively computing the roots of a polynomial $P_n(x)$ by factor-extraction methods.

DERIVATION EXERCISES

1. Derive the recursion formulas for Horner's nesting procedure to evaluate $P_n(x_0)$ by dividing $P_n(x)$ by $(x - x_0)$, noting that

$$P_n(x) = (x - x_0)P_{n-1}(x) + R$$

where $P_{n-1}(x)$ is a polynomial of degree $n - 1$ and R is zero or a constant. Note also that $P_n(x_0) = R$.

2. Determine a recursion relation for the Chebyshev polynomials $T_{k+1}(x)$, given that

$$T_0(x) = 1$$
$$T_1(x) = x$$
$$T_2(x) = 2x^2 - 1$$
$$T_3(x) = 4x^3 - 3x.$$

3. Assuming that h contains an inherent relative error r_h, write the expression for the error in the third term of the arithmetic sequence $\{t_k = a + kh\}$, where the elements t_k of the sequence are generated by (a) computing $t_k = a + kh$; (b) using the recursion formula $t_k = t_{k-1} + h$, where $t_0 = a$. (The value of a is assumed to be exact in this exercise.) The result should show that if h contains an inherent error, the use of the recursion formula in b can lead to large errors.

4. Assuming that x, a_0, a_1, a_2, a_3 have relative errors r_x, r_0, r_1, r_2, r_3, respectively, show that the error bound obtained for polynomial

$$P_3(x) = a_0x^3 + a_1x^2 + a_2x + a_3$$

by adding the terms

$$a_0 \cdot x \cdot x \cdot x, \quad a_1 \cdot x \cdot x, \quad a_2 \cdot x, \quad a_3$$

is greater than the error bound obtained by computing the polynomial using Horner's nesting procedure.

Let α_k and μ_k denote the relative errors in the kth addition and multiplication, respectively, each bounded by $5 \cdot 10^{-d}$.

McCracken and Dorn [16, pp. 60–62] show that Horner's procedure minimizes the effects of roundoff in the evaluation of a polynomial obtained by truncating a convergent series.

3

Solution of Non-linear Algebraic
and Transcendental Equations

3.0 Introduction

There are few non-linear equations (except for special cases) that can be solved by direct methods, i.e., methods which produce a solution in a predetermined and finite number of computational steps. To illustrate this, consider the following four *examples* of non-linear problems, noting the special ones that can be solved by direct methods.

1. Find a number x such that $x^K = N$, where N is a positive real number.
2. Calculate a value of x such that

$$a_0 x^n + a_1 x^{n-1} + \cdots + a_{n-1} x + a_n = 0.$$

3. If $y = f(x)$ is a continuous function of x such that $f(x_1) \cdot f(x_2) < 0$, find a value of x in (x_1, x_2) such that $f(x) = 0$.
4. Given values of M and e, find a value of x that satisfies the relation $M = x - e \sin x$.

An examination of each of these problems reveals the following:

Problem 1 can be solved by taking the logarithm of each side of the equation and dividing through by k to obtain

$$\log x = \frac{1}{k} \log N.$$

EXAMPLE. Given $N = 81$, $k = 4$, find x such that $x^4 = 81$. The solution is

$$\log x = \frac{1}{4} \log 81 = \left(\frac{1}{4}\right) 1.90849 = 0.47712.$$

$$x = 3.0$$

Problem 2 can be solved by formula when $n \leq 4$, as follows:

$$a_0 x + a_1 = 0$$

has solution (root) $x = \dfrac{-a_1}{a_0}$.

$$a_0 x^2 + a_1 x + a_2 = 0$$

can be solved by the quadratic formula

$$x = \frac{-a_1 \pm \sqrt{a_1{}^2 - 4a_0 a_2}}{2a_0}.$$

The equations $a_0 x^3 + a_1 x^2 + a_2 x + a_3 = 0$ and $a_0 x^4 + a_1 x^3 + a_2 x^2 + a_3 x + a_4 = 0$ can be solved by formula by such methods as Cardan's and Ferrari's; see Uspensky [33]. However, in general, no solution by formula exists for Problem 2 when $n > 4$.

Problem 3 can be solved by formula if $f(x)$ is a *linear* function of x. The solution in this case is accomplished by writing the equation of the straight line through the two points $(x_1, f(x_1))$ and $(x_2, f(x_2))$, and finding the point of intersection of this line with the x-axis; i.e., finding the value of x for which $f(x) = 0$. This equation has the form

$$\frac{f(x) - f(x_1)}{x - x_1} = \frac{f(x_2) - f(x_1)}{x_2 - x_1}.$$

Setting $f(x) = 0$ in this equation, we obtain the solution

$$x = \frac{x_1 f(x_2) - x_2 f(x_1)}{f(x_2) - f(x_1)}.$$

If $f(x)$ is a *non-linear* function of x, there is, in general, no formula for the solution as in the linear case.

Problem 4 is the classical equation of Kepler that relates the motion between any two heavenly bodies. This equation cannot be solved by formula. However, we can rewrite the equation in the form

$$f(x) \equiv (x - M) - e \sin x = 0$$

and plot the graphs of the line $y = x - M$ and the sine curve $y = e \sin x$.

The point at which these two graphs cross is the value of x that satisfies Kepler's equation.

Fortunately, each of these problems can be formulated as a *non-linear* equation of the form $f(x) = 0$, as follows:

1. $f(x) \equiv x^K - N = 0$.
2. $f(x) \equiv a_0 x^n + a_1 x^{n-1} + \cdots + a_{n-1} x + a_n = 0$
3. $f(x) \equiv$ some function of $x = 0$.
4. $f(x) \equiv x - M - e \sin x = 0$.

These four equations are typical examples of ãlgebraic and transcendental equations: 1 and 2 are algebraic equations, 4 is a transcendental equation, and 3 is either algebraic or transcendental, depending on the explicit form of the function $f(x)$.

A solution of the equation $f(x) = 0$ is a value of x, say α, for which the function is zero. That is, if $f(\alpha) \equiv 0$, α is a solution of the equation $f(x) = 0$. Such a solution α is called a *zero* of the function $f(x)$, or equivalently, a *root* of the equation $f(x) = 0$. Geometrically, a root of the equation $f(x) = 0$ is a value of x at which the graph of $f(x)$ crosses the x-axis.

Since each of the foregoing problems can be written in the form of an equation $f(x) = 0$, it seems reasonable that they can all be solved in the same way, without recourse to the specific *formula solutions* that pertain to any of them [such as logarithms for Problem 1]. It will be shown in this chapter that any algebraic or transcendental equation of the form $f(x) = 0$ can be solved by methods of *successive approximations*. Four representative methods of this type are presented in Secs. 3.1 and 3.2 for determining a *real* root of an algebraic or transcendental equation.

The Approximate Nature of Solutions

In classical methods, a root can be calculated by evaluating a formula. For example, the roots of the quadratic equation

$$f(x) \equiv ax^2 + bx + c = 0$$

are computed by the formulas

$$\alpha_1 = (-b + \sqrt{b^2 - 4ac})/2a$$
$$\alpha_2 = (-b - \sqrt{b^2 - 4ac})/2a$$

If α_1 and α_2 could be computed without roundoff error, $f(\alpha_1)$ and $f(\alpha_2)$ would be *exactly* zero. However, there are errors in computing the square roots, and roundoff errors usually occur, since only a finite number of digits can be carried in the computations; so that the computed values of α_1 and α_2 are only approximations of the true roots.

In methods of successive approximation, a root α is produced by an approximation method which generates a sequence $x_1, x_2, x_3, \ldots$ which (hopefully) converges to a root, and in general the value produced is not an exact root even if no roundoff error occurs. Then, in general, $f(x_{i+1})$ is not exactly zero, both because x_{i+1} is not an exact root and because roundoff error occurs in the evaluation of $f(x_{i+1})$. Because of this, it is not possible to algebraically compare $f(x)$ with 0. It is necessary to compare† $|x_{i+1} - x_i|$ with some small positive quantity ε. Then if $|x_{i+1} - x_i| \leq \varepsilon$, x_{i+1} is considered a root α of the equation $f(x) = 0$.

Outline

In this chapter representative methods of solution by successive approximation are developed for:

1. The solution of a single non-linear equation $f(x) = 0$. The types of methods presented include:

Simple, Sure-Fire Methods

Given interval (x_1, x_2) containing a root α, that root can be determined by methods which are guaranteed (neglecting roundoff) to converge; representative methods of this type are:

Method of successive bisection. Interval (x_1, x_2) is bisected at $\bar{x} = (x_1 + x_2)/2$, and a determination is made as to which half-interval contains a root; that half-interval replaces (x_1, x_2) and the procedure is repeated. The process is continued until $|x_2 - x_1| \leq \varepsilon$.

Method of inverse linear interpolation. According to this method, $f(x)$ is approximated by the line segment (chord) through points $(x_1, f(x_1))$ and $(x_2, f(x_2))$, and this chord intersects the x-axis at x'. The subinterval, either (x_1, x') or (x', x_2), containing a root replaces (x_1, x_2) and the procedure is repeated. This process is continued until convergence is achieved.

Sophisticated, Conditional-Convergence Methods

Given an estimate x_i of a root, an improved estimate x_{i+1} can be computed by iterative methods that converge, provided that specified conditions hold. Representative methods of this type are:

Newton-Raphson method—based on approximating $f(x)$ by a linear Taylor expansion of a function of a single variable

$$y(x) = f(x_i) + f'(x_i)(x - x_i).$$

Bailey's method—based on approximating $f(x)$ by a quadratic Taylor expansion of a function of a single variable

$$y(x) = f(x_i) + f'(x_i)(x - x_i) + \frac{f''(x_i)(x - x_i)^2}{2}.$$

† Or compare $(|x_{i+1} - x_i|/|x_i|)$ to ε.

2. The solution of systems of non-linear equations. Representative methods are:

Newton's method—based on linear Taylor expansions of functions of several variables.

Modified Newton method—based on quadratic Taylor expansions of functions of several variables.

3.1   Simple, Sure-Fire Methods for Solving Equations $f(x) = 0$

If x_1, x_2 are values of x such that $f(x_1)f(x_2) < 0$, where $f(x)$ is a real-valued function continuous throughout interval (x_1, x_2), then there exists at least one root α in (x_1, x_2) of the equation $f(x) = 0$. If the interval (x_1, x_2) is divided into two parts (equal or unequal), then a root must be contained in one of these two subintervals or at the point common to the two subintervals. If a root does not fall at this common point, then it can be determined which of the two subintervals contains a root. The subinterval containing a root can then be redesignated as interval (x_1, x_2), and this new interval can be divided into two parts. Again, if a root does not fall at the point common to the two parts of new interval (x_1, x_2), a determination is made as to which part contains a root. Continuing in this way, we can "shrink" the interval containing a root until a root is located. The methods of *successive bisection* and *inverse linear interpolation* are simple successive-approximation methods that locate a real root of an algebraic or transcendental equation by such a procedure.

The Method of Successive Bisection

This method is the simplest and most intuitive of the successive-approximation methods for locating a root of the equation $f(x) = 0$ in the interval (x_1, x_2) where $f(x_1)f(x_2) < 0$. In this method, the interval (x_1, x_2) is divided into *two equal parts*, i.e., *bisected*, at point $\bar{x} = (x_1 + x_2)/2$. If $|x_2 - x_1| \leq \varepsilon$, then $\bar{x}$ is a root. If not, then either $(x_1, \bar{x})$ or $(\bar{x}, x_2)$ contains a root. If $f(x_1)f(\bar{x}) \leq 0$, $(x_1, \bar{x})$ contains a root. If not, $(\bar{x}, x_2)$ contains a root. The subinterval that contains a root is then redesignated (x_1, x_2), and the process is repeated. Eventually, a root is located at the midpoint $\bar{x}$ of the current (x_1, x_2), or the interval (x_1, x_2) is sufficiently small that its midpoint is essentially a root, since the interval must contain a root. This simple iterative method is guaranteed to locate a root of the equation. However, the number of iterations required may be excessive.

Computational Summary of Method of Successive Bisection

Step 0. Define $f(x)$. Read x_1, x_2 = values of x such that $f(x_1)f(x_2) < 0$. Read ε = convergence term.

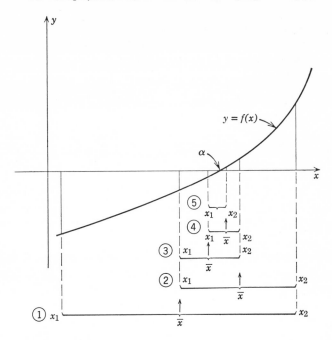

Figure 3-1. Illustration of method of successive bisection. Iteration numbers are encircled.

Step 1. Calculate $f(\bar{x})$, where $\bar{x} = (x_1 + x_2)/2$.

Step 2. If $|x_2 - x_1| \le \varepsilon$, $\bar{x}$ is a root α. Go to Step 4. If $|x_2 - x_1| > \varepsilon$, go to Step 3.

Step 3. If $f(x_1)f(\bar{x}) \le 0$, $(x_1, \bar{x})$ contains a root. Set $x_2 = \bar{x}$. Return to Step 1. If not, then $(\bar{x}, x_2)$ contains a root. Set $x_1 = \bar{x}$. Return to Step 1.

Step 4. Write root α.

Method of Inverse Linear Interpolation

Let x_1 and x_2 be values of x such that $f(x_1)f(x_2) < 0$, where $f(x)$ is a real-valued function that is continuous throughout interval (x_1, x_2). A simple successive-approximation method for determining a real root α of equation $f(x) = 0$ in interval (x_1, x_2) is the method of inverse linear interpolation. This method is derived by approximating the graph of $f(x)$ in interval (x_1, x_i), $(i = 2, 3, 4, \ldots)$, by the straight line through the points $(x_1, f(x_1))$ and $(x_i, f(x_i))$ and then determining the value x_{i+1} corresponding

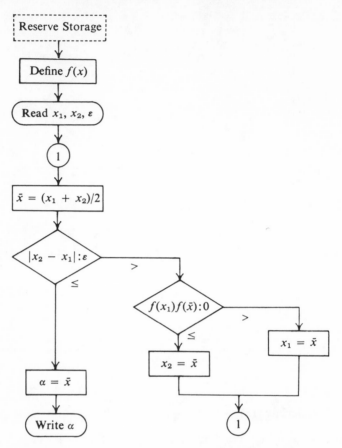

Figure 3-2. Flow chart of method of successive bisection.

to $y = 0$ on that line. Since a value of x is interpolated for, in terms of y, the method is called the method of inverse linear interpolation.†
Jennings [12, pp. 24–25] shows that the *Fourier conditions*

$$f(x_1)f(x_2) < 0$$
$$f(x_1)f''(x_1) > 0$$
$$f''(x) \neq 0, \qquad x_1 < x < x_2$$

are *sufficient* to guarantee convergence of the method of inverse linear interpolation. In the remainder of this discussion it is assumed that the

† Also called the method of chords, method of secants, or the method of false position.

function $f(x)$ under consideration satisfies these conditions, so that α is the only root of $f(x) = 0$ in (x_1, x_2).

Using the two-point formula for a straight line, we can write the equation of the chord through $(x_1, f(x_1))$ and $(x_i, f(x_i))$ in the form

(3.1.1)
$$\frac{f(x_i) - y}{x_i - x} = \frac{f(x_i) - f(x_1)}{x_i - x_1}.$$

This chord intersects the x-axis at $(x_{i+1}, 0)$, and the abscissa satisfies the recursion relation

(3.1.2)
$$x_{i+1} = x_i - \frac{(x_i - x_1)f(x_i)}{f(x_i) - f(x_1)}.$$

Further, this chord intersects the x-axis at a point such that its abscissa x_{i+1} is inside interval (x_1, x_i), so that $x_1 < x_3 < x_2$, $x_1 < x_4 < x_3, \ldots$, $x_1 < x_{i+1} < x_i, \ldots$. That is, the values $x_2, x_3, \ldots, x_i, x_{i+1}, \ldots$ form a monotonic decreasing sequence that is bounded below by root α, as depicted in Figure 3-3.

The sequence of successive approximations $x_2, x_3, \ldots$ converges to some limit α_0, where $\alpha_0 \geq \alpha$, since it is a monotonic sequence bounded below by α. Taking the limit of (3.1.2), we obtain

$$\alpha_0 = \alpha_0 - \frac{(\alpha_0 - x_1)f(\alpha_0)}{f(\alpha_0) - f(x_1)},$$

and since $\alpha_0 \neq x_1$, it follows that $f(\alpha_0) = 0$, so that $\alpha_0 = \alpha$, since α is the

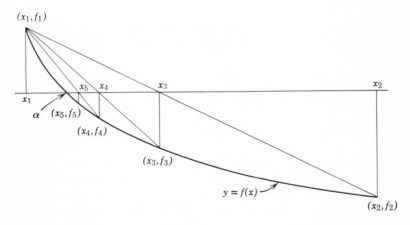

Figure 3-3. Illustration of method of inverse linear interpolation. Here f_i denotes $f(x_i)$.

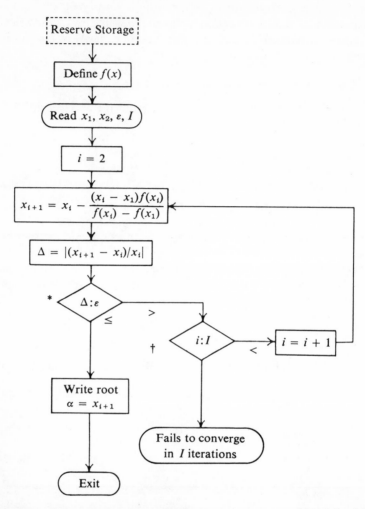

Figure 3-4. Flow chart for method of inverse linear interpolation. * Computation of root of $f(x) = 0$ with *relative* error less than ε. † In any successive-approximations routine it is advisable to limit the number of iterations by some maximum number, denoted here by I, where I is a positive integer. Although the method of inverse linear interpolation is guaranteed to converge, provided that the Fourier conditions are met, it is not always possible or feasible to check these conditions before using the method.

only root of $f(x) = 0$ in (x_1, x_2). That is, the sequence of successive approximations generated by (3.1.2) converges to the root α.

A number of variations of the method of inverse linear interpolation are given throughout the literature. See, for instance, Kunz [13], Hildebrand [10], Jennings [12], Henrici [7], Arden [1].

3.2 Sophisticated Conditional-Convergence Methods

The successive-bisection and the inverse-linear-interpolation methods are simple methods that are virtually guaranteed to locate a real root of the equation $f(x) = 0$ in the interval (x_1, x_2). However, these methods are quite slow in convergence. For this reason, more sophisticated methods are required, methods which converge to a solution more rapidly.

In this section, two representative methods, quite rapid in convergence, will be presented. These methods are based on *Taylor's expansion* of the function $f(x)$ about a point x_i, where x_i is an initial estimate of a root α of the equation $f(x) = 0$. Such an initial estimate might be obtained by a few iterations of one of the simple methods presented in Sec. 3.1, by graphical methods, or by the analytic methods described in various texts.

The first method to be discussed is that of Newton, sometimes referred to as the Newton-Raphson method. In this method, the function $f(x)$ is approximated by the line tangent to $f(x)$ at x_i, obtained by approximating $f(x)$ by a linear Taylor polynomial. The intersection of this tangent line with the x-axis is used as an improved approximation x_{i+1} of a root α.

The second method is that of Bailey.† In this method, the function $f(x)$ is approximated by a parabola that passes through the point $(x_i, f(x_i))$ with slope and curvature equal to the slope and curvature, respectively, of $f(x)$ at x_i (obtained by approximating $f(x)$ by a quadratic Taylor polynomial). The intersection of this parabola with the x-axis is used as an improved approximation x_{i+1} of a root α.

In either case, x_{i+1} can be calculated by an iterative algorithm of the form $x_{i+1} = F(x_i)$. Starting with initial estimate x_0, an improved estimate $x_1 = F(x_0)$ is calculated. From x_1 another improved estimate $x_2 = F(x_1)$ is calculated, and so forth. In this way, a *sequence* of successive approximations $x_1, x_2, x_3, \ldots$ is generated, which, under *specified conditions*, converges to a root α of the equation $f(x) = 0$.

Convergence, Divergence, or Failure to Converge

Iterative methods for solving an algebraic or transcendental equation $f(x) = 0$ generate a *sequence of successive approximations* of a true solution. Given an initial approximation x_0 of the solution, the particular

† H. S. Wall, "A Modification of Newton's Method," *Amer. Math. Monthly*, Vol. 55, 1948, pp. 90–94.

iterative method generates a sequence of successive approximations $x_1, x_2, x_3, \ldots$ using a recursive computational formula that can be expressed as an *algorithm* of the form $x_{i+1} = F(x_i)$. That is, the $(i + 1)$st approximation is obtained from the ith approximation by evaluating the *iteration function* $F(x)$ at $x = x_i$.

The properties of infinite sequences reviewed in Chapter 2 can be applied to the determination of whether or not a sequence of successive approximations converges to a root α of the equation $f(x) = 0$. In the following three categories, it is assumed that there exists an x_0 such that the sequence will converge to a root α (Case I). Unreasonable estimates of x_0 may yield a locally diverging sequence (Case II) or a sequence which fails to converge (Case III). In the latter two cases, a more reasonable estimate of x_0 must be supplied.

Case I. The sequence $\{x_{i+1}\}$ generated by algorithm $x_{i+1} = F(x_i)$ *converges* to a root α if and only if for every $\varepsilon > 0$ there exists an integer I such that $n > I$ and $m > I$ implies that $|x_n - x_m| < \varepsilon$. That is, the sequence $\{x_{i+1}\}$ converges to a root α ($\lim_{i \to \infty} x_{i+1} = \alpha$) if the Cauchy condition is satisfied.

Case II. The sequence $\{x_{i+1}\}$ generated by algorithm $x_{i+1} = F(x_i)$ is *diverging* if successive quantities $\Delta_{i+1} = |x_{i+1} - x_i|$ increase monotonically.

Case III. The sequence $\{x_{i+1}\}$ generated by algorithm $x_{i+1} = F(x_i)$ *fails to converge* if the Cauchy condition for convergence is *not* satisfied.

Cauchy's test cannot be applied directly, using a finite length of computer time. In practice, the behavior of a sequence of successive approximations is determined by the following test: Given an $\varepsilon > 0$ and a positive integer I (maximum number of iterations), if $|x_{i+1} - x_i| < \varepsilon$ for some $i < I$, the sequence is said to converge. If not, the sequence fails to converge or is diverging.

In the succeeding discussion, we will derive Newton's and Bailey's iterative method algorithms, and investigate the conditions required for the convergence of each of these methods.

The Newton-Raphson Method

Suppose that we are given an estimate x_i of a real root of the real equation

(3.2.1) $$f(x) = 0.$$

The equation of the line tangent to $f(x)$ at $x = x_i$ can be expressed (Exercise 3.1) as the linear Taylor polynomial

(3.2.2) $$y(x) = f(x_i) + f'(x_i)(x - x_i).$$

Let $(x_{i+1}, 0)$ denote the intersection of this tangent line with the x-axis.

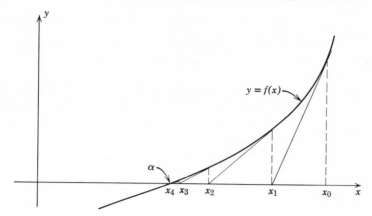

Figure 3-5. Illustration of the Newton-Raphson method.

This point is found by setting $y(x)$ equal to zero in (3.2.2) when $x = x_{i+1}$. Equation (3.2.2) then becomes

$$(3.2.3) \qquad 0 = f(x_i) + f'(x_i)(x_{i+1} - x_i).$$

Solving this equation for x_{i+1}, we obtain

$$(3.2.4) \qquad x_{i+1} = x_i - \frac{f(x_i)}{f'(x_i)}$$

which is the classical form of the Newton-Raphson iterative method. In effect, we are obtaining a refined approximation x_{i+1} of a root α of $f(x) = 0$ by approximating the graph of $f(x)$ by the line tangent to $f(x)$ at $x = x_i$.

The proof of convergence of Newton's method and the formula for determining the rate of convergence are facilitated by introducing an auxiliary function

$$(3.2.5) \qquad F(x) = x - \frac{f(x)}{f'(x)}, \qquad \text{with } F'(x) = \frac{f(x)f''(x)}{[f'(x)]^2}$$

which we call *the iteration function for Newton's method.*† This iteration function has two important properties: If α is a root of $f(x) = 0$, and if $f'(\alpha) \neq 0$, then these properties are

$$(3.2.6a) \qquad\qquad\qquad F(\alpha) = \alpha$$

$$(3.2.6b) \qquad\qquad\qquad F'(\alpha) = 0.$$

Property (a) is proved by simply evaluating (3.2.5) at $x = \alpha$, while property (b) is proved by evaluating the derivative of (3.2.5) at $x = \alpha$.

† Note that $F(x_i) = x_i - f(x_i)/f'(x_i) = x_{i+1}$.

The properties (a) and (b) of the iteration function $F(x)$ are illustrated in the following example.

EXAMPLE. Calculate $\sqrt{N}$, $N > 0$, by successive approximations, assuming $x_0 = N$.

Solution. This problem can be solved by finding a root of the equation $f(x) \equiv x^2 - N = 0$ by the Newton-Raphson iterative method. That is, given an approximation x_i of root α, we can compute approximation x_{i+1} by the recursive formula

$$x_{i+1} = x_i - \frac{f(x_i)}{f'(x_i)}$$

which for $f(x) = x^2 - N$ is

$$x_{i+1} = x_i - \left[\frac{x_i^2 - N}{2x_i}\right].$$

This formula can be simplified to the form

$$x_{i+1} = \frac{1}{2}\left(x_i + \frac{N}{x_i}\right)$$

so that the iteration function is

$$F(x) = \frac{1}{2}\left(x + \frac{N}{x}\right)$$

and its derivative is

$$F'(x) = \frac{1}{2}\left(1 - \frac{N}{x^2}\right).$$

It is obvious that $\alpha = \sqrt{N}$ is a root of the equation $x^2 - N = 0$, so that $F(\sqrt{N}) = \sqrt{N}$, and $F'(\sqrt{N}) = 0$. Figure 3-6 points out these properties.

Proof of Convergence of Newton's Method

To prove that under specified conditions the Newton-Raphson iterative formula

$$x_{i+1} = x_i - \frac{f(x_i)}{f'(x_i)}$$

generates (from initial estimate x_0) a convergent sequence $x_1, x_2, \ldots$ of successive approximations of root α of equation $f(x) = 0$, we examine the iteration function

$$F(x) = x - \frac{f(x)}{f'(x)}$$

and its derivative

$$F'(x) = \frac{f(x)f''(x)}{[f'(x)]^2}$$

for which $F(x_i) = x_{i+1}$, $F(\alpha) = \alpha$, and $F'(\alpha) = 0$.

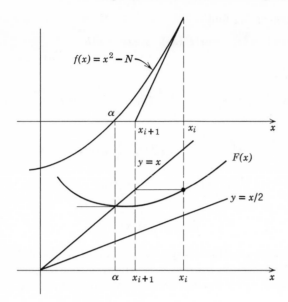

Figure 3-6. Note: $F(\alpha) = \alpha$
$$F'(\alpha) = 0.$$

The utility of this auxiliary (iteration) function $F(x)$ is illustrated in the proof of the following theorem:

Theorem. Let K be the largest magnitude of $F'(x)$ in the interval containing $x_0, x_1, x_2, \ldots$, and α. If $K < 1$, then the sequence $\{x_{i+1} = F(x_i)\}$ converges to α, where α is a root of the equation $f(x) = 0$, and $F(x) = x - f(x)/f'(x)$.

Proof. Using the Newton iteration formula $x_{i+1} = F(x_i)$, we compute

1. $$x_1 = F(x_0).$$

Then

2. $$x_1 - \alpha = F(x_0) - F(\alpha) \quad \text{because } F(\alpha) = \alpha$$

and

3. $$F(x_0) - F(\alpha) = (x_0 - \alpha)F'(\bar{x}_0) \quad \text{by mean value theorem}$$
$$\text{where } \bar{x}_0 \in (x_0, \alpha)$$

so that

4. $$x_1 - \alpha = (x_0 - \alpha)F'(\bar{x}_0) \quad \text{by equating 2 and 3.}$$

Taking absolute values, we get

5. $$|x_1 - \alpha| = |x_0 - \alpha|\,|F'(\bar{x}_0)| \le |x_0 - \alpha|K \quad \text{(from hypothesis).}$$

In the same manner, we find

6. $|x_2 - \alpha| = |x_1 - \alpha| \, |F'(\bar{x}_1)| \leq |x_1 - \alpha|K$
$$\leq |x_0 - \alpha|K^2 \quad \text{(from 5)}.$$

Continuing, we obtain

7. $|x_{i+1} - \alpha| = |x_i - \alpha| \, |F'(\bar{x}_i)| \leq |x_i - \alpha|K$
$$\leq |x_0 - \alpha|K^{i+1}.$$

From the relation

$$|x_{i+1} - \alpha| \leq |x_0 - \alpha|K^{i+1}, \qquad \text{where } K < 1$$

we see that

$$\lim_{i \to \infty} |x_{i+1} - \alpha| = 0.$$

That is,

(3.2.7) $$\lim_{i \to \infty} x_{i+1} = \alpha$$

so that the sequence $\{x_{i+1}\}$ converges to a root α of $f(x) = 0$. Hence, the algorithm defined by $x_{i+1} = F(x_i)$ generates a convergent sequence $\{x_{i+1}\}$ of successive approximations which converges to root α.

Rate of Convergence of Newton-Raphson Method

The rate of convergence of an iterative method can be determined by comparing successive values of the error term

(3.2.8) $$\delta_i = x_i - \alpha.$$

If we can find a relation between δ_{i+1} and δ_i, we can estimate how rapidly (or slowly) the algorithm converges to a root α of the equation $f(x) = 0$, provided the algorithm converges. Such a relation can be determined by expanding the iteration function $F(x)$ in a Taylor series about $x = \alpha$:

(3.2.9) $$F(x) = F(\alpha) + F'(\alpha)(x - \alpha) + \frac{F''(\alpha)}{2}(x - \alpha)^2 + \cdots.$$

For the Newton-Raphson algorithm, the iteration function $F(x)$ and its first two derivatives are

$$F(x) = x - \frac{f(x)}{f'(x)}$$

(3.2.10) $$F'(x) = \frac{f(x)f''(x)}{[f'(x)]^2}$$

$$F''(x) = \frac{(f')^2(ff''' + f'f'') - (ff'')2f'f''}{(f')^4}$$

and

(3.2.11) $$F(\alpha) = \alpha, \qquad F'(\alpha) = 0, \qquad F''(\alpha) = \frac{f''(\alpha)}{f'(\alpha)}.$$

Substituting the values (3.2.11) into series (3.2.9), truncating the series after terms of degree 2, and evaluating at $x = x_i$, we get

$$(3.2.12) \qquad F(x_i) \doteq \alpha + \frac{f''(\alpha)}{2f'(\alpha)}(x_i - \alpha)^2.$$

Using the relation $x_{i+1} = F(x_i)$ and definition (3.2.8), we find that

$$(3.2.13) \qquad \delta_{i+1} \doteq \frac{f''(\alpha)}{2f'(\alpha)} \delta_i^2.$$

From this relation, the *rate of convergence* is said to be *quadratic*.

Numerical Example of Newton-Raphson Method

Calculate $\sqrt[3]{5}$ by solving the equation $f(x) \equiv x^3 - 5 = 0$ by the Newton-Raphson method, using $x_0 = 5$ as initial estimate of a root.

Solution

$$f(x) = x^3 - 5$$
$$f'(x) = 3x^2$$
$$x_{i+1} = x_i - \frac{f(x_i)}{f'(x_i)}$$
$$= x_i - \frac{x_i^3 - 5}{3x_i^2}$$
$$= \frac{2x_i^3 + 5}{3x_i^2}$$

The successive approximations of the cube root of 5 generated by the Newton-Raphson method are listed in the following table, together with the value of $f(x_i)$, $i = 0, 1, 2, \ldots$.

x_i	$f(x_i)$
5.0	120.0
3.400	34.304
2.4108	9.011466
1.89365	1.793848
1.727271	0.153253
1.710149	0.001519
1.7099759	-0.00000041
1.7099759	-0.00000041

The last two approximations differ by less than 10^{-7}, so that the value 1.7099759 is an excellent approximation of $\sqrt[3]{5}$.

In manual computations with iterative methods of successive approximation the number of significant figures carried in each step can vary.

Usually, for manual computations only a few digits are carried in the early iterations, and the number of digits carried is increased as the solution is approached.

Another characteristic of iterative methods is that even if numerical errors are committed the method can still produce a valid result. This is because each iteration can be considered as though the current x_i were the initial estimate and the current x_{i+1} were an improved estimate. However, if the x_{i+1} produced as the result of an error is too far off, the iterative process may fail to converge from that point on.

Computational Summary for Newton-Raphson Method

Step 0. Input and definition. Read

x_1 = initial approximation of root of $f(x) = 0$.

ε = convergence term.

N = maximum number of iterations.

Step 1. Initialization. Set iteration counter $i = 1$. Set $\Delta_1 = c_1$ (c_1 = arbitrary large positive number).

Step 2. Compute successive approximations of root, using the Newton iterative formula†

$$x_{i+1} = x_i - \frac{f(x_i)}{f'(x_i)}.$$

Compute magnitude of correction in current iteration‡

$$\Delta_{i+1} = |x_{i+1} - x_i|.$$

Step 3. Test for convergence, divergence, failure to converge.

(a) Converged yet? If $\Delta_{i+1} \leq \varepsilon$ and $|f(x_{i+1})| \leq \varepsilon$, go to Step 4. If not, go to Step 3b.

(b) Diverging or converging? If $\Delta_{i+1} > \Delta_i$, select new x_1 and return to Step 1. If $\Delta_{i+1} \leq \Delta_i$, go to Step 3c.

(c) Maximum number of iterations exceeded? If $i \leq N$, set $i = i + 1$, and return to Step 2. If $i > N$, select new x_1 and return to Step 1.

Step 4. Output root α. Set $\alpha = x_{i+1}$. Write α.

Note. If a root has not been located after a number of starting values x_1 have been tried, the program should be stopped. Another program might then be used to locate a root.

A representative flow chart of the Newton-Raphson iterative method is given in Figure 3-7.

† The magnitude of $f'(x_i)$ should also be tested to insure against division by a near-zero number.

‡ Alternatively, we could define the magnitude of the relative correction

$$\Delta_{i+1} = \left| \frac{x_{i+1} - x_i}{x_{i+1}} \right|$$

provided $|x_{i+1}| > \varepsilon$.

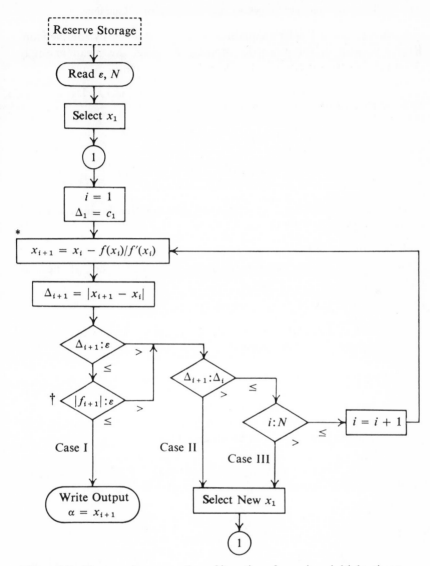

Figure 3-7. N = maximum number of iterations for a given initial estimate x_1. Case I: convergence. Case II: divergence. Case III: failure to converge in N iterations. c_1 = initial guess of error term (an arbitrary large positive number).

* Also test $|f'(x_i)| \leq \varepsilon$.

† Although $\Delta_{i+1} \leq \varepsilon$, x_{i+1} may not be a good approximation of root α (if graph of $f(x)$ is very nearly vertical at x_i). This pathological case can be guarded against by also checking that $|f(x_{i+1})| \leq \varepsilon$.

EXAMPLE. Solve Kepler's equation $M = E - e \sin E$, by the Newton-Raphson method, for eccentric anomaly E, given the mean anomaly $M = 0.8$ and eccentricity $e = 0.2$. Use $E_1 = M$ as initial estimate.

Solution. Define $f(E) = E - e \sin E - M$. Then $f'(E) = 1 - e \cos E$, and the Newton-Raphson iterative formula is

$$E_{i+1} = E_i - \frac{f(E_i)}{f'(E_i)} = E_i - \frac{E_i - e \sin E_i - M}{1 - e \cos E_i}$$

$$= \frac{M - E_i e \cos E_i + e \sin E_i}{1 - e \cos E_i} = F(E_i).$$

Successive approximations generated by this iterative formula are 0.8, 0.967, 0.96434, ...

Bailey's Iterative Method

Suppose that we are given an estimate x_i of a real root of the real equation

(3.2.14) $f(x) = 0.$

The equation of the *osculating parabola*† to $f(x)$ at $x = x_i$ can be expressed as the quadratic Taylor polynomial

(3.2.15) $y(x) = f(x_i) + f'(x_i)(x - x_i) + \dfrac{f''(x_i)}{2}(x - x_i)^2.$

Bailey's iterative method calculates (from initial estimate x_0) a sequence $x_1, x_2, \ldots$ of successive approximations to root α by approximating $f(x)$ by this osculating parabola and determining the intersection $(x_{i+1}, 0)$ of this parabola with the x-axis.

To calculate this intersection point we set $y(x) = 0$ when $x = x_{i+1}$ in equation (3.2.15), obtaining

$$0 = f(x_i) + f'(x_i)(x_{i+1} - x_i) + \frac{f''(x_i)}{2}(x_{i+1} - x_i)^2$$

(3.2.16)

$$= f(x_i) + (x_{i+1} - x_i)\left[f'(x_i) + \frac{f''(x_i)}{2}(x_{i+1} - x_i)\right].$$

This equation can be rewritten in the form

(3.2.17) $x_{i+1} = x_i - \left[\dfrac{f(x_i)}{f'(x_i) + \dfrac{f''(x_i)}{2}(x_{i+1} - x_i)} \right].$

† A parabola $y(x) = Ax^2 + Bx + C$ is said to be an osculating parabola of function $f(x)$ at x_i if the following three conditions are satisfied: (1) $y(x_i) = f(x_i)$; (2) $y'(x_i) = f'(x_i)$; (3) $y''(x_i) = f''(x_i)$.

Suppose we now let the coefficient of $f''(x_i)/2$ be computed by the Newton-Raphson iterative formula $x_{i+1} = x_i - f(x_i)/f'(x_i)$. That is, substitute $(-f(x_i)/f'(x_i))$ for $(x_{i+1} - x_i)$ in the term of the denominator of (3.2.17). The resulting formula

(3.2.18)
$$x_{i+1} = x_i - \left[\frac{f(x_i)}{f'(x_i) - \dfrac{f(x_i)f''(x_i)}{2f'(x_i)}} \right]$$

is known as Bailey's iterative formula for computing a refined approximation x_{i+1} of root α from approximation x_i.

The proof of convergence and the analysis of the rate of convergence for Bailey's method are simplified if an iteration function is defined in the same way that $F(x)$ was defined for Newton's method. For Bailey's method, the iteration function has the form

(3.2.19)
$$F(x) = x - \left[\frac{f(x)}{f'(x) - \dfrac{f(x)f''(x)}{2f'(x)}} \right]$$

so that $F(x)$ has the properties: $F(x_i) = x_{i+1}$, and $F(\alpha) = \alpha$.

Proof of convergence of Bailey's method. The proof of the convergence of Bailey's method is identical with the proof of convergence of Newton's method, with the $F(x)$ as defined in (3.2.19) replacing the $F(x)$ defined for Newton's method.

Rate of convergence of Bailey's method. The rate of convergence for Bailey's method is determined in the same way it was done for Newton's method. That is, the iteration function $F(x)$ defined in (3.2.19) is expanded in a Taylor series about $x = \alpha$ [where α is a root of equation $f(x) = 0$], as follows:

(3.2.20) $F(x) = F(\alpha) + F'(\alpha)(x - \alpha) + \dfrac{F''(\alpha)}{2}(x - \alpha)^2$

$$+ \frac{F'''(\alpha)}{6}(x - \alpha)^3 + \cdots.$$

For Bailey's method it can be shown† that

(3.2.21) $F'(\alpha) = F''(\alpha) = 0, \qquad F'''(\alpha) = \dfrac{3}{2} \dfrac{[f''(\alpha)]^2}{[f'(\alpha)]^2} - \dfrac{f'''(\alpha)}{f'(\alpha)}.$

By substituting (3.2.21) into (3.2.20), equation (3.2.20) can be written in the form (neglecting terms of order higher than 3)

(3.2.22) $F(x) - F(\alpha) \doteq K(x - \alpha)^3$

† J. F. Traub, "Comparison of Iterative Methods for the Calculation of nth Roots," *J. ACM*, Vol. 4, No. 3, 1961, pp. 143–145.

where

$$K \equiv F'''(\alpha).$$

Evaluating (3.2.22) at $x = x_{i+1}$, and using the relations $F(x_i) = x_{i+1}$ and $F(\alpha) = \alpha$, we find that

$$(3.2.23) \qquad x_{i+1} - \alpha \doteq K(x_i - \alpha)^3.$$

From this relation, the rate of convergence is said to be *cubic*. This would indicate that Bailey's method should converge faster than Newton's method, i.e., in fewer iterations. However, it should be noted that Bailey's method requires a greater number of computations per iteration than Newton's method. This fact somewhat offsets the advantage of converging in fewer iterations.

3.3   Solution of Systems of Non-linear Equations

The methods of Newton and Bailey can be extended to solve systems of non-linear equations, such as

$$(3.3.1) \qquad \begin{aligned} f(x, y) &= 0 \\ g(x, y) &= 0 \end{aligned}$$

provided that estimates (x_0, y_0) of the solution are available.

The derivation of these methods for solving systems of nonlinear equations are based on Taylor's expansion of functions of several variables. For the case of functions of two independent variables x and y, these Taylor expansions can be expressed in the form

$$f(x_0 + \delta x, y_0 + \delta y) = f(x_0, y_0) + f_x(x_0, y_0)\,\delta x + f_y(x_0, y_0)\,\delta y$$

$$+ \frac{1}{2}\,[f_{xx}(x_0, y_0)\,\delta x^2 + 2f_{xy}(x_0, y_0)\,\delta x\,\delta y$$

$$+ f_{yy}(x_0, y_0)\,\delta y^2] + \cdots$$

$$(3.3.2)$$

$$g(x_0 + \delta x, y_0 + \delta y) = g(x_0, y_0) + g_x(x_0, y_0)\,\delta x + g_y(x_0, y_0)\,\delta y$$

$$+ \frac{1}{2}\,[g_{xx}(x_0, y_0)\,\delta x^2 + 2g_{xy}(x_0, y_0)\,\delta x\,\delta y$$

$$+ g_{yy}(x_0, y_0)\,\delta y^2] + \cdots.$$

The extension of Newton's method to solve systems of nonlinear equations is based on approximating functions of several variables by *linear* Taylor polynomials. For example, to solve system (3.3.1), functions $f(x, y)$ and $g(x, y)$ are approximated by the Taylor expansions given in (3.3.2) *truncated* after terms of first degree. Similarly, the extension of Bailey's method to solve systems of nonlinear equations is based on approximating functions of several variables by *quadratic* Taylor polynomials, such as (3.3.2) *truncated* after terms of second degree in δx and δy. The extension of Bailey's method is referred to simply as a modified Newton method in this text. The extensions of Newton's and Bailey's methods are derived in the following pages and illustrated by numerical examples.

Newton's Method of Solving Systems of Non-linear Equations

Suppose that estimates (x_0, y_0) of the solution of equations (3.3.1) are known. If these estimates x_0, y_0 are incremented respectively by changes δx, δy, then *first-order approximations* of the resulting changes in $f(x, y)$ and $g(x, y)$ are given by the total differentials

(3.3.3)
$$\delta f = f_x(x_0, y_0)\,\delta x + f_y(x_0, y_0)\,\delta y$$
$$\delta g = g_x(x_0, y_0)\,\delta x + g_y(x_0, y_0)\,\delta y.$$

A solution of system (3.3.1) can be obtained by determining δx, δy such that the total differentials δf, δg satisfy the constraints

(3.3.4)
$$\delta f = -f(x_0, y_0)$$
$$\delta g = -g(x_0, y_0).$$

A set of equations linear in δx and δy can be found by substituting these constraints into (3.3.3)

(3.3.5)
$$-f(x_0, y_0) = f_x(x_0, y_0)\,\delta x + f_y(x_0, y_0)\,\delta y$$
$$-g(x_0, y_0) = g_x(x_0, y_0)\,\delta x + g_y(x_0, y_0)\,\delta y$$

and these linear equations can be solved for δx and δy.

If the functions f and g are evaluated at $(x_0 + \delta x, y_0 + \delta y)$ and expressed in a linear Taylor expansion, it follows from (3.3.5) that the right sides of the resulting equations are zero, i.e.,

$$f(x_0 + \delta x, y_0 + \delta y) \doteq f(x_0, y_0) + f_x(x_0, y_0)\,\delta x + f_y(x_0, y_0)\,\delta y = 0$$
$$g(x_0 + \delta x, y_0 + \delta y) \doteq g(x_0, y_0) + g_x(x_0, y_0)\,\delta x + g_y(x_0, y_0)\,\delta y = 0.$$
(3.3.6)

If these linear Taylor expansions are sufficiently accurate, it is clear that $(x_0 + \delta x, y_0 + \delta y)$ are fairly good approximations of the solutions of equations (3.3.1). If $|\delta x| > \varepsilon$ or if $|\delta y| > \varepsilon$, where ε is a small positive quantity, it is necessary to replace x_0 by $x_0 + \delta x$ and y_0 by $y_0 + \delta y$, and repeat the entire process. Usually a few iterations of this process will produce accurate values of the roots of (3.3.1), provided that the *original* estimates (x_0, y_0) are sufficiently close to the true solution.

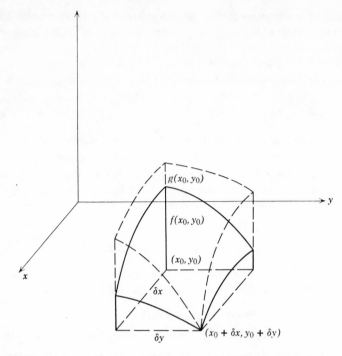

Figure 3-8. Newton's method for solving two non-linear equations:
$$f(x, y) = 0$$
$$g(x, y) = 0.$$

Determine δx and δy such that $\delta f = -f(x_0, y_0)$ and $\delta g = -g(x_0, y_0)$.
$$\delta f = f_x(x_0, y_0)\, \delta x + f_y(x_0, y_0)\, \delta y$$
$$\delta g = g_x(x_0, y_0)\, \delta x + g_y(x_0, y_0)\, \delta y.$$

An elegant proof of the convergence of Newton's method is given in Henrici [7, p. 106].

The extension of this method to the solution of n equations in n variables is obtained by replacing the two-variable expressions in the above equations by their n-variable counterparts.

Numerical Example of Newton's Method

Given solution estimate $(x_0, y_0) = (1, 1)$, determine a solution of the non-linear equations

$$f(x, y) \equiv x^2 + y^2 - 4 = 0$$
$$g(x, y) \equiv \frac{x^2}{9} + y^2 - 1 = 0.$$

In each iteration, calculate the differential corrections δx, δy by solving the linear equations:

$$\begin{bmatrix} f_x(x_0, y_0) & f_y(x_0, y_0) \\ g_x(x_0, y_0) & g_y(x_0, y_0) \end{bmatrix} \begin{bmatrix} \delta x \\ \delta y \end{bmatrix} = \begin{bmatrix} -f(x_0, y_0) \\ -g(x_0, y_0) \end{bmatrix}.$$

Let $x_0 + \delta x \to x_0$, $y_0 + \delta y \to y_0$. Continue to iterate until the differential corrections are negligible.

Solution

$$f_x(x, y) = 2x \qquad f_y(x, y) = 2y$$
$$g_x(x, y) = \frac{2}{9} x \qquad g_y(x, y) = 2y$$

Original $x_0 = 1$, $y_0 = 1$; $f(x_0, y_0) = -2.0$, $g(x_0, y_0) = 1/9$.

$$\begin{bmatrix} 2 & 2 \\ 2/9 & 2 \end{bmatrix} \begin{bmatrix} \delta x \\ \delta y \end{bmatrix} = \begin{bmatrix} -(-2.0) \\ -1/9 \end{bmatrix}$$

Solution

$$\delta x = 1.1875 \qquad x_0 + \delta x = 2.1875 \to x_0$$
$$\delta y = -0.1875 \qquad y_0 + \delta y = 0.8125 \to y_0$$

New $x_0 = 2.1875$, $y_0 = 0.8125$; $f(x_0, y_0) = 1.445$, $g(x_0, y_0) = 0.192$.

$$\begin{bmatrix} 4.375 & 1.625 \\ 0.486 & 1.625 \end{bmatrix} \begin{bmatrix} \delta x \\ \delta y \end{bmatrix} = \begin{bmatrix} -1.445 \\ -0.192 \end{bmatrix}$$

Solution

$$\delta x = -0.322 \qquad x_0 + \delta x = 1.8655 \to x_0$$
$$\delta y = -0.0218 \qquad y_0 + \delta y = 0.7907 \to y_0$$

New $x_0 = 1.8655$, $y_0 = 0.7909$; $f(x_0, y_0) = 0.1052$, $g(x_0, y_0) = 0.0119$.

$$\begin{bmatrix} 3.7310 & 1.5818 \\ 0.4146 & 1.5818 \end{bmatrix} \begin{bmatrix} \delta x \\ \delta y \end{bmatrix} = \begin{bmatrix} -0.1052 \\ -0.0119 \end{bmatrix}$$

Solution

$$\delta x = -0.0281 \qquad x_0 + \delta x = 1.8374 \to x_0$$
$$\delta y = -0.00015 \qquad y_0 + \delta y = 0.79055 \to y_0$$

New $x_0 = 1.8374$, $y_0 = 0.79055$, $f(x_0, y_0) = 0.001008$, $g(x_0, y_0) = 0.000084$.

Since f and g are approximately zero, the last values for x_0, y_0 are good approximations of the roots of the given equations.

It is interesting to note that this particular example can be stated as the following geometric problem: Find the first-quadrant intersection of the circle $x^2 + y^2 = 4$ and the ellipse $x^2/9 + y^2 = 1$.

A Modification of Newton's Method Using a Quadratic Taylor Expansion

Newton's method can be modified to take into account second derivatives. That is, if estimates x_0, y_0 are incremented respectively by δx, δy, then *second-order approximations* of the resulting changes in $f(x, y)$ and $g(x, y)$ are given by the expressions†

$$\delta f = f_x \, \delta x + f_y \, \delta y + \frac{1}{2} [f_{xx} \, \delta x^2 + 2f_{xy} \, \delta x \, \delta y + f_{yy} \, \delta y^2]$$

$$\delta g = g_x \, \delta x + g_y \, \delta y + \frac{1}{2} [g_{xx} \, \delta x^2 + 2g_{xy} \, \delta x \, \delta y + g_{yy} \, \delta y^2].$$

(3.3.7)

A solution of system (3.3.1) can be obtained by determining δx, δy such that δf and δg satisfy the constraints

(3.3.8)
$$\begin{aligned} \delta f &= -f(x_0, y_0) \\ \delta g &= -g(x_0, y_0). \end{aligned}$$

Equating (3.3.8) and (3.3.7) and factoring the right sides of the resulting expressions, we find that

$$-f(x_0, y_0) = \left[f_x + \frac{f_{xx} \, \delta x}{2} + \frac{f_{xy} \, \delta y}{2} \right] \delta x + \left[f_y + \frac{f_{yy} \, \delta y}{2} + \frac{f_{xy} \, \delta x}{2} \right] \delta y$$

$$-g(x_0, y_0) = \left[g_x + \frac{g_{xx} \, \delta x}{2} + \frac{g_{xy} \, \delta y}{2} \right] \delta x + \left[g_y + \frac{g_{yy} \, \delta y}{2} + \frac{g_{xy} \, \delta x}{2} \right] \delta y.$$

(3.3.9)

Now, if the δx, δy inside the brackets are approximated respectively by $-f/f_x$ and $-f/f_y$ in the first equation and by $-g/g_x$ and $-g/g_y$ in the second equation, system (3.3.9) can be written as a system of equations linear in δx and δy:

$$-f(x_0, y_0) = \left[f_x - \frac{f_{xx}f}{2f_x} - \frac{f_{xy}f}{2f_y} \right] \delta x + \left[f_y - \frac{f_{yy}f}{2f_y} - \frac{f_{xy}f}{2f_x} \right] \delta y$$

$$-g(x_0, y_0) = \left[g_x - \frac{g_{xx}g}{2g_x} - \frac{g_{xy}g}{2g_y} \right] \delta x + \left[g_y - \frac{g_{yy}g}{2g_y} - \frac{g_{xy}g}{2g_x} \right] \delta y.$$

(3.3.10)

These equations can then be directly solved for δx and δy, and $(x_0 + \delta x, y_0 + \delta y)$ should be an improved estimate of the solution of (3.3.1), depending on the accuracy of the quadratic Taylor expansion and the approximations of δx and δy by the quotients used in (3.3.10).

† The functions f and g and their partials are evaluated at (x_0, y_0).

If $|\delta x| > \varepsilon$ or if $|\delta y)| > \varepsilon$ the entire process is repeated after replacing the values of x_0, y_0 by the values of $x_0 + \delta x$, $y_0 + \delta y$.

Numerical Example of Modified Newton Method

Given solution estimate $(x_0, y_0) = (1, 1)$, determine a solution of the nonlinear equations

$$f(x, y) \equiv x^2 + y^2 - 4 = 0$$

$$g(x, y) \equiv \frac{x^2}{9} + y^2 - 1 = 0.$$

In each iteration, calculate the differential corrections δx, δy by solving the linear equations:

$$\begin{bmatrix} f_x - \dfrac{f_{xx}f}{2f_x} - \dfrac{f_{xy}f}{2f_y} & f_y - \dfrac{f_{yy}f}{2f_y} - \dfrac{f_{xy}f}{2f_x} \\[2mm] g_x - \dfrac{g_{xx}g}{2g_x} - \dfrac{g_{xy}g}{2g_y} & g_y - \dfrac{g_{yy}g}{2g_y} - \dfrac{g_{xy}g}{2g_x} \end{bmatrix} \begin{bmatrix} \delta x \\[2mm] \delta y \end{bmatrix} = \begin{bmatrix} -f(x_0, y_0) \\[2mm] -g(x_0, y_0) \end{bmatrix}.$$

Solution

$$\begin{aligned} f_x &= 2x & f_{xx} &= 2 & f_{xy} &= 0 \\ f_y &= 2y & f_{yy} &= 2 & \\ g_x &= \frac{2}{9}x & g_{xx} &= \frac{2}{9} & g_{xy} &= 0 \\ g_y &= 2y & g_{yy} &= 2 & \end{aligned}$$

Original $(x_0, y_0) = (1, 1)$, $f(x_0, y_0) = -2$, $g(x_0, y_0) = 1/9$.

$$\begin{bmatrix} 2 - \left(\dfrac{2}{2}\right)\left(\dfrac{-2}{2}\right) - 0 & 2 - \left(\dfrac{2}{2}\right)\left(\dfrac{-2}{2}\right) - 0 \\[2mm] \dfrac{2}{9} - \left(\dfrac{2/9}{2}\right)\left(\dfrac{1/9}{2/9}\right) - 0 & 2 - \left(\dfrac{2}{2}\right)\left(\dfrac{1/9}{2}\right) - 0 \end{bmatrix} \begin{bmatrix} \delta x \\[2mm] \delta y \end{bmatrix} = \begin{bmatrix} -(-2) \\[2mm] -1/9 \end{bmatrix}$$

Solution

$$\begin{aligned} \delta x &= 0.7916 & x_0 + \delta x &= 1.7916 \rightarrow x_0 \\ \delta y &= -0.125 & y_0 + \delta y &= 0.8750 \rightarrow y_0 \end{aligned}$$

New $(x_0, y_0) = (1.7916, 0.8750)$, $f(x_0, y_0) = -0.0246$, $g(x_0, y_0) = 0.1222$.

$$\begin{bmatrix} 3.5832 - \left(\dfrac{2}{2}\right)\left(\dfrac{-0.0246}{3.5832}\right) - 0 & 1.75 - \left(\dfrac{2}{2}\right)\left(\dfrac{-0.0246}{1.75}\right) - 0 \\[2mm] 0.3981 - \left(\dfrac{2/9}{2}\right)\left(\dfrac{0.1222}{0.3981}\right) - 0 & 1.75 - \left(\dfrac{2}{2}\right)\left(\dfrac{0.1222}{1.75}\right) - 0 \end{bmatrix} \begin{bmatrix} \delta x \\[2mm] \delta y \end{bmatrix}$$

$$= \begin{bmatrix} -(-0.0246) \\[4mm] -0.1222 \end{bmatrix}$$

Solution

$$\delta x = 0.0477 \qquad x_0 + \delta x = 1.8393 \rightarrow x_0$$
$$\delta y = -0.0831 \qquad y_0 + \delta y = 0.7919 \rightarrow y_0$$

New $(x_0, y_0) = (1.8393, 0.7919)$, $f(x_0, y_0) = 0.0101$, $g(x_0, y_0) = 0.0030$.
For this particular example, two iterations of the modified method have produced a solution that is almost as accurate as that produced by Newton's method in three iterations. In general, the modified method should converge faster and over a larger region than the unmodified Newton method. However, this improvement is obtained at the expense of additional computations in each iteration.

It should be noted that this modified method based on a *quadratic* Taylor expansion is an extension of Bailey's method for solving one nonlinear (algebraic or transcendental) equation in one unknown.

Summary

Solution of a Single Non-linear Equation

The method of successive bisection is a simple, highly reliable method that is guaranteed to converge unless the roundoff error in evaluating $f(x_1)f(\bar{x})$ causes selection of the wrong half-interval; this can be avoided by carrying extra figures (for manual computation) and by making the convergence term ε greater than the anticipated roundoff in evaluating $f(x_1)f(\bar{x})$. This method, which requires only one evaluation of $f(x)$ each iteration, is widely used to compute the eigenvalues of tridiagonal matrices as in Givens' and Householder's methods. The principal disadvantage is the slow convergence of the method.

The same remarks about simplicity, reliability, and roundoff apply to the method of inverse linear interpolation. The convergence of this technique is super-linear. Variations of the basic method appear in numerous numerical analysis texts. The method requires only one evaluation of $f(x)$ each iteration.

The Newton-Raphson method is the most widely used iterative technique for calculating a real root of a non-linear equation, because it offers an excellent compromise between efficiency and simplicity. This method requires the evaluation of $f(x)$ and $f'(x)$ each iteration, and converges quadratically provided specified conditions hold. The Birge-Vieta method (Chapter 4) employs the Newton-Raphson technique to calculate the roots of a polynomial.

Bailey's method† requires the evaluation of $f(x)$, $f'(x)$, and $f''(x)$ each

---

† H. S. Wall, "A modification of Newton's Method," *Amer. Math. Monthly*, Vol. 55, 1948, pp. 90–94.

iteration, which somewhat offsets the advantage provided by its cubic convergence. Since Bailey's method is based on a quadratic Taylor approximation, it should normally converge over a larger region than does the Newton-Raphson method.

If the evaluation of $f''(x)$ is difficult or time-consuming, the Newton-Raphson method is probably more efficient than Bailey's method. Likewise, if the evaluation of $f'(x)$ is difficult or time-consuming, the method of inverse linear interpolation is probably more efficient than Newton-Raphson.

COMPUTATIONAL EXERCISES

1. Calculate $\sqrt{43}$ by solving the equation $f(x) \equiv x^2 - 43 = 0$ with $x_1 = 0$, $x_2 = 43$, using
   (a) the method of successive bisection;
   (b) the method of inverse linear interpolation.

2. Calculate $\sqrt{43}$ by solving the equation $f(x) \equiv x^2 - 43 = 0$ with $x_0 = 43$, using
   (a) the Newton-Raphson method;
   (b) Bailey's method.

3. Locate a root in the interval $(2, 3)$ of $f(x) \equiv x^3 - 2x - 5 = 0$, using
   (a) the method of successive bisection;
   (b) the method of inverse linear interpolation.

4. Calculate a real root of $f(x) \equiv 1 - x - 2 \cos x = 0$, with $x_0 = 0$, using
   (a) the Newton-Raphson method;
   (b) Bailey's method.

5. Calculate a real root of $f(x) \equiv x^2 - e^x - 3 = 0$, with $x_0 = 0$, using
   (a) the Newton-Raphson method;
   (b) Bailey's method.

6. Calculate a real root of $f(x) \equiv x^3 - 18 = 0$, using the Newton-Raphson method.

7. Find the first-quadrant intersection of the graphs of the following two functions:

$$f(x, y) = x^2 + y^2 - 9$$

$$g(x, y) = \frac{x^2}{16} + \frac{y^2}{4} - 1$$

using
   (a) Newton's method;
   (b) Newton's modified method.

PROGRAMMING EXERCISES

1. Write a FORTRAN program to compute the cube root of a real positive number N, using the Newton-Raphson iterative method, with $x_0 = N$ as the initial estimate.

2. Write a FORTRAN program to compute the cube root of a real positive number N, using Bailey's iterative method, with $x_0 = N$ as the initial estimate.

3. Write a FORTRAN program to calculate a real root of the equation $f(x) = 0$, using the method of successive bisection. Note that the explicit form of $f(x)$ must be defined in the program.

4. Write a FORTRAN program to calculate a real root of the equation $f(x) = 0$, using the method of inverse linear interpolation. Note that the explicit form of the function $f(x)$ must be defined in the program.

4

Polynomials and Zeros

of Polynomials

4.0 Introduction

The nth-degree polynomial of the form

$$(4.0.1) \qquad P_n(x) = a_0 x^n + a_1 x^{n-1} + a_2 x^{n-2} + \cdots + a_{n-1} x + a_n$$

is one of the workhorses of the mathematical stable. Polynomials can be easily multiplied and divided by other polynomials. They can be easily differentiated, integrated, and evaluated. By allowing the polynomial to take forms other than (4.0.1), we are able to derive the various finite-difference forms and ordinate-form interpolating polynomials, which are in turn used extensively in the derivation of formulas for numerical differentiation and integration. In this chapter, we will concern ourselves only with polynomials of the form of (4.0.1).

Polynomials occur extensively throughout the scientific disciplines, especially in connection with differential equations. The characteristic equation of a system of n first-order constant-coefficient linear differential equations can be expressed in the form of a polynomial equation. Equivalently, the characteristic (auxiliary) equation of an nth-order linear differential equation is a polynomial equation of degree n. In this connection, the matrix eigen problem and the determination of the roots of a polynomial equation are mathematically one and the same problem. However,

101

the matrix eigen problem is usually solved without reducing the characteristic equation to explicit polynomial form (See Sec. 5.10).

The roots of polynomial equations of degree 1, 2, 3, and 4 can be determined by classical (closed-form) methods. However, these methods are quite involved for the third- and fourth-degree equations. For this reason, and so that we may solve polynomial equations of higher degree, numerical methods of successive approximation are used extensively for the determination of the roots of polynomial equations. Before developing such numerical methods, we will briefly discuss some basic theorems and relations concerning polynomials.

Fundamental Theorem of Algebra (FTA)

Every algebraic (polynomial) equation with complex coefficients has at least one real or complex root. A detailed proof of this theorem is given in Appendix I of Uspensky [34].

Division Algorithm

If $P(x)$ and $F(x)$ are polynomials in x and $F(x) \neq 0$, then polynomials $Q(x)$ and $R(x)$ can be found which satisfy the relation

$$(4.0.2) \qquad P(x) = Q(x)F(x) + R(x)$$

where either $R(x) = 0$, or degree $R(x) <$ degree $F(x)$.

EXAMPLE. $x^3 + x^2 - 10x + 8 = (x^2 + 4x + 2)(x - 3) + 14$

where $P(x) = x^3 + x^2 - 10x + 8$
  $F(x) = x - 3$
  $Q(x) = x^2 + 4x + 2$
  $R(x) = 14.$

Remainder Theorem

The remainder obtained in dividing $P(x)$ by $(x - \alpha)$ is the value of $P(\alpha)$.

Proof. Dividing $P(x)$ by $(x - \alpha)$ we obtain

$$(4.0.3) \qquad P(x) = (x - \alpha)Q(x) + R$$

where $R = 0$ or $R = $ constant, since degree $R < 1$ by the division algorithm. Evaluating (4.0.3) for $x = \alpha$, we get

$$(4.0.4) \qquad P(\alpha) = (\alpha - \alpha)Q(\alpha) + R$$

so that
$$R = P(\alpha) \quad \text{and} \quad P(x) = (x - \alpha)Q(x) + P(\alpha).$$

Factor Theorem

Every polynomial equation of the form†

$$(4.0.5) \qquad P_n(x) \equiv x^n + a_1 x^{n-1} + a_2 x^{n-2} + \cdots + a_{n-1} x + a_n = 0$$

has at most n distinct roots α_i $(i = 1, 2, \ldots, n)$.

By the Fundamental Theorem of Algebra (FTA) there exists at least one root of the algebraic equation (4.0.5). If α_1 is such a root, i.e., if $P_n(\alpha_1) = 0$, then by the remainder theorem

$$(4.0.6) \qquad \begin{aligned} P_n(x) &= (x - \alpha_1) P_{n-1}(x) + P_n(\alpha_1) \\ &= (x - \alpha_1) P_{n-1}(x). \end{aligned}$$

Similarly, there exists at least one root, say α_2, of $P_{n-1}(x) = 0$, so that

$$(4.0.7) \qquad P_{n-1}(x) = (x - \alpha_2) P_{n-2}(x).$$

This process is continued until we obtain

$$(4.0.8) \qquad P_1(x) = (x - \alpha_n).$$

Now, by successive substitution of the relation

$$(4.0.9) \qquad P_{n-i}(x) = (x - \alpha_{i+1}) P_{n-(i+1)}(x)$$

into (4.0.6), we obtain

$$(4.0.10) \qquad P_n(x) = (x - \alpha_1)(x - \alpha_2) \ldots (x - \alpha_n).$$

Equation (4.0.10) states that an nth-degree polynomial $P_n(x)$ can be factored into n linear factors $(x - \alpha_i)$, where the α_i $(i = 1, 2, \ldots, n)$ are roots of the corresponding polynomial equation. It should be noted that the α_i need not be distinct. It can be proved by the principle of finite induction that this factorization is unique except for the order in which the factors appear. (See for example [Birkhoff and MacLane, 35].)

This chapter presents iterative methods of the factor-extraction (deflation) type for successively computing all the zeros of a real-coefficient polynomial. A representative method of each of the following categories is presented.

1. *Extraction of linear factor $(x - \alpha)$ from $P_n(x)$.* The Birge-Vieta method is a combination of the process of synthetic division [used to compute values of $P_n(x_i)$ and $P_n{}'(x_i)$] *and* the Newton-Raphson iterative

† $\bar{P}_n(x) \equiv \bar{a}_0 x^n + \bar{a}_1 x^{n-1} + \bar{a}_2 x^{n-2} + \cdots + \bar{a}_{n-1} x + \bar{a}_n = 0$ is a more general form of the nth-degree polynomial. However, to avoid the zero subscript which cannot be handled by most FORTRAN processors, we use the form (4.0.5), which is obtained from the above by dividing through by $\bar{a}_0$. It can easily be shown that both polynomials have the same roots if $a_k = \bar{a}_k / \bar{a}_0$.

technique (used to compute a sequence x_{i+1} of successive approximations of each root α). When this sequence converges to a real zero α, the corresponding linear factor $(x - \alpha)$ is extracted by synthetic division. The $(n - 1)$st-degree (deflated) polynomial $P_{n-1}(x) \equiv P_n(x)/(x - \alpha)$ then replaces $P_n(x)$. The procedure is repeated until all zeros of the original polynomial have been computed.

2. *Extraction of a quadratic factor* $(x^2 + rx + s)$ *from* $P_n(x)$. The Bairstow method is a combination of the process of synthetic division by a quadratic term *and* Newton's method for solving a system of two nonlinear equations. When the sequence of quadratic terms converges to a quadratic factor, that factor is extracted, and the deflated polynomial $P_{n-2}(x) \equiv P_n(x)/(x^2 + rx + s)$ replaces $P_n(x)$. The procedure is repeated until all quadratic factors of the original polynomial have been computed. The *zeros* of the polynomial are computed from the quadratic factors by the quadratic formula.

4.1 The Birge-Vieta Iterative Method

The Birge-Vieta method is a straightforward algorithm for calculating the real roots of a polynomial equation.† We will consider herein an Nth-degree polynomial equation of the form

$$(4.1.1) \quad P_N(x) = x^N + a_1 x^{N-1} + a_2 x^{N-2} + \cdots + a_{N-1}x + a_N = 0$$

where the coefficients a_k are real, and the roots α_k are assumed to be real.

The factor theorem (Sec. 4.0) states that an Nth-degree polynomial

$$P_N(x) = x^N + a_1 x^{N-1} + a_2 x^{N-2} + \cdots + a_{N-1}x + a_N$$

can be expressed as the product of N linear factors $(x - \alpha_j)$, where the α_j are the roots of the polynomial equation $P_N(x) = 0$. That is, we can express $P_N(x)$ in the form

$$P_N(x) = (x - \alpha_1)(x - \alpha_2)\ldots(x - \alpha_N).$$

The Birge-Vieta method calculates $P_N(x_i)$ and its derivative $P_N'(x_i)$ by simple recursive formulas, and solves for a root of equation (4.1.1) by the Newton-Raphson iterative formula

$$(4.1.2) \qquad x_{i+1} = x_i - \frac{P_N(x_i)}{P_N'(x_i)}.$$

† Complex roots of a real- or complex-coefficient polynomial equation can be calculated by analogous recursion formulas using a complex variable z in place of real variable x, and by modifying the Newton-Raphson iterative formula, as required, to eliminate division by a complex number. See Jennings [12, p. 29, Exercise 3].

When a root $\alpha_1 = x_{i+1}$ of $P_N(x) = 0$ has been computed, $P_N(x)$ is replaced by the $(N-1)$st-degree polynomial $P_{N-1}(x)$, where $P_{N-1}(x) = P_N(x)/(x - \alpha_1)$. A root α_2 of $P_{N-1}(x) = 0$, and hence also a root of $P_N(x) = 0$, is then calculated using the same technique by which α_1 was calculated. $P_{N-1}(x)$ is then replaced by the $(N-2)$nd-degree polynomial $P_{N-2}(x) = P_{N-1}(x)/(x - \alpha_2)$. Continuing this process, we can obtain all roots $\alpha_1, \alpha_2, \ldots, \alpha_N$ of $P_N(x) = 0$. It should be noted that formula (4.1.2) must be modified in the event of *multiple roots*, since, in general, $P_N'(\alpha) = P_N''(\alpha) = \cdots = P_N^{(m-1)}(\alpha) = 0$, for root α of multiplicity m.

As Henrici [7, p. 86] notes, the coefficients of succeeding polynomials $P_{N-1}(x), P_{N-2}(x), \ldots$ are affected by the accumulation of rounding error in the process, so that it is advisable to recheck each root by using it as a starting value in Newton's iteration (4.1.2) for the original polynomial $P_N(x)$.

Computation of $P_n(x_i)$ and $P_n'(x_i)$ by Synthetic Division

Let n denote the degree of the polynomial from which a linear factor $(x - \alpha)$, where α is one of the roots α_j of $P_N(x)$, is to be extracted. That is, $n = N - k$, where k linear factors $(x - \alpha_j)$ $(j = 1, k)$, have been previously extracted from the original polynomial $P_N(x)$.

If $P_n(x)$ is divided by a *trial* linear term $(x - x_i)$, we obtain

$$(4.1.3) \qquad P_n(x) = (x - x_i)P_{n-1}(x) + p_n$$

where $P_{n-1}(x)$ is a polynomial of degree $n - 1$, and p_n is a constant remainder. Evaluation of (4.1.3) at $x = x_i$ shows that

$$(4.1.4) \qquad P_n(x_i) = p_n.$$

The value of $P_n'(x_i)$ could be obtained by differentiating (4.1.3) and evaluating the result at $x = x_i$, i.e.,

$$(4.1.5) \qquad P_n'(x) = (x - x_i)P_{n-1}'(x) + P_{n-1}(x)$$

from which it follows that

$$(4.1.6) \qquad P_n'(x_i) = P_{n-1}(x_i).$$

Instead of actually performing this differentiation to get $P_n'(x_i)$, we can divide $P_{n-1}(x)$ by $(x - x_i)$ to obtain

$$(4.1.7) \qquad P_{n-1}(x) = (x - x_i)P_{n-2}(x) + \bar{p}_n$$

where $P_{n-2}(x)$ is the quotient polynomial of degree $n - 2$ and $\bar{p}_n$ is the constant remainder. Evaluating (4.1.7) at $x = x_i$, we find that

$$(4.1.8) \qquad P_{n-1}(x_i) = \bar{p}_n.$$

Equating (4.1.6) and (4.1.8), we obtain

(4.1.9) $P_n'(x_i) = \bar{p}_n$.

Therefore, the values $P_n(x_i)$ and $P_n'(x_i)$, required in the Newton-Raphson iterative formula for computing x_{i+1}, can be obtained by dividing $P_n(x)$ by $(x - x_i)$ and $P_{n-1}(x)$ by $(x - x_i)$.

Recursion formulas for computing $P_n(x_i)$ and $P_n'(x_i)$ can be derived as follows.

Write (4.1.3) in expanded form

$$x^n + a_1 x^{n-1} + a_2 x^{n-2} + \cdots + a_{n-1} x + a_n$$
$$= (x - x_i)(x^{n-1} + p_1 x^{n-2} + p_2 x^{n-3} + \cdots + p_{n-2} x + p_{n-1}) + p_n.$$

Carrying out the indicated multiplication and equating coefficients of like powers, we obtain

$$
\begin{aligned}
a_1 &= p_1 - x_i & p_1 &= a_1 + x_i \\
a_2 &= p_2 - p_1 x_i & p_2 &= a_2 + p_1 x_i \\
a_3 &= p_3 - p_2 x_i & p_3 &= a_3 + p_2 x_i \\
&\;\;\vdots & &\;\;\vdots \\
a_k &= p_k - p_{k-1} x_i & p_k &= a_k + p_{k-1} x_i \\
&\;\;\vdots & &\;\;\vdots \\
a_{n-1} &= p_{n-1} - p_{n-2} x_i & p_{n-1} &= a_{n-1} + p_{n-2} x_i \\
a_n &= p_n - p_{n-1} x_i & p_n &= a_n + p_{n-1} x_i.
\end{aligned}
$$

The formulas for computing the coefficients p_k of the quotient polynomial $P_{n-1}(x)$ and the remainder p_n can be grouped into the single recursion formula

(4.1.10)
$$
\begin{aligned}
p_1 &= a_1 + x_i & &\text{(starter)} \\
p_k &= a_k + p_{k-1} x_i \quad (k = 2, n) & &\text{(repeater)}
\end{aligned}
$$

so that $P_n(x_i) \equiv p_n$ can be computed simply by calculating p_k $(k = 1, n)$ by these recursion formulas.

Similarly, recursion formulas for computing $P_n'(x_i)$ can be derived by writing (4.1.7) in expanded form as

$$x^{n-1} + p_1 x^{n-2} + p_2 x^{n-3} + \cdots + p_{n-2} x + p_{n-1}$$
$$= (x - x_i)(x^{n-2} + \bar{p}_2 x^{n-3} + \bar{p}_3 x^{n-4} + \cdots + \bar{p}_{n-2} x + \bar{p}_{n-1}) + \bar{p}_n.$$

Carrying out the indicated multiplication and equating coefficients of like powers of x, we obtain

$$p_1 = \bar{p}_2 - x_i \qquad\qquad \bar{p}_2 = p_1 + x_i = p_1 + \bar{p}_1 x_i$$
$$p_2 = \bar{p}_3 - \bar{p}_2 x_i \qquad\qquad \bar{p}_3 = p_2 + \bar{p}_2 x_i$$
$$\vdots \qquad\qquad\qquad\qquad \vdots$$
$$p_{k-1} = \bar{p}_k - \bar{p}_{k-1} x_i \qquad\qquad \bar{p}_k = p_{k-1} + \bar{p}_{k-1} x_i$$
$$\vdots \qquad\qquad\qquad\qquad \vdots$$
$$p_{n-2} = \bar{p}_{n-1} - \bar{p}_{n-2} x_i \qquad\qquad \bar{p}_{n-1} = p_{n-2} + \bar{p}_{n-2} x_i$$
$$\vdots \qquad\qquad\qquad\qquad \vdots$$
$$p_{n-1} = \bar{p}_n - \bar{p}_{n-1} x_i \qquad\qquad \bar{p}_n = p_{n-1} + \bar{p}_{n-1} x_i$$

We define $\bar{p}_1 = 1$ to introduce symmetry in the formulas.

These formulas for computing the coefficients $\bar{p}_k$ of the quotient polynomial $P_{n-2}(x)$ and the remainder $\bar{p}_n$ can be grouped into the single recursion formula

(4.1.11)
$$\bar{p}_1 = 1 \qquad\qquad\qquad \text{(starter)}$$
$$\bar{p}_k = p_{k-1} + \bar{p}_{k-1} x_i \quad (k = 2, n) \quad \text{(repeater)}$$

where $\bar{p}_1 = 1$ has been introduced to provide symmetry in the recursion formulas.

Calculating a Zero of $P_n(x)$ by Newton-Raphson Iteration

Given an estimate x_i of a single root α of the polynomial equation $P_n(x) = 0$, we compute $P_n(x_i)$ and $P_n'(x_i)$ by simply evaluating the recursive formulas (4.1.10) and (4.1.11), respectively. An improved value x_{i+1} of the approximation of root α can then be calculated by the Newton-Raphson iterative formula

(4.1.12)
$$x_{i+1} = x_i - \frac{P_n(x_i)}{P_n'(x_i)}.$$

Let ε denote the convergence term determined by the degree of accuracy required in the computation of the roots of the polynomial equation. If $|x_{i+1} - x_i| > \varepsilon$, replace x_i by improved root estimate x_{i+1}, i.e., increment iteration index i by 1. Under certain conditions† a convergent sequence $\{x_{i+1}\}$ of successive approximations of solution α is generated by repeated application of (4.1.12). If successive differences $|x_{i+1} - x_i|$ are decreasing in magnitude, the Newton-Raphson iteration is continued, until $|x_{i+1} - x_i| \leq \varepsilon$, for some iteration i. When this convergence condition is achieved, we have calculated a root $\alpha = x_{i+1}$ of equation (4.1.1). If sequence $\{x_{i+1}\}$ diverges or fails to converge, a new initial approximation of the root must be selected.

† Sequence $\{x_{i+1}\}$ is convergent if $|F'(x_i)| < 1$ for $x_1, x_2, x_3, \ldots, \alpha$ where $x_{i+1} = x_i - P_n(x_i)/P_n'(x_i)$, and $F(x_i) = x_{i+1}$.

Once we have calculated a root α, where α is one of the roots α_j of $P_n(x) = 0$, we can extract its corresponding linear factor $(x - \alpha)$ from $P_n(x)$, obtaining polynomial $P_{n-1}(x)$ whose coefficients are simply the p_k calculated by recursion formula (4.1.10) for $x_i = \alpha$, since division of $P_n(x)$ by *factor* $(x - \alpha)$ gives a zero remainder, i.e.,

$$P_n(x) = (x - \alpha)P_{n-1}(x) + 0.$$

The process of calculating the next root is accomplished by replacing the coefficients a_k of $P_n(x)$ by the coefficients p_k of $P_{n-1}(x)$ and replacing n by $n - 1$, and repeating the same procedure.

Computation of Roots by Birge-Vieta Method

Calculate root of

$$P_n(x) = x^n + a_1 x^{n-1} + a_2 x^{n-2} + \cdots + a_{n-1}x + a_n$$

by Birge-Vieta method, using estimate $x_i = -a_n/a_{n-1}$.

For computations, write equations (4.1.10) and (4.1.11), in *synthetic-division form* as follows:

$x_i/$	1	a_1	a_2	a_3	$\ldots$	a_{n-2}	a_{n-1}	a_n
+		x_i	$x_i p_1$	$x_i p_2$	$\ldots$	$x_i p_{n-3}$	$x_i p_{n-2}$	$x_i p_{n-1}$
=	1	p_1	p_2	p_3	$\ldots$	p_{n-2}	p_{n-1}	$p_n \equiv P_n(x_i)$
+		$x_i \bar{p}_1$	$x_i \bar{p}_2$	$x_i \bar{p}_3$	$\ldots$	$x_i \bar{p}_{n-2}$	$x_i \bar{p}_{n-1}$	
=	$\bar{p}_1$	$\bar{p}_2$	$\bar{p}_3$	$\bar{p}_4$	$\ldots$	$\bar{p}_{n-1}$	$\bar{p}_n \equiv P_n'(x_i)$	

Then compute improved estimate x_{i+1} of root by Newson-Raphson iterative formula

$$x_{i+1} = x_i - \frac{P_n(x_i)}{P_n'(x_i)}.$$

NUMERICAL EXAMPLE. Calculate root of $P_3(x) \equiv x^3 - 11x^2 + 32x - 22 = 0$ using $x_0 = -(-22)/32 = 0.6875$.

Solution

$x_0 = 0.6875 /$	1.0	-11.0	32.0	-22.0
+		0.6875	-7.0898	17.1258
=	1.0	-10.3125	24.9102	$-4.8742 = P_3(0.6875)$
+		0.6875	-6.6172	
=	1.0	-9.6250	$18.2930 = P_3'(0.6875)$	

$$x_1 = x_0 - \frac{P_3(x_0)}{P_3'(x_0)} = 0.6875 - \frac{-4.8742}{18.2930} = 0.9540$$

$x_1 = 0.954$ /

$$
\begin{array}{rrrr}
1.0 & -11.0 & 32.0 & -22.0 \\
+ & 0.954 & -9.5839 & 21.3850 \\
\hline
= 1.0 & -10.046 & 22.4161 & -0.6150 = P_3(0.954) \\
+ & 0.954 & -8.6738 \\
\hline
= 1.0 & -9.092 & 13.7423 = P_3{}'(0.954)
\end{array}
$$

$$
x_2 = x_1 - \frac{P_3(x_1)}{P_3{}'(x_1)} = 0.954 - \frac{-0.6150}{13.7423} = 0.9988
$$

$x_2 = 0.9988$ /

$$
\begin{array}{rrrr}
1.0 & -11.0 & 32.0 & -22.0 \\
+ & 0.9988 & -9.9892 & 21.9844 \\
\hline
= 1.0 & -10.0012 & 22.0108 & -0.0156 = P_3(0.9988) \\
+ & 0.9988 & -8.9916 \\
\hline
= 1.0 & -9.0024 & 13.0192 = P_3{}'(0.9988)
\end{array}
$$

$$
x_3 = x_2 - \frac{P_3(x_2)}{P_3{}'(x_2)} = 0.9988 - \frac{-0.0156}{13.0192} = 1.0000
$$

Thus $x_3 = 1.0000$. It can be shown that $x_3 = 1.0$ is a root of the given polynomial, i.e., $P_3(1.0) = 0.0$. Dividing $P_3(x)$ by $(x - 1)$, the quadratic $x^2 - 10x + 22$ is found. The roots of this are then computed by the quadratic formula; these roots are $5 + \sqrt{3}$, $5 - \sqrt{3}$.

Computational Summary for Birge-Vieta Method

Step 0. Input and Initialization. Read parameters N (degree), M (maximum number of iterations), ε (convergence term), x_0 (initial estimate of root), h (if $|P_n'(x_0)| \le \varepsilon$, offset x_0 by increment h), and coefficients a_k ($k = 1, N$) of $P_N(x)$. Set root counter $j = 1$.

Step 1. Calculate degree n of current polynomial $P_n(x)$, where $n = N + 1 - j$. Set initial estimate x_0 of root α_j. Reset Newton-Raphson iteration counter $m = 1$.

Step 2. (a) Calculate nested terms $p_k(x_0)$†

$$p_1(x_0) = x_0 + a_1$$
$$p_k(x_0) = p_{k-1}(x_0)x_0 + a_k \qquad (k = 2, n).$$

(b) Calculate derivatives $\bar{p}_k(x_0)$

$$\bar{p}_1(x_0) = 1$$
$$\bar{p}_k(x_0) = x_0\bar{p}_{k-1}(x_0) + p_{k-1}(x_0) \qquad (k = 2, n).$$

Step 3. ‡Calculate improved estimate $\bar{x}$ of root α_j by Newton-Raphson.

$$\bar{x} = x_0 - \frac{P_n(x_0)}{P_n'(x_0)}, \qquad \text{where } P_n(x_0) = p_n(x_0)$$
$$P_n'(x_0) = \bar{p}_n(x_0).$$

Test convergence of root α_j.

If $|\bar{x} - x_0| > \varepsilon$, test iteration counter m. If $m \le M$, set $m = m + 1$, set $x_0 = \bar{x}$, return to Step 2. If $m > M$, go to "failure to converge" exit.
If $|\bar{x} - x_0| \le \varepsilon$, root $\alpha_j = \bar{x}$, go to Step 4.

Step 4. Replace $P_n(x)$ by

$$P_{n-1}(x) = x^{n-1} + p_1(\alpha_j)x^{n-2} + \cdots + p_{n-2}(\alpha_j)x + p_{n-1}(\alpha_j).$$

That is, replace a_k by $p_k(\alpha_j)$ ($k = 1, n - 1$).

Step 5. Have first $N - 1$ roots been computed?
If $j < N - 1$, set $j = j + 1$, and return to Step 1.
If $j = N - 1$, set $j = j + 1$, and go to Step 6.

Step 6. Calculate Nth (last) root of original equation by solving linear equation $x + a_1 = 0$, i.e., $\alpha_N = -a_1$.

Step 7. Output. Write out roots α_j ($j = 1, N$) of $P_N(x) = 0$.

---

† The terms p_k and $\bar{p}_k$ of (4.1.10) and (4.1.11) are written here as $p_k(x_0)$ and $\bar{p}_k(x_0)$, respectively, since these terms are functions of x_0.

‡ Should also test: If $|P_n'(x_0)| \le \varepsilon$

```
      PROGRAM BVIETA
C     BIRGE-VIETA METHOD FOR CALCULATING ROOTS OF POLYNOMIAL EQS.
      DIMENSION A(20), ROOT(20), P(20), PP(20)
C     INPUT AND INITIALIZATION
      READ (5,100) EPS, XZERO, H, NC, MC
      READ (5,101) (A(K), K = 1,NC)
      NCM1 = NC - 1
      J = 1
C     CALCULATE DEGREE N OF CURRENT POLYNOMIAL
    1 N = NC + 1 - J
      NM1 = N - 1
      M = 1
      XO = XZERO
C     ENTRY POINT FOR REFINED ESTIMATE OF ROOT
    2 P(1) = XO + A(1)
      PP(1) = 1.0
      DO 10 K = 2,N
      P(K) = XO * P(K-1) + A(K)
   10 PP(K) = XO * PP(K-1) + P(K-1)
C     CALCULATE NEXT APPROXIMATION OF ROOT J BY NEWTON-RAPHSON
      IF (ABS(PP(N)) - EPS) 15,15,18
   15 XO = XO + H
      GO TO 2
   18 XBR = XO - P(N)/PP(N)
   19 IF(ABS (XBR - XO) - EPS) 20, 20, 50
   20 ROOT (J) = XBR
C     REPLACE DEGREE N POLYNOMIAL BY DEGREE N - 1 POLYNOMIAL
      DO 30  K = 1,NM1
   30 A(K) = P(K)
      IF (J - NCM1) 35,40,40
   35 J = J+1
      GO TO 1
C     CALCULATE LAST ROOT
   40 ROOT(NC) = -A(1)
C     WRITE OUTPUT
      WRITE(6,110) (ROOT(J), J=1,NC)
   79 STOP
C     TEST NEWTON-RAPHSON ITERATION COUNTER M
   50 IF (M-MC) 55, 55, 60
   55 M = M+1
      XO = XBR
      GO TO 2
   60 PRINT 113
      WRITE (6,110) (ROOT(K), K=1,J)
      GO TO 79
  100 FORMAT (F10.8, 2F10.5, 2I2)
  101 FORMAT (F19.9)
  110 FORMAT (2X, F19.9)
  113 FORMAT (21H0 FAILURE TO CONVERGE)
      END
```

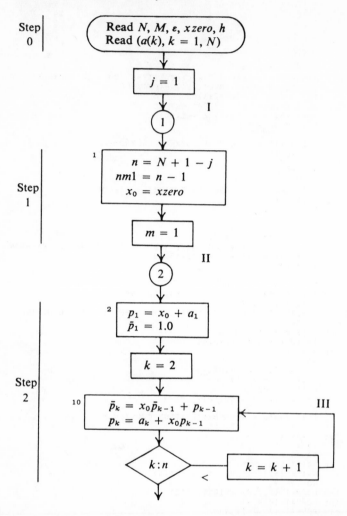

Figure 4-1. Micro flow chart for Birge-Vieta method. Assumptions: no complex roots; no close roots; no multiple roots. Loops: I, successive roots; II, iteration loop for a root; III, calculate p_k, $\bar{p}_k$ $(k = 2, n)$ for x_0; IV, replace $P_n(x)$ by $P_{n-1}(x)$.

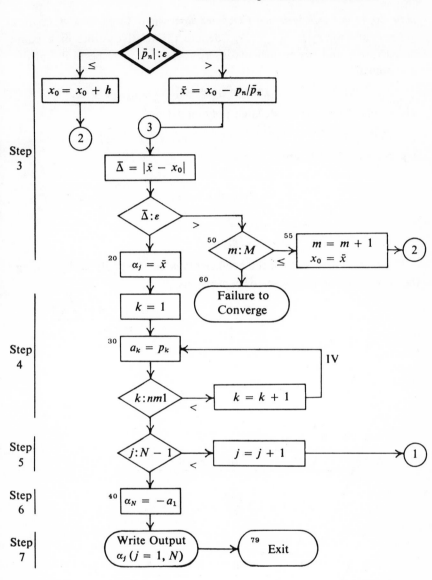

How Errors in Coefficients and Computed Root Affect Coefficients of $P_{n-1}(x)$

Let r_γ and $r_{a_1}, r_{a_2}, \ldots, r_{a_{n-1}}, r_{a_n}$ denote the relative errors in a computed root γ and the coefficients $a_1, a_2, \ldots, a_{n-1}, a_n$, respectively, of the polynomial

$$P_n(x) = x^n + a_1 x^{n-1} + a_2 x^{n-2} + \cdots + a_{n-1}x + a_n.$$

The coefficients p_k of the deflated polynomial,

$$P_{n-1}(x) = x^{n-1} + p_1 x^{n-2} + p_2 x^{n-3} + \cdots + p_{n-2}x + p_{n-1}$$

expressed in (4.1.10), are

$$p_1 = \gamma + a_1$$
$$p_2 = (p_1 \cdot \gamma) + a_2$$
$$p_3 = (p_2 \cdot \gamma) + a_3$$
$$\vdots$$
$$p_{n-1} = (p_{n-2} \cdot \gamma) + a_{n-1}.$$

The relative errors r_{p_k} in the coefficients p_k can be analyzed by writing the corresponding relative-error expressions

$$r_{p_1} = r_\gamma \frac{\gamma}{\gamma + a_1} + r_{a_1} \frac{a_1}{\gamma + a_1} + \alpha_1 = \frac{r_\gamma \gamma + r_{a_1}a_1 + \alpha_1(\gamma + a_1)}{\gamma + a_1}$$

$$ N \quad\; N \quad\;\; 0 \qquad N \quad\;\; 0 \qquad\quad R$$

$$r_{p_2} = (r_{p_1}\cdot 1 + r_\gamma \cdot 1 + \mu_2) \frac{p_1 \gamma}{p_2} + r_{a2} \frac{a_2}{p_2} + \alpha_2$$

$$ N \;\; 0 \;\; N \;\; 0 \qquad R$$
$$ N \qquad\qquad 0 \quad\;\; N \;\; 0 \qquad R$$

$$r_{p_3} = (r_{p_2}\cdot 1 + r_\gamma \cdot 1 + \mu_3) \frac{p_2 \gamma}{p_3} + r_{a3} \frac{a_3}{p_3} + \alpha_3$$

$$ N \;\; 0 \;\; N \;\; 0 \quad R$$
$$ N \qquad\qquad 0 \quad\;\; N \;\; 0 \qquad R$$

$$\vdots \quad \vdots$$

$$r_{p_{n-1}} = (r_{p_{n-2}}\cdot 1 + r_\gamma \cdot 1 + \mu_{n-1}) \frac{p_{n-2}\gamma}{p_{n-1}} + r_{a_{n-1}} \frac{a_{n-1}}{p_{n-1}} + \alpha_{n-1}$$

$$ N \;\; 0 \;\; N \;\; 0 \qquad R$$
$$ N \qquad\qquad 0 \qquad\; N \quad\;\; 0 \qquad R.$$

Multiplying the kth equation through by p_k, we obtain the error expressions (where ε_k denotes the error in p_k)

$$\varepsilon_{p_1} = r_\gamma\gamma + r_{a_1}a_1 + \alpha_1 p_1$$
$$\varepsilon_{p_2} = (r_{p_1} + r_\gamma + \mu_2)p_1\gamma + r_{a_2}a_2 + \alpha_2 p_2$$
$$\varepsilon_{p_3} = (r_{p_2} + r_\gamma + \mu_3)p_2\gamma + r_{a_3}a_3 + \alpha_3 p_3$$
$$\vdots$$

Note. The μ_k and α_k denote the relative roundoff in the multiplication and addition, respectively, in the kth expression.

$$\varepsilon_{p_{n-1}} = (r_{p_{n-2}}1 + r_\gamma 1 + \mu_{n-1})p_{n-2}\gamma + r_{a_{n-1}}a_{n-1} + \alpha_{n-1}p_{n-1}$$

The bounds for the errors in the coefficients p_k are

$$|\varepsilon_{p_1}| \le |r_\gamma\gamma| + |r_{a_1}a_1| + |\alpha_1 p_1|$$
$$|\varepsilon_{p_2}| \le |r_{p_1}p_1\gamma| + |r_\gamma p_1\gamma| + |\mu_2 p_1\gamma| + |r_{a_2}a_2| + |\alpha_2 p_2|$$
$$|\varepsilon_{p_3}| \le |r_{p_2}p_2\gamma| + |r_\gamma p_2\gamma| + |\mu_3 p_2\gamma| + |r_{a_3}a_3| + |\alpha_3 p_3|$$
$$\vdots$$

Clearly, then, the bounds for the errors ε_{p_k} in the coefficients p_k of the deflated polynomial $P_{n-1}(x)$ decrease as the magnitude of γ decreases. Hence, it is advantageous to extract the roots of smallest magnitude first in a factor-deflation process such as the Birge-Vieta method.

4.2   The Lin-Bairstow Iterative Method

The Lin-Bairstow method is an iterative procedure for calculating the roots (real or complex) of a real-coefficient polynomial equation while requiring only the manipulation of real numbers in the computations. The method is based on the successive extraction of quadratic factors $F_m(x)$ ($m = 1, 2, \ldots$) from the original polynomial of degree N and from succeeding factor polynomials of degree $N - 2m$. Each quadratic factor is determined by an iterative differential-correction procedure.

The factor theorem states that the Nth-degree polynomial

$$(4.2.1) \qquad P_N(x) = x^N + a_1 x^{N-1} + a_2 x^{N-2} + \cdots + a_{N-1}x + a_N$$

can be expressed as the product of N linear factors, in the form

$$(4.2.2) \qquad\qquad P_N(x) = (x - \alpha_1)(x - \alpha_2)\ldots(x - \alpha_N)$$

where the α_j ($j = 1, 2, \ldots, N$), not necessarily distinct, are the roots of the polynomial equation

$$(4.2.3) \qquad\qquad\qquad P_N(x) = 0.$$

Obviously, then, a polynomial of even degree $N = 2k$, where k is a positive integer, can be expressed as a product of k quadratic factors, as can be seen from equation (4.2.2) by grouping the linear factors pairwise, and considering the product of each pair as a quadratic factor of the polynomial. Since complex roots of a real-coefficient polynomial equation occur in conjugate pairs, α and $\bar{\alpha}$, the complex conjugates should be paired so that all quadratic products will have only real coefficients.

Therefore we can express the polynomial $P_N(x)$, where $N = 2k$, in the form

$$(4.2.4) \qquad P_N(x) = \prod_{m=1}^{k} F_m(x)$$

where

$$(4.2.5) \qquad F_m(x) = x^2 + r_m x + s_m \qquad (m = 1, 2, \ldots, k).$$

Similarly, a polynomial of odd degree $N = 2k + 1$ can be expressed as a product of k quadratic factors and a single linear factor, as follows:

$$(4.2.6) \qquad P_N(x) = (x - \alpha_{2k+1}) \prod_{m=1}^{k} F_m(x)$$

where $F_m(x)$ is a quadratic factor as defined in (4.2.5), and α_{2k+1} is the real root corresponding to the single linear factor of the polynomial.

The roots of a polynomial of arbitrary degree N can therefore be calculated by determining successively the quadratic factors of that polynomial, and further (in the case of an odd-degree polynomial) by finally determining the single linear factor. The Lin-Bairstow method is an efficient algorithm for calculating the zeros of a polynomial by successive extraction of its quadratic factors.

EXAMPLES. N even (polynomial with all complex zeros):

$$P_4(x) = [x - (a_1 + b_1 i)][x - (a_1 - b_1 i)][x - (a_2 + b_2 i)][x - (a_2 - b_2 i)]$$
$$= (x^2 - 2a_1 x + a_1{}^2 + b_1{}^2)(x^2 - 2a_2 x + a_2{}^2 + b_2{}^2).$$

N even (polynomial with mixed real and complex zeros):

$$P_4(x) = [x - \alpha_1][x - \alpha_2][x - (a + bi)][x - (a - bi)]$$
$$= [x^2 - (\alpha_1 + \alpha_2)x + \alpha_1 \alpha_2][x^2 - 2ax + a^2 + b^2].$$

N odd (polynomial with mixed real and complex zeros):

$$P_5(x) = [x - (a + bi)][x - (a - bi)](x - \alpha_1)(x - \alpha_2)(x - \alpha_3)$$
$$= [x^2 - 2ax + a^2 + b^2][x^2 - (\alpha_1 + \alpha_2)x + \alpha_1 \alpha_2](x - \alpha_3).$$

N odd (polynomial with all real zeros):

$$P_5(x) = (x - \alpha_1)(x - \alpha_2)(x - \alpha_3)(x - \alpha_4)(x - \alpha_5)$$
$$= [x^2 - (\alpha_1 + \alpha_2)x + \alpha_1 \alpha_2][x^2 - (\alpha_3 + \alpha_4)x + \alpha_3 \alpha_4](x - \alpha_5).$$

Determining a Quadratic Factor of a Polynomial

Let n denote the degree of the polynomial to be factored in a given stage of the Lin-Bairstow method. That is, $n = N - 2k$ ($k = 0, 1, \ldots\ldots$), where k quadratic factors have been already extracted from original polynomial $P_N(x)$.

If $P_n(x)$ is divided by a trial quadratic factor $F(x) = x^2 + rx + s$, where r and s are arbitrary real constants, we obtain

(4.2.7) $P_n(x) = F(x)P_{n-2}(x) + Rx + S.$

In expanded form, this equation can be written as

$$x^n + a_1 x^{n-1} + a_2 x^{n-2} + \cdots + a_{n-1}x + a_n$$
$$= (x^2 + rx + s)(x^{n-2} + b_1 x^{n-3} + b_2 x^{n-4} + \cdots + b_{n-3}x + b_{n-2})$$
(4.2.8) $+ Rx + S.$

Note that changes in r or s will cause changes in the coefficients b_k of the quotient polynomial $P_{n-2}(x)$ and in the coefficients R and S of the remainder term. Hence, let us consider r and s as independent "variables," and express the b_k, R, and S as functions of these "variables." We denote these functions by $b_k(r, s)$, $R(r, s)$, and $S(r, s)$, respectively.

The requirement that $F(x)$ be a factor of $P_n(x)$ imposes the constraints that

(4.2.9) $$R(r, s) = 0$$
$$S(r, s) = 0.$$

The problem of extracting a quadratic factor from a polynomial can therefore be solved by calculating the roots of equations (4.2.9).

Calculating Roots of $R(r, s) = 0 = S(r, s)$ (See Newton's Method, Sec. 3.3)

Suppose that initial estimates r_0, s_0 of the roots of system of equations (4.2.9) are known. If these initial values are incremented respectively by small changes δr and δs, then first-order approximations of the resulting changes† in the functions $R(r, s)$ and $S(r, s)$, respectively, are given by the total differentials

(4.2.10) $$\delta R = R_r\, \delta r + R_s\, \delta s$$
$$\delta S = S_r\, \delta r + S_s\, \delta s$$

where R_r, R_s, S_r, S_s denote the partial derivatives of R and S with respect to r and s, at the current values of r and s.

---

† The exact changes $\overline{\delta R}$ and $\overline{\delta S}$ are

$$\overline{\delta R} = R(r + \delta r, s + \delta s) - R(r, s) = R_r \delta r + R_s \delta s + \cdots.$$
$$\overline{\delta S} = S(r + \delta r, s + \delta s) - S(r, s) = S_r \delta r + S_s \delta s + \cdots.$$

Given root estimates r_0, s_0, we can calculate $R(r_0, s_0)$ and $S(r_0, s_0)$ by dividing $P_n(x)$ by $x^2 + r_0 x + s_0$, obtaining the coefficients R and S of the remainder term, as

(4.2.11)
$$R = R(r_0, s_0) \neq 0$$
$$S = S(r_0, s_0) \neq 0.$$

We must therefore determine δr and δs such that the total differentials δR and δS satisfy the constraints

(4.2.12)
$$R(r_0, s_0) + \delta R = 0$$
$$S(r_0, s_0) + \delta S = 0.$$

That is, we impose the constraints that

(4.2.13)
$$\delta R = -R(r_0, s_0)$$
$$\delta S = -S(r_0, s_0).$$

Equations (4.2.10) and (4.2.13) together give us the following relations:

(4.2.14)
$$R_r \, \delta r + R_s \, \delta s = -R(r_0, s_0)$$
$$S_r \, \delta r + S_s \, \delta s = -S(r_0, s_0).$$

Equations (4.2.14) are called the differential-correction equations. The solutions δr and δs of these equations are referred to as the differential corrections.

Having calculated the differential corrections δr, δs, we can evaluate the first-order Taylor series expansions

(4.2.15)
$$R(r_0 + \delta r, s_0 + \delta s) \doteq R(r_0, s_0) + R_r \, \delta r + R_s \, \delta s$$
$$S(r_0 + \delta r, s_0 + \delta s) \doteq S(r_0, s_0) + S_r \, \delta r + S_s \, \delta s$$

By (4.2.14) we see that the right sides of (4.2.15) are zero. Hence,

$$R(r_0 + \delta r, s_0 + \delta s) \doteq 0$$
$$S(r_0 + \delta r, s_0 + \delta s) \doteq 0.$$

That is, $r_1 = r_0 + \delta r$ and $s_1 = s_0 + \delta s$ are first-order approximations of the zeros of functions $R(r, s)$ and $S(r, s)$.

Using solution approximations r_1, s_1 we can calculate refined solution approximations r_2, s_2 by the same procedure defined above for using r_0, s_0 to calculate r_1, s_1. Similarly, r_3, s_3 can be calculated from r_2, s_2. In this manner, a convergent† sequence $(r_0, s_0), (r_1, s_1), (r_2, s_2), (r_3, s_3), \ldots$ of successive solution approximations for equations (4.2.9) can be generated. If, for some iteration k, the conditions

(4.2.16)
$$|r_{k+1} - r_k| \leq \varepsilon$$
$$|s_{k+1} - s_k| \leq \varepsilon$$

† Provided (r_0, s_0) are sufficiently close to true zeros (r^+, s^+) which simultaneously satisfy $R(r^+, s^+) = 0$, $S(r^+, s^+) = 0$.

are simultaneously satisfied for some preassigned convergence criterion $\varepsilon > 0$, then the differential-correction routine is said to have converged, and we have

(4.2.17)
$$R(r_{k+1}, s_{k+1}) = 0$$
$$S(r_{k+1}, s_{k+1}) = 0.$$

Denoting r_{k+1} and s_{k+1} by r^+ and s^+, respectively, we see that the quadratic

$$x^2 + r^+x + s^+$$

is a factor of polynomial $P_n(x)$, satisfying

(4.2.18) $P_n(x) = (x^2 + r^+x + s^+)(x^{n-2} + b_1{}^+x^{n-3} + b_2{}^+x^{n-4}$
$$+ \cdots + b_{n-3}^+x + b_{n-2}^+)$$
$$= F(x)P_{n-2}(x)$$

and $P_{n-2}(x)$ is the polynomial to be factored at the next stage.

Calculation of Coefficients R and S of the Remainder

As noted earlier, division of $P_n(x)$ by $F(x) = x^2 + rx + s$ gives a quotient $P_{n-2}(x)$ and a remainder $Rx + S$. This section is devoted to the derivation of *recursive formulas* for calculating R and S for use in the differential-correction equations.

Dividing $P_n(x)$ by $F(x)$, we obtain

$$x^n + a_1 x^{n-1} + a_2 x^{n-2} + \cdots + a_{n-1}x + a_n$$
$$= (x^2 + rx + s)(x^{n-2} + b_1 x^{n-3} + b_2 x^{n-4} + \cdots + b_{n-3}x + b_{n-2})$$
(4.2.19)
$$+ Rx + S.$$

Carrying out the indicated multiplications and equating coefficients of like powers of x, we find

(4.2.20)
$$a_1 = b_1 + r$$
$$a_2 = b_2 + rb_1 + s$$
$$a_3 = b_3 + rb_2 + sb_1$$
$$\vdots$$
$$a_k = b_k + rb_{k-1} + sb_{k-2}$$
$$\vdots$$
$$a_{n-1} = R + rb_{n-2} + sb_{n-3}$$
$$a_n = S + sb_{n-2}.$$

If we now introduce the recursion formula

(4.2.21) $b_k = a_k - rb_{k-1} - sb_{k-2}$ $(k = 3, 4, \ldots, n)$

where

$$b_1 = a_1 - r$$
$$b_2 = a_2 - rb_1 - s$$

are used to start the recursion, we obtain

(4.2.22)

$$b_1 = a_1 - r$$
$$b_2 = a_2 - rb_1 - s$$
$$b_3 = a_3 - rb_2 - sb_1$$
$$\vdots$$
$$b_k = a_k - rb_{k-1} - sb_{k-2}$$
$$\vdots$$
$$b_{n-1} = a_{n-1} - rb_{n-2} - sb_{n-3}$$
$$b_n = a_n - rb_{n-1} - sb_{n-2}.$$

Comparing the last two equations of (4.2.20) with the corresponding equations of (4.2.22), we see that

(4.2.23)

$$R = b_{n-1}$$
$$S = b_n + rb_{n-1}.$$

Note that R and S are the functions $R(r, s)$ and $S(r, s)$ required in the differential-correction equations (4.2.14). The remaining four terms R_r, R_s, S_r, S_s required in the differential-correction equations are obtained by taking the partial derivatives of R and S in (4.2.23), with respect to r and s. Thus we find

(4.2.24)

$$R_r = \frac{\partial b_{n-1}}{\partial r}$$
$$R_s = \frac{\partial b_{n-1}}{\partial s}$$
$$S_r = \frac{\partial b_n}{\partial r} + r \frac{\partial b_{n-1}}{\partial r} + b_{n-1}$$
$$S_s = \frac{\partial b_n}{\partial s} + r \frac{\partial b_{n-1}}{\partial s}.$$

Recursion formulas for these partial derivatives are derived in the following section.

Calculation of R_r, R_s, S_r, S_s

To simplify the notation in the derivation of the recursion formulas for the partial derivatives, we define

(4.2.25)

$$p_k = \frac{\partial b_k}{\partial r}$$
$$q_k = \frac{\partial b_k}{\partial s}$$

Differentiating the b_k $(k = 1, 2, \ldots, n)$ of (4.2.22) with respect to r and s, respectively, we find

$$
\begin{aligned}
p_1 &= -1 & q_1 &= 0 \\
p_2 &= r - b_1 & q_2 &= -1 \\
p_3 &= -b_2 - rp_2 - sp_1 & q_3 &= -b_1 - rq_2 - sq_1 \\
&\;\;\vdots & &\;\;\vdots \\
p_k &= -b_{k-1} - rp_{k-1} - sp_{k-2} & q_k &= -b_{k-2} - rq_{k-1} - sq_{k-2} \\
&\;\;\vdots & &\;\;\vdots \\
p_{n-1} &= -b_{n-2} - rp_{n-2} - sp_{n-3} & q_{n-1} &= -b_{n-3} - rq_{n-2} - sq_{n-3} \\
p_n &= -b_{n-1} - rp_{n-1} - sp_{n-2} & q_n &= -b_{n-2} - rq_{n-1} - sq_{n-2}.
\end{aligned}
$$

(4.2.26)

Substituting the simpler notation of (4.2.25) into (4.2.24) we obtain

(4.2.27)
$$
\begin{aligned}
R_r &= p_{n-1} \\
R_s &= q_{n-1} \\
S_r &= p_n + rp_{n-1} + b_{n-1} \\
S_s &= q_n + rq_{n-1}.
\end{aligned}
$$

We now have recursion formulas for calculating the six terms required in the solution of the differential-correction equations. That is, R and S can be calculated by (4.2.22) and (4.2.23), and R_r, R_s, S_r, S_s can be calculated by (4.2.26) and (4.2.27).

Notes. 1. The differential-correction technique described herein for calculating zeros of $R(r, s)$ and $S(r, s)$ simultaneously is simply Newton's method of solving two non-linear equations in two unknowns. An elegant proof of the convergence, for sufficiently close initial estimates, of Newton's method is given in Henrici [7, p. 106].

2. The coefficients of the deflated polynomials $P_{N-2}(x)$, $P_{N-4}(x)$, $P_{N-6}(x)$, ... are not exact because of the accumulation of roundoff, since remainder coefficients R, S at any stage of the deflation are only approximately zero. For this reason, the almost exact factor $x^2 + rx + s$, in stage m, should be used as starting values for extraction of a more exact factor from the original polynomial $P_N(x)$.

3. For most cases, the Bairstow method can be started using initial estimates $r_0{}^0 = 0$, $s_0{}^0 = 0$. The procedure may, however, have to "search" until it finds values of r and s which lead to convergence. Wilkinson [31, p. 67] indicates that it is essential that the quadratic factors having the smaller values of r and s be extracted first. This is done to minimize the bounds for the errors in succeeding deflated polynomials.

Example of Computation of Roots by Bairstow Method

The number of computations required in each iteration of the Bairstow method can be reduced by using the relation $q_{k+1} = p_k$ (see Exercise 4.2), and the differential-correction equations (4.2.14) can be simplified by this relation to the form

$$(4.2.28) \qquad \begin{bmatrix} p_{n-1} & p_{n-2} \\ p_n + b_{n-1} & p_{n-1} \end{bmatrix} \begin{bmatrix} \delta r \\ \delta s \end{bmatrix} = \begin{bmatrix} -b_{n-1} \\ -b_n \end{bmatrix}.$$

For manual computations, the terms in the differential-correction equations can be computed by writing equations (4.2.22) and (4.2.26) in a synthetic-division-by-quadratic form as follows:†

1		1	a_1	a_2	$a_3 \ldots$	a_{n-3}	a_{n-2}	a_{n-1}	a_n
$-r$			$-r$	$-rb_1$	$-rb_2 \ldots$	$-rb_{n-4}$	$-rb_{n-3}$	$-rb_{n-2}$	$-rb_{n-1}$
	$-s$			$-s$	$-sb_1 \ldots$	$-sb_{n-5}$	$-sb_{n-4}$	$-sb_{n-3}$	$-sb_{n-2}$
1		1	b_1	b_2	$b_3 \ldots$	b_{n-3}	b_{n-2}	$\boxed{b_{n-1}}$	$\boxed{b_n}$
$-r$			rp_1	rp_2	$rp_3 \ldots$	rp_{n-3}	rp_{n-2}	rp_{n-1}	
	$-s$			sp_1	$sp_2 \ldots$	sp_{n-4}	sp_{n-3}	sp_{n-2}	
		$-p_1$	$-p_2$	$-p_3$	$-p_4 \ldots$	$\boxed{-p_{n-2}}$	$\boxed{-p_{n-1}}$	$\boxed{-p_n}$	

NUMERICAL EXAMPLE. Calculate roots of $P_4(x) \equiv x^4 - 7x^3 + 13x^2 + 45x - 50 = 0$ using $(r, s) = (0.1, 0.1)$.

1			1	-7	13	45	-50
	-0.1			-0.1	0.71	-1.36	-4.44
		-0.1			-0.10	0.71	-1.36
1			1	-7.1	13.61	44.35	-55.80
-0.1				-0.1	0.72	-1.42	
		-0.1			-0.10	0.72	
			1	-7.2	14.23	44.65	

$$\begin{bmatrix} -14.23 & 7.2 \\ 0.70 & -14.23 \end{bmatrix} \begin{bmatrix} \delta r \\ \delta s \end{bmatrix} = \begin{bmatrix} -44.35 \\ 55.80 \end{bmatrix}$$

† The circled terms in the synthetic-division array are the terms required for the differential-correction equations.

Solution

$$\delta r = 1.16 \qquad r + \delta r = 1.26 \rightarrow r$$
$$\delta s = -3.86 \qquad s + \delta s = -3.76 \rightarrow s$$

1			1	−7	13	45	−50
	−1.26			−1.26	10.41	−72.03	73.19
		3.76			3.76	−31.06	214.96
1			1	−8.26	57.17	−58.09	238.15
	−1.26			−1.26	11.99	−91.88	
		3.76			3.76	−35.80	
			1	−9.52	72.92	−185.77	

$$\begin{bmatrix} -72.92 & 9.52 \\ 127.68 & -72.92 \end{bmatrix} \begin{bmatrix} \delta r \\ \delta s \end{bmatrix} = \begin{bmatrix} 58.09 \\ -238.15 \end{bmatrix}$$

Solution

$$\delta r = -0.48 \qquad r + \delta r = 0.78 \rightarrow r$$
$$\delta s = 2.43 \qquad s + \delta s = -1.33 \rightarrow s$$

1			1	−7	13	45	−50
	−0.78			−0.78	6.07	−15.91	−14.62
		1.33			1.33	−10.35	27.13
1			1	−7.78	20.40	18.74	−37.49
	−0.78			−0.78	6.68	−22.16	
		1.33			1.33	−11.38	
			1	−8.56	28.41	−14.80	

$$\begin{bmatrix} -28.41 & 8.56 \\ 33.54 & -28.41 \end{bmatrix} \begin{bmatrix} \delta r \\ \delta s \end{bmatrix} = \begin{bmatrix} -18.74 \\ 37.49 \end{bmatrix}$$

Solution

$$\delta r = 0.41 \qquad r + \delta r = 1.19 \rightarrow r$$
$$\delta s = -0.84 \qquad s + \delta s = -2.17 \rightarrow s$$

1			1	−7	13	45	−50
	−1.19			−1.19	9.75	−29.65	2.88
		2.17			2.17	−17.77	54.08
1			1	−8.19	24.92	−2.42	6.96
	−1.19			−1.19	11.16	−45.52	
		2.17			2.17	−20.35	
			1	−9.38	38.25	−68.29	

$$\begin{bmatrix} -38.25 & 9.38 \\ 65.87 & -38.25 \end{bmatrix} \begin{bmatrix} \delta r \\ \delta s \end{bmatrix} = \begin{bmatrix} 2.42 \\ -6.96 \end{bmatrix}$$

Solution

$$\delta r = -0.03 \qquad r + \delta r = \quad 1.16 \rightarrow r$$
$$\delta s = \quad 0.13 \qquad s + \delta s = -2.04 \rightarrow s$$

<pre>
1   1   -7   13   45   -50
   -1.16   -1.16   9.47   -28.43   0.09
          2.04 _   2.04   -16.65   50.00
         ───
                  1   -8.16   24.51   -0.08=b_{n-1} 0.09=b_n
</pre>

$\underbrace{\qquad\qquad\qquad\qquad}$

Coefficients of $R = b_{n-1}$ $S = b_n + rb_{n-1}$

quotient polynomial $= 0.09 + 1.16(-0.08)$

$\doteq 0.0$

    Since the remainder terms $R = -0.08$ and $S \doteq 0.0$ are essentially zero, the quadratic $x^2 + 1.16x - 2.04$ is essentially a factor of $P_4(x)$. The other quadratic factor is the quotient polynomial $x^2 - 8.16x + 24.51$. The roots of these quadratics can then be calculated by the quadratic formula; these roots are 0.963, -2.12, $4.079 \pm i \cdot 2.80$.

Computational Summary for Lin-Bairstow Method

Step 0. Input and Initialization. Read parameters: degree $= N$, initial values (r_0, s_0), convergence criteria $= \varepsilon$. Read coefficients a_k $(k = 1, N)$ of $P_N(x)$. Set index $m = 0$ ($m =$ number of quadratic factors extracted). Set index $j = 1$ ($j =$ root-pair counter).

Step 1. Calculate degree $n = N - 2m$ of current polynomial. Reset Newton iteration counter $L = 0$. Reset (r, s) to initial values (r_0, s_0).

Step 2. Test degree n. If $n > 2$, go to Step 3. If $n = 2$, go to Step 2*b*. If $n < 2$, go to Step 2*a*.

(a) Calculate root α_j of linear equation $x + a_1 = 0$; go to Step 9.

(b) Calculate root pair α_j, α_{j+1} of $x^2 + a_1 x + a_2 = 0$; go to Step 9.

Step 3. Divide $P_n(x)$ by $x^2 + rx + s$, and compute R, S.

$$b_1 = a_1 - r$$
$$b_2 = a_2 - rb_1 - s$$
$$b_k = a_k - rb_{k-1} - sb_{k-2} \qquad (k = 3, 4, \ldots, n).$$

Then

$$R = b_{n-1}, \qquad S = b_n + rb_{n-1}.$$

Step 4. Calculate partial derivatives R_r, R_s, S_r, S_s.

$$p_1 = -1 \qquad\qquad\qquad q_1 = 0$$
$$p_2 = -b_1 - r(-1) \qquad\quad q_2 = -1$$
$$p_k = -b_{k-1} - rp_{k-1} - sp_{k-2} \quad q_k = -b_{k-2} - rq_{k-1} - sq_{k-2} \quad (k = 3, n)$$

Then

$$R_r = p_{n-1} \qquad\qquad\qquad R_s = q_{n-1}$$
$$S_r = p_n + rp_{n-1} + b_{n-1} \qquad S_s = q_n + rq_{n-1}.$$

Step 5. Solve differential-correction equations for δr, δs.

$$R_r\, \delta r + R_s\, \delta s = -R$$
$$S_r\, \delta r + S_s\, \delta s = -S.$$

Step 6. Calculate improved values of roots of $R(r, s) = 0 = S(r, s)$.

$$r = r + \delta r$$
$$s = s + \delta s.$$

Step 7. Test for convergence of differential corrections.

(a) If both $|\delta r| \le \varepsilon$ and $|\delta s| \le \varepsilon$, calculate root pair α_j, α_{j+1} of quadratic $x^2 + rx + s = 0$. Go to Step 8.

(b) If either $|\delta r| > \varepsilon$ or $|\delta s| > \varepsilon$, test index L. If $L \le$ LMAX, increment L by 1, and go to Step 3. If $L >$ LMAX, go to "failure to converge" exit.

Step 8. Replace $P_n(x)$ by $P_{n-2}(x)$, i.e., replace a_k by b_k $(k = 1, n - 2)$. Increment quadratic factor counter m by 1. Increment root-pair counter j by 2. Return to Step 1.

Step 9. Write output. Write out coefficients a_k and roots α_j.

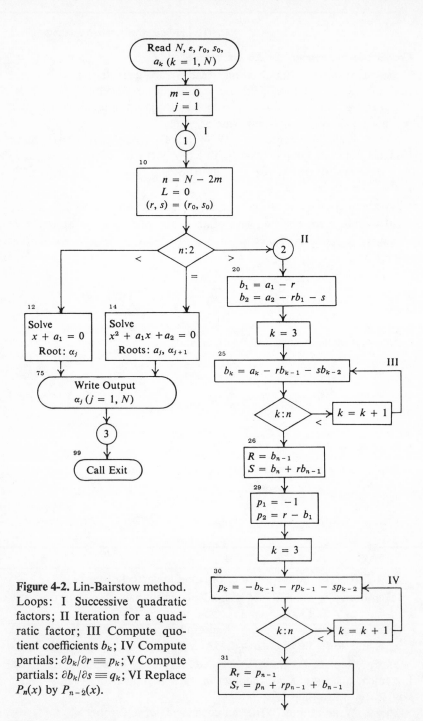

Figure 4-2. Lin-Bairstow method. Loops: I Successive quadratic factors; II Iteration for a quadratic factor; III Compute quotient coefficients b_k; IV Compute partials: $\partial b_k/\partial r \equiv p_k$; V Compute partials: $\partial b_k/\partial s \equiv q_k$; VI Replace $P_n(x)$ by $P_{n-2}(x)$.

126

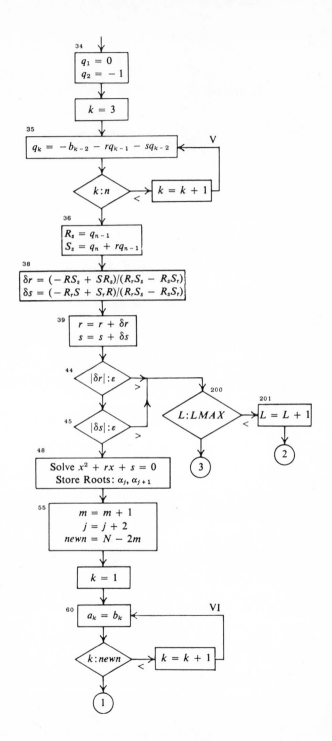

34

$$q_1 = 0$$
$$q_2 = -1$$

$$k = 3$$

35

$$q_k = -b_{k-2} - rq_{k-1} - sq_{k-2}$$ V

$$k:n$$ $$k = k + 1$$
$<$

36

$$R_s = q_{n-1}$$
$$S_s = q_n + rq_{n-1}$$

38

$$\delta r = (-RS_s + SR_s)/(R_rS_s - R_sS_r)$$
$$\delta s = (-R_rS + S_rR)/(R_rS_s - R_sS_r)$$

39

$$r = r + \delta r$$
$$s = s + \delta s$$

44

$$|\delta r| : \varepsilon$$ $>$

200

$$L : LMAX$$ $<$ 201 $$L = L + 1$$

45

$$|\delta s| : \varepsilon$$ $>$

3

2

48

Solve $x^2 + rx + s = 0$
Store Roots: α_j, α_{j+1}

55

$$m = m + 1$$
$$j = j + 2$$
$$newn = N - 2m$$

$$k = 1$$

60

$$a_k = b_k$$ VI

$$k:newn$$ $$k = k + 1$$
$<$

1

```
      PROGRAM BAIRSTOW
      DIMENSION A(20),B(20),P(20),Q(20),RTREAL(20),RTIMAG(20)
C     NC = DEGREE OF ORIGINAL POLYNOMIAL EQUATION
C     EPS = CONVERGENCE TERM
C     RZERO, SZERO ARE INITIAL VALUES OF R,S
      READ (5,100) NC, EPS,RZERO, SZERO
      READ (5,101) (A(K), K=1,NC)
      M = 0
      J = 1
C     ENTRY POINT FOR INITIAL VALUES RZERO, SZERO
   10 R = RZERO
      S = SZERO
      L = 0
      N = NC - 2 * M
      IF (N-2) 12, 14, 20
C     CALCULATE ROOT CORRESPONDING TO LINEAR FACTOR (IF ANY)
   12 RTREAL (J) = - A(1)
      GO TO 75
C     CALCULATE ROOT PAIR CORRESPONDING TO LAST QUADRATIC
   14 IF (A(1) * A(1) - 4.0 * A(2)) 16,17,18
   16 RADTRM = - (A(1) * A(1) - 4.0 * A(2))
      RAD = SQRT (RADTRM)
      RTREAL (J) = - A(1)/2.0
      RTREAL (J+1) = -A(1)/2.0
      RTIMAG(J) = RAD/2.0
      RTIMAG (J+1) = -RAD/2.0
      GO TO 75
   17 RTREAL (J) = -A(1)/2.0
      RTREAL (J+1) = -A(1)/2.0
      RTIMAG (J) = 0.0
      RTIMAG (J+1) = 0.0
      GO TO 75
   18 RAD = SQRT (A(1) * A(1) - 4.0 * A(2))
      RTREAL (J) = (-A(1) + RAD)/2.0
      RTREAL (J+1) = (-A(1) - RAD)/2.0
      RTIMAG (J) = 0.0
      RTIMAG (J+1) = 0.0
      GO TO 75
C     ENTRY POINT FOR REFINED VALUES OF APPROX R,S
   20 B(1) = A(1) - R
      B(2) = A(2) - R * B(1) - S
      DO 25 K = 3,N
   25 B(K) = A(K) - R * B(K-1) - S * B(K-2)
      RC = B(N-1)
      SC = B(N) + R * B(N-1)
C     CALCULATE PARTIAL DERIVATIVES OF B (K) W.R.T. R
      P(1) = -1.0
      P(2) = R - B(1)
      DO 30 K = 3,N
   30 P(K) = -B(K-1) - R * P(K-1) - S * P(K-2)
      RCR = P(N-1)
      SCR = P(N) + R * P(N-1) + B(N-1)
C     CALCULATE PARTIAL DERIVATIVES OF B(K) W.R.T. S
      Q(1) = 0.0
      Q(2) = -1.0
      DO 35 K = 3,N
```

```
 35 Q(K) = -B(K-2) - R * Q(K-1) - S * Q(K-2)
    RCS = Q(N-1)
    SCS = Q(N) + R * Q(N-1)
C   SOLVE DIFFERENTIAL-CORRECTION EQUATIONS
    DENOM = RCR * SCS - RCS * SCR
    RNUM = -RC * SCS + SC * RCS
    SNUM = -RCR * SC + SCR * RC
    DELR = RNUM/DENOM
    DELS = SNUM/DENOM
C   CALCULATE IMPROVED VALUES OF R,S
    R = R + DELR
    S = S + DELS
C   TEST FOR CONVERGENCE OF DIFFERENTIAL CORRECTIONS
    IF (ABS (DELR) - EPS) 45, 45, 200
 45 IF (ABS (DELS) - EPS) 48, 48, 200
C   TEST ITERATION COUNTER L
200 IF (L-100) 201, 201, 202
201 L = L +1
    GO TO 20
202 PRINT 113, N
    GO TO 99
C   CALCULATE ROOT PAIR CORRESPONDING TO QUADRATIC F(M)
 48 IF (R * R - 4,0 * S) 50,51,52
 50 RADTRM = - (R * R - 4.0 * S)
    RAD = SQRT(RADTRM)
    RTREAL(J) = -R/2.0
    RTREAL(J+1) = -R/2.0
    RTIMAG (J) = RAD/2.0
    RTIMAG (J+1) = -RAD/2.0
    GO TO 55
 51 RTREAL(J) = -R/2.0
    RTREAL (J+1) = -R/2.0
    RTIMAG (J) = 0.0
    RTIMAG (J+1) = 0.0
    GO TO 55
 52 RAD = SQRT (R * R - 4.0 * S)
    RTREAL (J) = (-R + RAD)/2.0
    RTREAL (J+1) = (-R - RAD)/2.0
    RTIMAG (J) = 0.0
    RTIMAG (J+1) = 0.0
    GO TO 55
C   REPLACEPOLYNOMIAL P(N) BY P(N-2) AND INCREMENT COUNTERS
 55 M = M+1
    J = J+2
    NEWN = NC - 2 + M
    DO 60 K = 1, NEWN
 60 A(K) = B(K)
    GO TO 10
 75 WRITE (6,110) (RTREAL (J), RTIMAG (J), J=1, NC)
 99 STOP
100 FORMAT (I2, F8.6, 2F10.5)
101 FORMAT (F10.5)
110 FORMAT (2X, 2F19.9)
113 FORMAT (28H0 FAILS TO CONVERGE FOR N = I2)
    END
```

Summary

The Birge-Vieta method, under specified conditions, converges quadratically to a linear factor of $P_n(x)$. Since the coefficients of deflated polynomials are degraded by the cumulative effects of roundoff, each approximate root calculated from the current $P_n(x)$ can be used as an initial estimate, and an improved root can be calculated by the Newton-Raphson method on the original polynomial $P_N(x)$. The Birge-Vieta method can be easily extended to compute complex zeros of a polynomial by using complex arithmetic in the same recursion formulas for synthetic division, and by modifying the Newton-Raphson formula to eliminate division by a complex number.

However, the quadratically convergent Bairstow method is more efficient for complex zeros because it simultaneously computes a complex zero and its conjugate, requiring only real arithmetic.

Neither the Birge-Vieta nor the Bairstow method is guaranteed to converge from arbitrary starting values. For this reason, it may be advisable to use some other starting method, such as the Bernoulli method or a generalization thereof called the quotient-difference algorithm. A combination Bernoulli-Birge-Vieta method is presented in Ralston-Wilf [24, p. 233]. Henrici [8, p. 162] presents an excellent derivation of the quotient-difference algorithm.

If all the zeros of a polynomial are known to be real, then the method of successive bisection (for computing initial approximation) can be combined with Birge-Vieta (for refining the zeros) into a simple, reliable program. The initial interval (x_1, x_2) for the successive-bisection method can be chosen as $(-M, M)$, where $M = (|a_0| + |a_1| + \cdots + |a_n|)/|a_0|$. This is based on the theorem: All zeros of a real-coefficient polynomial $P_n(x) = a_0 x^n + a_1 x^{n-1} + \cdots + a_n$ have a modulus less than $(|a_0| + |a_1| + \cdots + |a_n|)/|a_0|$. [21.]

If complex zeros exist, the quotient-difference algorithm (for obtaining initial approximations of the zeros corresponding to a quadratic factor) can be combined with Bairstow's method (for refining the quadratic factors) into a reliable, efficient program.

SHARE routines SDA 3314 and 3315 are available as efficiently programmed examples of the Birge-Vieta and Bairstow methods, respectively.

COMPUTATIONAL EXERCISES

1. Calculate a root of

$$P_3(x) \equiv x^3 - 0.5x^2 - 8x + 7.5 = 0$$

by the Birge-Vieta method, using

$$x_0 = \frac{-a_3}{a_2} = \frac{-7.5}{-8}.$$

2. Calculate a root of

$$P_4(x) \equiv x^4 + 2x^3 - 9.25x^2 - 12.5x + 18.75 = 0$$

using Bailey's iterative formula (see Programming exercise 6 of this chapter), with $x_0 = 0$.

Note. The values of $P_n(x_i)$, $P_n'(x_i)$, $P_n''(x_i)$ can be computed using the synthetic-division array

$x_i /$	1	a_1	a_2	a_3	$\ldots$	a_{n-2}	a_{n-1}	a_n
		x_i	$x_i p_1$	$x_i p_2$	$\ldots$	$x_i p_{n-3}$	$x_i p_{n-2}$	$x_i p_{n-1}$
	1	p_1	p_2	p_3	$\ldots$	p_{n-2}	p_{n-1}	p_n
		$x_i \bar{p}_1$	$x_i \bar{p}_2$	$x_i \bar{p}_3$	$\ldots$	$x_i \bar{p}_{n-2}$	$x_i \bar{p}_{n-1}$	
	$\bar{p}_1$	$\bar{p}_2$	$\bar{p}_3$	$\bar{p}_4$	$\ldots$	$\bar{p}_{n-1}$	$\bar{p}_n$	
		$x_i \bar{\bar{p}}_2/2$	$x_i \bar{\bar{p}}_3/2$	$x_i \bar{\bar{p}}_4/2$	$\ldots$	$x_i \bar{\bar{p}}_{n-1}/2$		
	$\bar{\bar{p}}_2/2$	$\bar{\bar{p}}_3/2$	$\bar{\bar{p}}_4/2$	$\bar{\bar{p}}_5/2$	$\ldots$	$\bar{\bar{p}}_n/2$		

3. Calculate a quadratic factor of

$$P_4(x) = x^4 - 7x^3 + 13x^2 + 45x - 50$$

by Bairstow's method using initial $(r, s) = (-7, 21)$.

DERIVATION EXERCISES

1. Derive the Birge-Vieta recursion formula (4.1.10) by writing polynomial $P_n(x)$ in nested form, as in Sec. 2.3, and determining the relation between successive nested terms, where the terms are denoted p_k ($k = 1, 2, \ldots$), from innermost to outermost. Then derive formula (4.1.11) by differentiating this recursion formula.

2. Show that in Bairstow's method the partial derivatives $p_k \equiv \partial b_k/\partial r$ and $q_k = \partial b_k/\partial s$ satisfy the relation

$$q_{k+1} = p_k \qquad (k = 1, n - 1).$$

3. Show that the differential-correction equations (4.2.14) in Bairstow's method can be reduced to the form

$$\begin{bmatrix} p_{n-1} & p_{n-2} \\ p_n + b_{n-1} & p_{n-1} \end{bmatrix} \begin{bmatrix} \delta r \\ \delta s \end{bmatrix} = \begin{bmatrix} -b_{n-1} \\ -b_n \end{bmatrix}$$

by using the relation $q_{k+1} = p_k$ ($k = 1, n - 1$).

4. Derive recursion formulas for calculating roots of a polynomial equation by Bailey's method [Sec. 3.2].

Hint: Use fact that $P_n''(x_i) = 2P_{n-2}(x_i)$, where

$$P_n(x) = (x - x_i)P_{n-1}(x) + p_n, \qquad P_n(x_i) = p_n$$
$$P_{n-1}(x) = (x - x_i)P_{n-2}(x) + \bar{p}_n, \qquad P_{n-1}(x_i) = \bar{p}_n$$
$$P_{n-2}(x) = (x - x_i)P_{n-3}(x) + \bar{\bar{p}}_n, \qquad P_{n-2}(x_i) = \bar{\bar{p}}_n.$$

5. Derive recursion formulas for the coefficients of the quotient and remainder polynomials resulting from division of $P_n(x)$ by $D_m(x)$, where

$$P_n(x) = x^n + a_1 x^{n-1} + a_2 x^{n-2} + \cdots + a_{n-1}x + a_n$$

and

$$D_m(x) = x^m + d_1 x^{m-1} + d_2 x^{m-2} + \cdots + d_{m-1}x + d_m.$$

Assume that $m \geq 1$ and $n > m$.

Hint: Categorize the formulas into three cases as follows:

$$
\begin{array}{ll}
\text{Case I} & n - m < m \\
\text{Case II} & n - m = m \\
\text{Case III} & n - m > m.
\end{array}
$$

PROGRAMMING EXERCISES

1. Modify the Birge-Vieta program to use $x_0 = -a_n/a_{n-1}$ as an initial estimate of the root.

2. Modify the Birge-Vieta program to use α_i as an initial estimate of the root α_{i+1}.

3. Modify the Birge-Vieta method for (a) multiple roots; (b) close roots. (a) Hint. If α is a root of multiplicity m, $P(\alpha) = P'(\alpha) = \cdots = P^{(m-1)}(\alpha) = 0$. (b) See Macon [14, p. 34] for modification of the Newton iterative formula.

4. Modify the Birge-Vieta program for complex roots. See Jennings [12, p. 23, exercise 2].

5. Modify the Birge-Vieta program to conserve storage by storing successive p values in a single location and successive $\bar{p}$ values in a single location. (Caution: $\bar{p}$ must be computed before p in each iteration if this is done.)

6. Write a flow chart and a FORTRAN program to compute the real roots of polynomial equations, using Bailey's iterative formula

$$x_{i+1} = x_i - \left[\dfrac{P_n(x_i)}{P_n'(x) - \dfrac{P_n(x_i)P_n''(x_i)}{2P_n'(x_i)}} \right].$$

Hint: Use the nested form of the polynomial to find the recursion formula to calculate $P_n(x_i)$; then differentiate twice to get the recursion formulas for $P_n'(x_i)$ and $P_n''(x_i)$.

7. Modify the Bairstow program to conserve storage and reduce computations by using the relation $q_{k+1} = p_k$ ($k = 1, n - 1$), and the simplified form of the differential correction equations

$$
\begin{bmatrix} p_{n-1} & p_{n-2} \\ p_n + b_{n-1} & p_{n-1} \end{bmatrix} \begin{bmatrix} \delta r \\ \delta s \end{bmatrix} = \begin{bmatrix} -b_{n-1} \\ -b_n \end{bmatrix}.
$$

8. Modify the Bairstow program to solve non-linear equations

$$
R(r, s) = 0
$$
$$
S(r, s) = 0
$$

by modified Newton method (Sec. 3.3). The recursion formulas for the second partials can be obtained by taking the partials of p_k with respect to r, denoted $p_k{}^r$. The partials with respect to s of p_k, denoted $p_k{}^s$, satisfy the relation $p_{k+1}^s = p_k{}^r$.

5

Matrix Methods and Systems of Linear Algebraic Equations

5.0 Introduction

In almost every scientific discipline there are problems that give rise to a system of linear algebraic equations of the form

$$a_{11}x_1 + a_{12}x_2 + \cdots + a_{1n}x_n - b_1 = 0$$
$$a_{21}x_1 + a_{22}x_2 + \cdots + a_{2n}x_n - b_2 = 0$$
$$\vdots$$
$$a_{n1}x_1 + a_{n2}x_2 + \cdots + a_{nn}x_n - b_n = 0.$$

For example, an electrical network of resistive components can be represented in the form above, where the coefficients a_{ij} are real numbers representing resistance values, the real unknowns x_i representing values of direct current, and the real constants b_i representing voltage sources. In polynomial curve fitting, the coefficients of a least-squares polynomial are determined by solving such a system of linear algebraic equations. Partial differential equations can be replaced by linear difference equations of this form and solved by methods of linear algebra. These examples illustrate but a few of the many applied problems that give rise to linear algebraic systems of equations. In this chapter we will discuss the elements of the classical theory of linear equations, and develop numerical methods for solving such systems.

A solution of this system of equations is a set of values $(x_1, x_2, \ldots, x_n)$ which simultaneously satisfies (i.e., reduces to an identity) each equation of the system. Two systems of equations in n variables $x_1, x_2, \ldots, x_n$ are said to be *equivalent* if every solution of one system is also a solution of the other system, and vice versa.

In the remainder of this section we will illustrate an algebraic and a geometric approach to the solution of two equations in two unknowns; this simple example points out the need for matrix methods in the solution of systems of linear equations. Such methods are developed in the remaining sections of this chapter.

The solution (x, y) of the system of equations

(5.0.1a) $$a_{11}x + a_{12}y - b_1 = 0$$

(5.0.1b) $$a_{21}x + a_{22}y - b_2 = 0$$

is defined as the number pair (x, y) that simultaneously satisfies (reduces to an identity) each equation of the system. A solution, if it exists, can be obtained by formal algebraic operations as follows:†

First, isolate x in each equation of system (5.0.1), obtaining

$$x = \frac{b_1 - a_{12}y}{a_{11}}$$

$$x = \frac{b_2 - a_{22}y}{a_{21}}.$$

Now, since the x coordinates of the solution must be the same for each equation of system (5.0.1), we equate the above expressions for x

$$\frac{b_1 - a_{12}y}{a_{11}} = \frac{b_2 - a_{22}y}{a_{21}}$$

and solve for y, obtaining

(5.0.2) $$y = \frac{-a_{21}b_1 + a_{11}b_2}{a_{11}a_{22} - a_{21}a_{12}}.$$

Finally, substituting this value of y into (5.0.1a) [or (5.0.1b)]

$$a_{11}x + a_{12}\left[\frac{-a_{21}b_1 + a_{11}b_2}{a_{11}a_{22} - a_{21}a_{12}}\right] - b_1 = 0$$

† This formal algebraic approach requires that both $a_{11} \neq 0$ and $a_{21} \neq 0$, which is not really true, as we will see later in the elimination methods. We use this approach, however, to illustrate how simple systems of equations might be solved by perhaps the most obvious, if not necessarily the best, approach.

and solving for x, we get

(5.0.3) $$x = \frac{a_{22}b_1 - a_{12}b_2}{a_{11}a_{22} - a_{21}a_{12}}.$$

The x and y of formulas (5.0.3) and (5.0.2), respectively, are called the coordinates of the solution (x, y) of system (5.0.1); the fact that this number pair (x, y) simultaneously satisfies each equation of system (5.0.1) can be demonstrated by direct substitution. Examination of either (5.0.2) or (5.0.3) indicates that the solution is not defined if the quantity $a_{11}a_{22} - a_{21}a_{12} = 0$.

Geometrically, the solution (x, y) of system (5.0.1) can be interpreted as the point $P(x, y)$ common to lines L_1 and L_2, represented by equations (5.0.1a) and (5.0.1b), respectively.

Now, consider the geometric problem of determining the point $P(x, y)$ common to lines L_1 and L_2, represented by

(5.0.4a) $$L_1 \equiv a_{11}x + a_{12}y - b_1 = 0$$

(5.0.4b) $$L_2 \equiv a_{21}x + a_{22}y - b_2 = 0.$$

The linear combination, $L_3 = k_1L_1 + k_2L_2$, where k_1 and k_2 are arbitrary constants (not both zero), of lines L_1, L_2 is also a line. To prove this fact, let us express this linear combination in the form

(5.0.5) $$k_1[a_{11}x + a_{12}y - b_1] + k_2[a_{21}x + a_{22}y - b_2] = 0.$$

A number pair (x, y) that simultaneously satisfies each equation of system (5.0.4) will also satisfy (5.0.5), for arbitrary constants k_1 and k_2, since each term in brackets is zero for an (x, y) that satisfies both (5.0.4a) and (5.0.4b).

Collecting terms in (5.0.5), we get

(5.0.6) $$(k_1a_{11} + k_2a_{21})x + (k_1a_{12} + k_2a_{22})y - (k_1b_1 + k_2b_2) = 0$$

which is an equation of the form $ax + by - c = 0$. Therefore, the linear combination $L_3 = k_1L_1 + k_2L_2$ is in fact a straight line. Also, L_3 *contains the point $P(x, y)$ common to L_1 and L_2 since (5.0.5) is satisfied by that number pair (x, y), corresponding to point P,* which simultaneously satisfies each equation of system (5.0.4).

If k_1 and k_2 are allowed to take on any combination of values, then a family of lines through $P(x, y)$, the point of intersection of L_1, L_2, is generated. We denote this family of lines $F(L_1, L_2)$.

To find the point $P(x, y)$ common to L_1 and L_2, replace L_2 by a member L_3 of family $F(L_1, L_2)$, choosing k_1 and k_2 such that the x coefficient in equation (5.0.6) is zero; that is, such that

(5.0.7) $$k_1a_{11} + k_2a_{21} = 0.$$

The constraint of equation (5.0.7) is satisfied if we choose $k_1 = -a_{21}$ and $k_2 = a_{11}$. Substituting these values for k_1, k_2 into (5.0.6) we get

(5.0.8) $L_3 \equiv (-a_{21}a_{11} + a_{11}a_{21})x + (-a_{21}a_{12} + a_{11}a_{22})y$
$$- (-a_{21}b_1 + a_{11}b_2) = 0$$

which has zero coefficient of x. Solving (5.0.8) for y, we get

(5.0.9) $$y = \frac{-a_{21}b_1 + a_{11}b_2}{a_{11}a_{22} - a_{21}a_{12}} = K$$

which is the equation of the horizontal line through the point $P(x, y)$ common to L_1 and L_2.

Now, the system of equations

(5.0.10a) $L_1 \equiv a_{11}x + a_{12}y - b_1 = 0$

(5.0.10b) $L_3 \equiv 0x + (a_{11}a_{22} - a_{21}a_{12})y - (-a_{21}b_1 + a_{11}b_2) = 0$

has the same solution as system (5.0.4) since L_1 and L_3 have the same point of intersection as L_1 and L_2 (see Figure 5-1).

The y coordinate of the solution is found directly by solving equation (5.0.10b), which is simply the y given in (5.0.9):

$$y = \frac{-a_{21}b_1 + a_{11}b_2}{a_{11}a_{22} - a_{21}a_{12}}.$$

Substituting this value of y into (5.0.10a), and solving for x, we find

(5.0.11) $$x = \frac{a_{22}b_1 - a_{12}b_2}{a_{11}a_{22} - a_{21}a_{12}}.$$

Note that the formulas for x and y given in (5.0.11) and (5.0.9) are the same as (5.0.3) and (5.0.2), respectively. Thus, we see that the algebraic and geometric solutions of the system (5.0.1) give the same result.

However, both the algebraic and geometric approaches to the solution of the system of linear algebraic equations are tedious, even for the simple

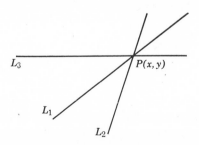

Figure 5-1.

case of two equations in two unknowns illustrated earlier. Let us see how we can simplify the solution of the system (5.0.1) of linear algebraic equations by introducing the concept of matrices.

First, write the coefficients of x and y in system (5.0.1) in an array $\begin{bmatrix} a_{11} & a_{12} \\ a_{21} & a_{22} \end{bmatrix}$, the unknowns in an array $\begin{bmatrix} x \\ y \end{bmatrix}$, and the constant terms in an array $\begin{bmatrix} b_1 \\ b_2 \end{bmatrix}$. Denote these arrays by A, X, B, respectively. Such a rectangular array of elements is called a *matrix*. A set of elements in a horizontal of the matrix is called a *row* of the matrix. A set of elements in a vertical of the matrix is called a *column* of the matrix. The number of rows and columns of a matrix are the row and column *dimensions* of the matrix.

Consider the *augmented* matrix

$$(5.0.12) \qquad \begin{bmatrix} a_{11} & a_{12} & b_1 \\ a_{21} & a_{22} & b_2 \end{bmatrix}$$

obtained by the juxtaposition of the elements of A and B. Let R_1 and R_2 denote row 1 and row 2 of augmented matrix (5.0.12), which we will call the matrix-equivalent of system (5.0.1). Conversely, system (5.0.1) will be called the system-equivalent to matrix (5.0.12).

Suppose we "reduce" matrix (5.0.12) by replacing row R_2 by the linear combination $R_2' = a_{11}R_2 - a_{21}R_1$. The reduced matrix with rows R_1 and R_2' is

$$(5.0.13) \quad \begin{bmatrix} a_{11} & a_{12} & b_1 \\ a_{11}a_{21} - a_{21}a_{11} & a_{11}a_{22} - a_{21}a_{12} & a_{11}b_2 - a_{21}b_1 \end{bmatrix}.$$

The system-equivalent of (5.0.13) is the set of linear equations

$$(5.0.14) \quad \begin{aligned} a_{11}x + \qquad\qquad a_{12}y - \qquad\qquad b_1 &= 0 \\ 0x + (a_{11}a_{22} - a_{21}a_{12})y - (a_{11}b_2 - a_{21}b_1) &= 0 \end{aligned}$$

which has solution (x, y) given by

$$(5.0.15) \quad y = \frac{a_{11}b_2 - a_{21}b_1}{a_{11}a_{22} - a_{21}a_{12}}$$

$$x = \frac{1}{a_{11}} [b_1 - a_{12}y] = \frac{1}{a_{11}} \left[b_1 - \frac{a_{12}(a_{11}b_2 - a_{21}b_1)}{a_{11}a_{22} - a_{21}a_{12}} \right]$$

$$(5.0.16) \quad x = \frac{a_{22}b_1 - a_{12}b_2}{a_{11}a_{22} - a_{21}a_{12}}.$$

Hence, the matrix method solution (x, y) given in (5.0.16) and (5.0.15) is identical with the algebraic and geometric solutions.

We will go further into matrix methods of solving simultaneous linear algebraic equations after we obtain a better understanding of matrix operations, as described in succeeding sections.

The first half of this chapter is devoted to the basic matrix methods required for solving many computational problems. The development of topics is presented in the following order:

1. *Basic matrix definitions and nomenclature.*

2. *Basic matrix operations*—The arithmetic operations of matrix addition, subtraction, and multiplication are included.

3. *Elementary row (column) operations and elementary matrices*—These provide the theoretical basis for computing the determinant and inverse of a square matrix by the elimination methods.

4. *Determinants*—The determinant of a square matrix can be computed by reducing the original matrix to either a triangular or diagonal matrix, using a sequence of properly selected row operations. The value of the determinant equals the product of the diagonal elements of the resulting matrix.

5. *Inversion of a square matrix*—The elimination methods are the most widely used methods for computing the inverse of a square matrix whose determinant is non-zero. Such methods are *direct methods*, i.e., they compute an inverse in a predetermined number of computational steps. (This inverse is exact if the computations can be performed without roundoff.) In practice, roundoff error occurs in each stage of the elimination process, and builds up as the elimination progresses; the effects of this roundoff can be minimized by using *pivotal* elimination. Since the row operations used in computing the inverse are the same as those used in computing the determinant, the inverse and determinant can be computed concurrently by the elimination methods.

6. *Improving an approximate inverse by Hotelling*—If the inverse of a matrix is sensitive to small changes in the elements of the original matrix, the matrix is said to be *ill-conditioned*. In such cases, the accumulative effects of roundoff error can be devastating, in that the computed inverse bears little resemblance to the actual inverse. Hotelling's iterative procedure can be used effectively to improve the approximate inverse of a matrix, provided that the norm of the first error matrix is less than 1.

5.1   Matrix Definitions and Nomenclature

A matrix A is defined as an $I \times J$ array of elements a_{ij}, where the subscripts i and j denote the row and column position, respectively, of the element in the array. That is, the $I \times J$ matrix A, denoted also by (a_{ij}), is the rectangular array

$$A = \begin{bmatrix} a_{11} & a_{12} & \cdots & a_{1J} \\ a_{21} & a_{22} & \cdots & a_{2J} \\ \cdot & \cdot & \cdot & \cdot \\ a_{I1} & a_{I2} & \cdots & a_{IJ} \end{bmatrix}.$$

The number of rows I and columns J are called the dimensions of A. A *row* matrix (also called a row vector) is a $1 \times J$ matrix. A *column* matrix (also called a column vector) is an $I \times 1$ matrix.

Square Matrices

Matrices in which the number of rows equals the number of columns are called square matrices. The number of rows (or columns) of a square matrix is called the *order* of the matrix. Such matrices are encountered in practice in a number of diverse scientific disciplines. The diagonal of elements a_{ii} is called the main diagonal of the square matrix.

Particular types of square matrices that are of special interest include the following:

An upper-triangular matrix is a square matrix that has all zero elements *below* the main diagonal, i.e., $a_{ij} = 0$ whenever $i > j$.

A lower-triangular matrix is a square matrix that has all zero elements *above* the main diagonal, i.e., $a_{ij} = 0$ whenever $i < j$.

A diagonal matrix is a square matrix with non-zero elements only on the main diagonal, i.e., $a_{ij} = 0$ whenever $i \neq j$.

An identity matrix, denoted by I, is a diagonal matrix with each main-diagonal element equal to unity.

5.2 Basic Matrix Operations

In this section the basic matrix operations of addition, subtraction, and multiplication are defined. Recursion formulas for elements of the sum, difference, and product matrices are derived. Finally, flow charts (Figures 5-2 and 5-3) and FORTRAN programs for these basic matrix operations are given.

Matrix Addition

Matrices A and B can be added (subtracted) provided that they have the same dimensions. The sum (difference) of two $I \times J$ matrices A and B is obtained by adding (subtracting) the corresponding elements of A and B. That is, if

(5.2.1)
$$\begin{aligned} C &= A + B \\ D &= A - B \end{aligned}$$

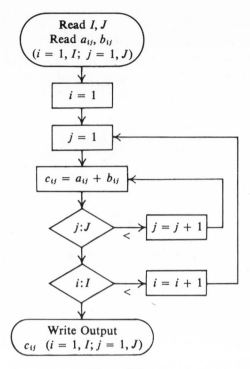

Figure 5-2.

then the elements of C and D are defined, respectively, as

$$(5.2.2) \quad \begin{array}{ll} c_{ij} = a_{ij} + b_{ij} & (i = 1, 2, \ldots, I; j = 1, 2, \ldots, J) \\ d_{ij} = a_{ij} - b_{ij} & (i = 1, 2, \ldots, I; j = 1, 2, \ldots, J). \end{array}$$

Figure 5-2 is a representative flow chart for the process of matrix addition.

The corresponding FORTRAN program for matrix addition is given below.

```
      PROGRAM MATADD
C     PROGRAM FOR MATRIX SUM C (IXJ) = A(IXJ) + B(IXJ)
      DIMENSION A(20,20), B(20,20), C(20,20)
      READ (5,100) IC, JC
      READ (5,101) ((A(I,J),B(I,J), I=1,IC), J=1,JC)
      DO 5 I = 1, IC
      DO 2 J = 1, JC
    2 C(I,J) = A(I,J) + B(I,J)
    5 CONTINUE
      WRITE (6,110) ((C(I,J), I=1,IC), J=1,JC)
      STOP
  100 FORMAT (2I2)
  101 FORMAT (2F19.9)
  110 FORMAT (2X, F19.9)
      END
```

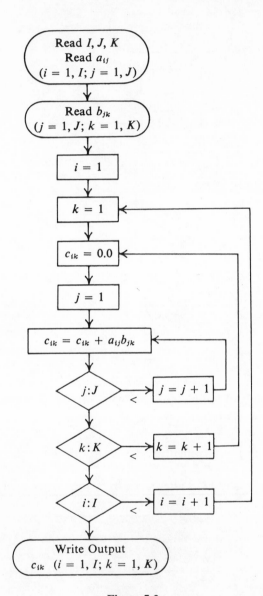

Figure 5-3.

Matrix Multiplication

The product $C = AB$ of two matrices A and B is defined if the number of columns of A equals the number of rows of B. The product of $A(I \times J)$† and $B(J \times K)$ is the matrix $C(I \times K)$ with elements c_{ik} defined by

$$(5.2.3) \qquad c_{ik} = \sum_{j=1}^{J} a_{ij} b_{jk}.$$

Figure 5-3 is a representative flow chart for calculating the elements c_{ik} of the matrix product $C = AB$.

```
        PROGRAM MATMUL
  C     PROGRAM FOR MATRIX PRODUCT C(IXK) = A(IXJ) X B(JXK)
        DIMENSION A (20, 20), B(20, 20), C(20, 20)
        READ (5,100) IC, JC, KC
        READ (5, 101) ((A(I,J), I=1,IC), J=1,JC)
        READ (5,101) ((B(J,K), J=1,JC), K=1,KC)
        DO 15 I = 1, IC
        DO 13 K = 1, KC
        C(I, K) = 0.0
        DO 11 J = 1, JC
     11 C(I,K) = C(I,K) + A(I,J) * B(J,K)
     13 CONTINUE
     15 CONTINUE
        WRITE (6,110) ((C(I,K), I=1, IC), K=1, KC)
        STOP
    100 FORMAT (3I2)
    101 FORMAT (F19.9)
    110 FORMAT (2X, E19.9)
        END
```

5.3 Elementary Row (Column) Operations and Elementary Matrices

In the *elimination* methods for calculating determinants and matrix inverses or for solving a system of linear algebraic non-homogeneous equations, the original square matrix is reduced to a simpler matrix (either the identity matrix or an upper-triangular matrix) by a sequence of properly chosen *row operations* of the following types:

(a) Multiplication of a row by a non-zero scalar c;

(b) The interchange of two rows;

(c) Replacement of row i by the sum of row i and c times row k, where c is any scalar, and $k \neq i$.

Perlis [22] shows that an elementary row operation on matrix A can be accomplished by premultiplying A by an elementary matrix E, where E is obtained by performing the same row operation on the identity matrix I. That is, the elementary matrices $E = [E_i(c), E_{ik}, E_{ik}(c)]$ are defined as follows:

$E_i(c)$ = identity matrix with row i multiplied by scalar c.

E_{ik} = identity matrix with rows i and k interchanged.

$E_{ik}(c)$ = identity matrix with row i replaced by the sum of row i and c times row k.

† The notation $A(I \times J)$ denotes a matrix A of I rows and J columns.

The following are examples of 3×3 elementary matrices.

$$E_2(c) = \begin{bmatrix} 1 & 0 & 0 \\ 0 & c & 0 \\ 0 & 0 & 1 \end{bmatrix}, \quad E_{13} = \begin{bmatrix} 0 & 0 & 1 \\ 0 & 1 & 0 \\ 1 & 0 & 0 \end{bmatrix}, \quad E_{12}(c) = \begin{bmatrix} 1 & c & 0 \\ 0 & 1 & 0 \\ 0 & 0 & 1 \end{bmatrix}.$$

The following examples illustrate that *premultiplication* of matrix A by an elementary matrix E effects the same row operation on matrix A as was performed on I to produce E; that is,

$A' = \quad E_i(c) \cdot A$ = matrix A with row i multiplied by scalar c.

$A' = \quad\quad E_{ik} \cdot A$ = matrix A with rows i and k interchanged.

$A' = E_{ik}(c) \cdot A$ = matrix A with row i replaced by the sum of row i and c times row k.

In the examples, we will use 3×3 matrices

$$A = \begin{bmatrix} a_{11} & a_{12} & a_{13} \\ a_{21} & a_{22} & a_{23} \\ a_{31} & a_{32} & a_{33} \end{bmatrix}, \quad A' = \begin{bmatrix} a_{11}' & a_{12}' & a_{13}' \\ a_{21}' & a_{22}' & a_{23}' \\ a_{31}' & a_{32}' & a_{33}' \end{bmatrix}$$

where A' is obtained by premultiplying A by an elementary matrix.

Note. Row i denotes ith row of A, while row i' denotes ith row of A'.

EXAMPLE *1.* $A' = E_1(c) \cdot A$

$$\begin{bmatrix} a_{11}' & a_{12}' & a_{13}' \\ a_{21}' & a_{22}' & a_{23}' \\ a_{31}' & a_{32}' & a_{33}' \end{bmatrix} = \begin{bmatrix} c & 0 & 0 \\ 0 & 1 & 0 \\ 0 & 0 & 1 \end{bmatrix} \begin{bmatrix} a_{11} & a_{12} & a_{13} \\ a_{21} & a_{22} & a_{23} \\ a_{31} & a_{32} & a_{33} \end{bmatrix} = \begin{bmatrix} ca_{11} & ca_{12} & ca_{13} \\ a_{21} & a_{22} & a_{23} \\ a_{31} & a_{32} & a_{33} \end{bmatrix}$$

so that

$$\left. \begin{aligned} a_{1j}' &= ca_{1j} \\ a_{2j}' &= a_{2j} \\ a_{3j}' &= a_{3j} \end{aligned} \right\} (j = 1, 3), \quad \text{i.e.,} \quad \left\{ \begin{aligned} \text{Row } 1' &= c \cdot \text{Row } 1 \\ \text{Row } 2' &= \quad \text{Row } 2 \\ \text{Row } 3' &= \quad \text{Row } 3. \end{aligned} \right.$$

In general, if $A' = E_i(c) \cdot A$, where A and A' are $n \times n$ matrices, then

$$\left. \begin{aligned} a_{ij}' &= ca_{ij} \\ a_{kj}' &= a_{kj} \quad (k \neq i) \end{aligned} \right\} (j = 1, n), \quad \text{i.e.,} \quad \left\{ \begin{aligned} \text{Row } i' &= c \cdot \text{Row } i \\ \text{Row } k' &= \quad \text{Row } k \quad (k \neq i). \end{aligned} \right.$$

EXAMPLE 2. $A' = E_{12} \cdot A$

$$\begin{bmatrix} a_{11}' & a_{12}' & a_{13}' \\ a_{21}' & a_{22}' & a_{23}' \\ a_{31}' & a_{32}' & a_{33}' \end{bmatrix} = \begin{bmatrix} 0 & 1 & 0 \\ 1 & 0 & 0 \\ 0 & 0 & 1 \end{bmatrix} \begin{bmatrix} a_{11} & a_{12} & a_{13} \\ a_{21} & a_{22} & a_{23} \\ a_{31} & a_{32} & a_{33} \end{bmatrix} = \begin{bmatrix} a_{21} & a_{22} & a_{23} \\ a_{11} & a_{12} & a_{13} \\ a_{31} & a_{32} & a_{33} \end{bmatrix}$$

so that

$$\left.\begin{array}{l} a_{1j}' = a_{2j} \\ a_{2j}' = a_{1j} \\ a_{3j}' = a_{3j} \end{array}\right\} (j = 1, 3), \qquad \text{i.e.,} \quad \left\{\begin{array}{l} \text{Row } 1' = \text{Row } 2 \\ \text{Row } 2' = \text{Row } 1 \\ \text{Row } 3' = \text{Row } 3. \end{array}\right.$$

In general, if $A' = E_{ik} \cdot A$, where A and A' are $n \times n$ matrices, then

$$\left.\begin{array}{l} a_{ij}' = a_{kj} \\ a_{kj}' = a_{ij} \\ a_{mj}' = a_{mj} \end{array}\right\} (j = 1, n; m \neq i, k), \qquad \text{i.e.,} \quad \left\{\begin{array}{l} \text{Row } i' = \text{Row } k \\ \text{Row } k' = \text{Row } i \\ \text{Row } m' = \text{Row } m \quad (m \neq i, k). \end{array}\right.$$

EXAMPLE 3. $A' = E_{13}(c) \cdot A$

$$\begin{bmatrix} a_{11}' & a_{12}' & a_{13}' \\ a_{21}' & a_{22}' & a_{23}' \\ a_{31}' & a_{32}' & a_{33}' \end{bmatrix} = \begin{bmatrix} 1 & 0 & c \\ 0 & 1 & 0 \\ 0 & 0 & 1 \end{bmatrix} \begin{bmatrix} a_{11} & a_{12} & a_{13} \\ a_{21} & a_{22} & a_{23} \\ a_{31} & a_{32} & a_{33} \end{bmatrix}$$

$$= \begin{bmatrix} a_{11} + ca_{31} & a_{12} + ca_{32} & a_{13} + ca_{33} \\ a_{21} & a_{22} & a_{23} \\ a_{31} & a_{32} & a_{33} \end{bmatrix}$$

so that

$$\left.\begin{array}{l} a_{1j}' = a_{1j} + ca_{3j} \\ a_{2j}' = a_{2j} \\ a_{3j}' = a_{3j} \end{array}\right\} (j = 1, 3), \qquad \text{i.e.,} \quad \left\{\begin{array}{l} \text{Row } 1' = \text{Row } 1 + c \cdot \text{Row } 3 \\ \text{Row } 2' = \text{Row } 2 \\ \text{Row } 3' = \text{Row } 3. \end{array}\right.$$

In general, if $A' = E_{ik}(c) \cdot A$ where A and A' are $n \times n$ matrices, then

$$\left.\begin{array}{l} a_{ij}' = a_{ij} + ca_{kj} \\ a_{mj}' = a_{mj} \quad (m \neq i) \end{array}\right\} (j = 1, n), \qquad \text{i.e.,} \quad \left\{\begin{array}{l} \text{Row } i' = \text{Row } i + c \cdot \text{Row } k \\ \text{Row } m' = \text{Row } m \quad (m \neq i). \end{array}\right.$$

5.4 Determinants

The solution of systems of linear algebraic equations is rooted in the theory of determinants. In this section, we will develop the essential properties of determinants needed to give a better understanding of the solution of systems of linear equations.

Every square matrix A has an associated *scalar quantity* called a determinant, which we denote by $|A|$. The determinant of a 2×2 matrix A is $|A| = a_{11}a_{22} - a_{21}a_{12}$, while the determinant of a 3×3 matrix A is $|A| = a_{11}a_{22}a_{33} - a_{11}a_{32}a_{23} + a_{12}a_{23}a_{31} - a_{12}a_{21}a_{33} + a_{13}a_{21}a_{32} - a_{13}a_{31}a_{22}$.

In general, the determinant of an $n \times n$ matrix A is *defined* as follows: first, generate the $n!$ possible products $a_{1j_1} a_{2j_2} \ldots a_{nj_n}$, where the column indices $j_1, j_2, \ldots, j_n$ are the numbers $1, 2, \ldots, n$ in some permutation; next, multiply each such product by $(-1)^t$ where t is the number of transpositions† required to rearrange the sequence of integers $j_1, j_2, \ldots, j_n$ in the natural order $1, 2, \ldots, n$; and, lastly, add the $n!$ products $(-1)^t a_{1j_1} a_{2j_2} \ldots a_{nj_n}$ to obtain the value of the determinant.

The determinant of matrix A can also be written as the original array of elements a_{ij} enclosed by two vertical lines, i.e., as

$$(5.4.1) \qquad |A| = \begin{vmatrix} a_{11} & a_{12} & \cdots & a_{1n} \\ a_{21} & a_{22} & \cdots & a_{2n} \\ \cdot & \cdot & \cdot \cdot \cdot & \cdot \\ a_{n1} & a_{n2} & \cdots & a_{nn} \end{vmatrix}.$$

It should be emphasized that the determinant is a *single scalar quantity* obtained from the elements of matrix A by the procedure defined above.

There are other forms of expressing the determinant of a matrix; these forms are based on minors and cofactors which are defined as follows:

A *minor* of order $n - 1$ is the $(n - 1) \times (n - 1)$ determinant obtained by deleting one row and one column from the original $n \times n$ array (a_{ij}).

Every element a_{ij} of a matrix A has an associated *cofactor* A_{ij} which is equal to $(-1)^{i+j}$ times the minor obtained by deleting row i and column j from the original array (a_{ij}).

With these definitions, we can express the determinant in terms of the elements of row i and their corresponding cofactors in the form

$$(5.4.2) \qquad |A| = \sum_{j=1}^{n} a_{ij} A_{ij} \qquad (i = 1, 2, \ldots, n)$$

or in terms of the elements of column j and their cofactors in the form

$$(5.4.3) \qquad |A| = \sum_{i=1}^{n} a_{ij} A_{ij} \qquad (j = 1, 2, \ldots, n).$$

---

† A transposition is the interchange of the order of two of the integers $j_1, j_2, \ldots, j_n$. The number of transpositions required to rearrange sequence $j_1, j_2, \ldots, j_n$ in natural order $1, 2, \ldots, n$ is *not* unique, but $(-1)^t$ is unique for a given sequence. See, for example, Perlis [22, p. 70].

EXAMPLE. The determinant of a 3×3 matrix A can be expressed in terms of the elements of row 1 and cofactors thereof as

$$|A| = \begin{vmatrix} a_{11} & a_{12} & a_{13} \\ a_{21} & a_{22} & a_{23} \\ a_{31} & a_{32} & a_{33} \end{vmatrix} = a_{11}A_{11} + a_{12}A_{12} + a_{13}A_{13}$$

$$= a_{11}(-1)^2 \begin{vmatrix} a_{22} & a_{23} \\ a_{32} & a_{33} \end{vmatrix} + a_{12}(-1)^3 \begin{vmatrix} a_{21} & a_{23} \\ a_{31} & a_{33} \end{vmatrix} + a_{13}(-1)^4 \begin{vmatrix} a_{21} & a_{22} \\ a_{31} & a_{32} \end{vmatrix}.$$

Properties of Determinants

The following is a list of properties of determinants that we will find useful in developing numerical methods of evaluating the determinant and in deriving Cramer's rule for the solution of a system of linear algebraic equations.

P1. If two rows (or two columns) of a matrix are identical, then its determinant is zero.

P2. If the elements a_{ij} of row i are multiplied by the cofactors A_{kj} of a different row k, then

$$\sum_{j=1}^{n} a_{ij}A_{kj} = 0 \qquad (k \neq i).$$

P3. If matrix C equals the product of matrices A and B, then $|C| = |A||B|$.

P4. The determinant of a triangular matrix equals the product of the diagonal elements of the array.

P5. The determinant of a diagonal matrix equals the product of the diagonal elements of the array.

P6. If row i of matrix A is replaced by row $(b_1 \quad b_2 \quad \ldots \quad b_n)$, then the determinant of the resulting matrix A_i can be expressed in the form

$$|A_i| = b_1 A_{i1} + b_2 A_{i2} + \cdots + b_n A_{in}.$$

P7. If column j of matrix A is replaced by column $(b_1 \quad b_2 \quad \ldots \quad b_n)$, then the determinant of the resulting matrix A_j can be expressed in the form

$$|A_j| = b_1 A_{1j} + b_2 A_{2j} + \cdots + b_n A_{nj}.$$

The evaluation of a determinant of an $n \times n$ matrix A by forming the sum of the products $(-1)^t a_{1j_1}a_{2j_2}\ldots a_{nj_n}$ or by expanding the determinant in terms of elements and cofactors is inefficient when n is greater than 3 or 4. There are more practical, efficient methods (such as the elimination methods) for calculating the value of the determinant of large-order matrices. The elimination methods reduce the original square array of elements a_{ij} to a triangular or a diagonal array by a sequence of operations on the rows of the array.

The evaluation of a determinant by these elimination methods is based on the following *three rules of transformation for determinants*:

R1. If matrix B is formed by multiplying each element of the ith row of A by a scalar c, then $|B| = c|A|$.

Proof. Expand $|B|$ in terms of elements of row i and cofactors:

$$
\begin{aligned}
|B| &= (ca_{i1})A_{i1} + (ca_{i2})A_{i2} + \cdots + (ca_{in})A_{in} \\
&= c[a_{i1}A_{i1} + a_{i2}A_{i2} + \cdots + a_{in}A_{in}] \\
&= c|A| \hspace{4cm} \text{by (5.4.2).}
\end{aligned}
$$

R2. If matrix B is formed by interchanging row i and row k of A, then $|B| = -|A|$.

Proof. Matrix B can be obtained from matrix A by premultiplying A by elementary matrix E_{ik}, where E_{ik} is the identity matrix with row i and row k interchanged.

Then

$$ B = E_{ik}A $$

and by property P3, it follows that

$$ |B| = |E_{ik}||A|. $$

It can be shown that $|E_{ik}| = -1$, so that $|B| = -|A|$.

R3. If matrix B is formed by adding c times row k of A to row i of A, then $|B| = |A|$.

Proof. Expand $|B|$ by terms of row i and cofactors:

$$
\begin{aligned}
|B| &= (a_{i1} + ca_{k1})A_{i1} + (a_{i2} + ca_{k2})A_{i2} + \cdots + (a_{in} + ca_{kn})A_{in} \\
&= [a_{i1}A_{i1} + a_{i2}A_{i2} + \cdots + a_{in}A_{in}] \\
&\quad + c[a_{k1}A_{i1} + a_{k2}A_{i2} + \cdots + a_{kn}A_{in}].
\end{aligned}
$$

The expression in the second pair of brackets is zero by property P2, so that

$$ |B| = |A| + c0 = |A|. $$

5.5   The Inverse of a Square Matrix

For every square matrix A whose determinant is non-zero† there exists an inverse matrix, denoted A^{-1}, which satisfies the relations

$$ (5.5.1) \hspace{3cm} AA^{-1} = A^{-1}A = I $$

where I is the identity matrix (diagonal matrix with $a_{ii} = 1$). *As we will*

---

† If a square matrix has a non-zero determinant, it is said to be non-singular. If the matrix has a zero determinant, the matrix is said to be singular.

show in Sec. 5.6, a system of n linear non-homogeneous algebraic equations in n variables can be solved by calculating the inverse of the matrix of coefficients.

To formally define the inverse matrix, we first need to introduce the concept of the adjoint matrix of A. This adjoint matrix, denoted Adj(A), is defined as the transpose of the matrix of cofactors of A; that is,

$$(5.5.2) \quad \text{Adj}(A) = \begin{bmatrix} A_{11} & A_{12} & \cdots & A_{1n} \\ A_{21} & A_{22} & \cdots & A_{2n} \\ \cdot & \cdot & \cdot & \cdot \\ A_{n1} & A_{n2} & \cdots & A_{nn} \end{bmatrix}^T = \begin{bmatrix} A_{11} & A_{21} & \cdots & A_{n1} \\ A_{12} & A_{22} & \cdots & A_{n2} \\ \cdot & \cdot & \cdot & \cdot \\ A_{1n} & A_{2n} & \cdots & A_{nn} \end{bmatrix}$$

where A_{ij} is the cofactor of element a_{ij} as defined in Sec. 5.4.

The inverse A^{-1} of matrix A can be defined in terms of the determinant of A and the adjoint of A as follows

$$(5.5.3) \qquad\qquad A^{-1} = \frac{1}{|A|} \, \text{Adj}(A).$$

EXAMPLE. Using definition (5.5.3), compute the inverse of matrix A where

$$A = \begin{bmatrix} 9 & 9 & 8 \\ 9 & 8 & 7 \\ 8 & 7 & 6 \end{bmatrix}.$$

Solution

$$|A| = 9 \begin{vmatrix} 8 & 7 \\ 7 & 6 \end{vmatrix} - 9 \begin{vmatrix} 9 & 7 \\ 8 & 6 \end{vmatrix} + 8 \begin{vmatrix} 9 & 8 \\ 8 & 7 \end{vmatrix} = 1$$

$$\text{Adj}(A) = \begin{bmatrix} -1 & 2 & -1 \\ 2 & -10 & 9 \\ -1 & 9 & -9 \end{bmatrix}^T = \begin{bmatrix} -1 & 2 & -1 \\ 2 & -10 & 9 \\ -1 & 9 & -9 \end{bmatrix}$$

so that

$$A^{-1} = \frac{1}{1} \begin{bmatrix} -1 & 2 & -1 \\ 2 & -10 & 9 \\ -1 & 9 & -9 \end{bmatrix} = \begin{bmatrix} -1 & 2 & -1 \\ 2 & -10 & 9 \\ -1 & 9 & -9 \end{bmatrix}.$$

Matrix Inversion by Elimination

If a non-singular matrix A can be reduced to the identity matrix I by premultiplying A by a sequence of properly-chosen elementary matrices

E_k of types $E_i(c)$, E_{ik}, $E_{ik}(c)$, then premultiplication of identity matrix I by the same sequence of elementary matrices will produce A^{-1}, the inverse matrix of A.

Proof. Suppose we find elementary matrices E_k $(k = 1, K)$ such that

$$E_K E_{K-1} \ldots E_3 E_2 E_1 A = I.$$

Postmultiplication of each side of the above equation by A^{-1} gives us

$$E_K E_{K-1} \ldots E_3 E_2 E_1 A A^{-1} = I A^{-1}$$

so that, applying the associative law of multiplication, we get the desired result:

$$E_K E_{K-1} \ldots E_3 E_2 E_1 I = A^{-1}.$$

The problem confronting us is *how* to find the required sequence of elementary matrices which reduce A to the identity I. If we examine the problem in terms of row operations (corresponding to premultiplication by elementary matrices), we can immediately see the required sequence of operations. First, let us summarize the *correspondence* among *premultiplication by elementary matrices*, *row operations*, and *recursion formulas for the matrix elements*; these are summarized in Table 5-1.

Table 5-1

Premultiplication	Row Operations†	Recursion Formulas‡
$A' = E_i(c) \cdot A$	$R_i{}' = c R_i$ $R_k{}' = R_k \quad (k \neq i)$	$a_{ij}{}' = c a_{ij}$ $a_{kj}{}' = a_{kj} \quad (k \neq i)$
$A' = E_{ik} \cdot A$	$R_i{}' = R_k$ $R_k{}' = R_i$ $R_m{}' = R_m \quad (m \neq i, k)$	$a_{ij}{}' = a_{kj}$ $a_{kj}{}' = a_{ij}$ $a_{mj}{}' = a_{mj} \quad (m \neq i, k)$
$A' = E_{ik}(c) \cdot A$	$R_i{}' = R_i + c R_k$ $R_m{}' = R_m \quad (m \neq i)$	$a_{ij}{}' = a_{ij} + c a_{kj}$ $a_{mj}{}' = a_{mj} \quad (m \neq i)$

The elimination method of calculating the inverse of an $n \times n$ nonsingular matrix

$$A = \begin{bmatrix} a_{11} & a_{12} & a_{13} & \ldots & a_{1n} \\ a_{21} & a_{22} & a_{23} & \ldots & a_{2n} \\ a_{31} & a_{32} & a_{33} & \ldots & a_{3n} \\ \cdot & \cdot & \cdot & \cdot & \cdot & \cdot \\ a_{n1} & a_{n2} & a_{n3} & \ldots & a_{nn} \end{bmatrix}$$

† $R_i{}'$ denotes row i of A', while R_i denotes row i of A.
‡ For the recursion formulas, $j = 1, 2, \ldots, n$.

is accomplished by n *stages* of computations, each stage consisting of two distinct steps. The kth stage ($k = 1, 2, \ldots, n$) of the elimination process is accomplished as follows:

Step 1. Normalization of element k, k by multiplying row k by the reciprocal of element k, k, provided the reciprocal exists.†

Step 2. Zeroing the off-diagonal elements of column k by replacing row i ($i \neq k$) by suitable linear combination of row i and row k.

The first stage of the elimination process produces a matrix $A^{(1)}$, the superscript denoting the stage number, of the form

$$A^{(1)} = \begin{bmatrix} 1 & a_{12}^{1} & a_{13}^{1} & \ldots & a_{1n}^{1} \\ 0 & a_{22}^{1} & a_{23}^{1} & \ldots & a_{2n}^{1} \\ 0 & a_{32}^{1} & a_{33}^{1} & \ldots & a_{3n}^{1} \\ & \cdot & \cdot & \cdot & \cdot \\ 0 & a_{n2}^{1} & a_{n3}^{1} & \ldots & a_{nn}^{1} \end{bmatrix}.$$

The row operations on A which produce $A^{(1)}$ are

$$R_1^1 = \frac{R_1}{a_{11}}$$

provided $a_{11} \neq 0$.

$$R_i^1 = R_i - a_{i1}R_1^1 \qquad (i \neq 1).$$

That is, the elements a_{ij}^1 of $A^{(1)}$ are calculated from the elements of A by the recursion formulas

$$\left. \begin{array}{l} a_{1j}^1 = \dfrac{a_{1j}}{a_{11}} \\[2mm] a_{ij}^1 = a_{ij} - a_{i1}a_{1j}^1 \qquad (i \neq 1) \end{array} \right\} (j = 1, n).$$

The second stage of the elimination process produces matrix $A^{(2)}$ from matrix $A^{(1)}$, where

$$A^{(2)} = \begin{bmatrix} 1 & 0 & a_{13}^{2} & \ldots & a_{1n}^{2} \\ 0 & 1 & a_{23}^{2} & \ldots & a_{2n}^{2} \\ 0 & 0 & a_{33}^{2} & \ldots & a_{3n}^{2} \\ & \cdot & \cdot & \cdot & \cdot \\ 0 & 0 & a_{n3}^{2} & \ldots & a_{nn}^{2} \end{bmatrix}.$$

† If element k, k is zero, then its reciprocal is not defined. In this case, row k must be interchanged with some row i which has non-zero element i, k. In practice, row k is interchanged with row *imax*, where element *imax*, k is the element of maximum magnitude in column k, on or below the main diagonal.

The row operations on $A^{(1)}$ which produce $A^{(2)}$ are

$$R_2{}^2 = \frac{R_2{}^1}{a_{22}{}^1}$$

provided $a_{22}{}^1 \neq 0$.

$$R_i{}^2 = R_i{}^1 - a_{i2}{}^1 R_2{}^2 \qquad (i \neq 2).$$

That is, the elements of $A^{(2)}$ are calculated from the elements of $A^{(1)}$ by the recursion formulas

$$\left.\begin{array}{l} a_{2j}{}^2 = \dfrac{a_{2j}{}^1}{a_{22}{}^1} \\[2mm] a_{ij}{}^2 = a_{ij}{}^1 - a_{i2}{}^1 a_{2j}{}^2 \qquad (i \neq 2) \end{array}\right\} (j = 1, n).$$

In general, the kth stage ($k = 1, 2, \ldots, n$) of the elimination process produces a matrix

$$A^{(k)} = \begin{bmatrix} 1 & 0 & 0 & \ldots & 0 & a_{1,k+1}^{k} & \ldots & a_{1n}{}^{k} \\ 0 & 1 & 0 & \ldots & 0 & a_{2,k+1}^{k} & \ldots & a_{2n}{}^{k} \\ 0 & 0 & 1 & \ldots & 0 & a_{3,k+1}^{k} & \ldots & a_{3n}{}^{k} \\ \cdot & \cdot & \cdot & \cdot & \cdot & \cdot & \cdot & \cdot \\ 0 & 0 & 0 & \ldots & 0 & a_{n,k+1}^{k} & \ldots & a_{nn}{}^{k} \end{bmatrix}$$

where

$$R_k{}^k = \frac{R_k^{k-1}}{a_{kk}^{k-1}}$$

provided $a_{kk}^{k-1} \neq 0$.

$$R_i{}^k = R_i^{k-1} - a_{ik}^{k-1} R_k{}^k \qquad (i \neq k).$$

That is, the elements of $A^{(k)}$ are calculated from the elements of $A^{(k-1)}$ by the recursion formulas

$$(5.51.1) \qquad \left.\begin{array}{l} a_{kj}{}^k = \dfrac{a_{kj}^{k-1}}{a_{kk}^{k-1}} \\[2mm] a_{ij}{}^k = a_{ij}^{k-1} - a_{ik}^{k-1} a_{kj}{}^k \qquad (i \neq k) \end{array}\right\} (j = 1, n)$$

for $k = 1, 2, \ldots, n$, provided we define $a_{ij} = a_{ij}{}^0$. (Likewise, we must define $b_{ij} = b_{ij}{}^0$, where $B \equiv I$.)

It is evident then that the matrix $A^{(n)}$ produced in stage n of the reduction process will be the identity matrix I. Now, the same sequence of row operations (applied to A, $A^{(1)}$, $A^{(2)}$, $\ldots$) is concurrently applied to matrix $I \equiv B$, $B^{(1)}$, $B^{(2)}$, $\ldots$, to produce finally $B^{(n)} = A^{-1}$.

If we denote by $r_i{}^k$ the ith row of matrix $B^{(k)}$, the row operations required to produce $B^{(k)}$ are

$$r_k{}^k = \frac{r_k^{k-1}}{a_{kk}^{k-1}}$$

$$r_i{}^k = r_i^{k-1} - a_{ik}^{k-1} r_k{}^k \qquad (i \neq k).$$

The elements of $B^{(k)}$ can therefore be calculated from the elements of $B^{(k-1)}$ by the recursion formulas

$$(5.51.2) \qquad \left. \begin{aligned} b_{kj}{}^{k} &= \frac{b_{kj}^{k-1}}{a_{kk}^{k-1}} \\ b_{ij}{}^{k} &= b_{ij}^{k-1} - a_{ik}^{k-1} b_{kj}{}^{k} \quad (i \neq k) \end{aligned} \right\} (j = 1, n).$$

The process of inversion by elimination and the process for a specific numerical example are illustrated in succeeding pages.

Note that formulas (5.51.1) and (5.51.2) are the basic recursion formulas for inversion by elimination, the former reducing matrix A to the identity I, the latter reducing I $\equiv$ B to the matrix A^{-1}.

Computation of the Determinant

Concurrently with the computation of the inverse, we can calculate the determinant of matrix A. The determinant is obtained by simply computing the cumulative product of elements a_{kk}^{k-1} ($k = 1, n$), in stage k of the elimination process. When the elimination process has been completed, the value of the cumulative product equals the determinant; that is,

$$|A| = a_{11} a_{22}{}^{1} a_{33}{}^{2} \ldots a_{kk}^{k-1} \ldots a_{nn}^{n-1}.$$

The determinant of A can be computed concurrently with the calculation of the inverse matrix by simply applying the rules of determinant transformation (p. 148). That is, since the elimination method reduces matrix A to the identity matrix I by a sequence of elementary row operations, we know the relation between the determinants of successive matrices A, $A^{(1)}$, $A^{(2)}$, $A^{(3)}$, $A^{(4)}$.

Only row operations 1 and 2 change the value of the determinant, so that we can ignore row operation 3 in the calculation of the value of the determinant. Let us assume for the moment that matrix A can be reduced to I without using row operations of type 2 (row interchange). Then, we can apply rule R1 at each stage of the elimination process to obtain the following relations (for $n = 4$):

$$|A^{(1)}| = \frac{1}{a_{11}} |A|$$

$$|A^{(2)}| = \frac{1}{a_{22}{}^{1}} |A^{(1)}| = \frac{1}{a_{22}{}^{1}} \frac{1}{a_{11}} |A|$$

$$|A^{(3)}| = \frac{1}{a_{33}{}^{2}} |A^{(2)}| = \frac{1}{a_{33}{}^{2}} \frac{1}{a_{22}{}^{1}} \frac{1}{a_{11}} |A|$$

$$|A^{(4)}| = \frac{1}{a_{44}{}^{3}} |A^{(3)}| = \frac{1}{a_{44}{}^{3}} \frac{1}{a_{33}{}^{2}} \frac{1}{a_{22}{}^{1}} \frac{1}{a_{11}} |A|$$

and since $A^{(4)} = I$, it follows that $|A^{(4)}| = 1$. Equating these two expressions for $|A^{(4)}|$, we find that

$$|A| = a_{11}a_{22}{}^1a_{33}{}^2a_{44}{}^3.$$

Now, if m row interchanges (row operations type 2) are required in the reduction of A to I, we apply rule R2 to obtain

$$|A| = (-1)^m a_{11}a_{22}{}^1a_{33}{}^2a_{44}{}^3.$$

In the computer program, the determinant of A is computed by initially setting $|A| = 1$ and then forming the above cumulative product, where in stage k factor $\pm a_{kk}^{k-1}$ multiplies the previous value of the cumulative product which was initially equal to 1.

The fact that the determinant of A can be computed concurrently with the calculation of the inverse matrix in the elimination methods is a valuable bonus. The simplicity and efficiency of the elimination methods, coupled with this extra feature, make these methods one of the most popular types for matrix inversion.

Illustration of Matrix Inversion by Elimination for $n = 4$

Reduce the 4×4 matrix A to the identity matrix I by a sequence of properly-chosen row operations.† Then the same sequence of row operations will reduce I to the inverse of matrix A.

$$A = \begin{bmatrix} a_{11} & a_{12} & a_{13} & a_{14} \\ a_{21} & a_{22} & a_{23} & a_{24} \\ a_{31} & a_{32} & a_{33} & a_{34} \\ a_{41} & a_{42} & a_{43} & a_{44} \end{bmatrix} \quad B = \begin{bmatrix} b_{11} & b_{12} & b_{13} & b_{14} \\ b_{21} & b_{22} & b_{23} & b_{24} \\ b_{31} & b_{32} & b_{33} & b_{34} \\ b_{41} & b_{42} & b_{43} & b_{44} \end{bmatrix} = \begin{bmatrix} 1 & 0 & 0 & 0 \\ 0 & 1 & 0 & 0 \\ 0 & 0 & 1 & 0 \\ 0 & 0 & 0 & 1 \end{bmatrix}$$

Stage 1. Eliminate column 1 of A except element 1, 1.

$$A^{(1)} = \begin{bmatrix} 1 & a_{12}{}^1 & a_{13}{}^1 & a_{14}{}^1 \\ 0 & a_{22}{}^1 & a_{23}{}^1 & a_{24}{}^1 \\ 0 & a_{32}{}^1 & a_{33}{}^1 & a_{34}{}^1 \\ 0 & a_{42}{}^1 & a_{43}{}^1 & a_{44}{}^1 \end{bmatrix} \quad B^{(1)} = \begin{bmatrix} b_{11}{}^1 & b_{12}{}^1 & b_{13}{}^1 & b_{14}{}^1 \\ b_{21}{}^1 & b_{22}{}^1 & b_{23}{}^1 & b_{24}{}^1 \\ b_{31}{}^1 & b_{32}{}^1 & b_{33}{}^1 & b_{34}{}^1 \\ b_{41}{}^1 & b_{42}{}^1 & b_{43}{}^1 & b_{44}{}^1 \end{bmatrix}$$

$$R_1{}^1 = \frac{R_1}{a_{11}} \qquad a_{1j}{}^1 = \frac{a_{1j}}{a_{11}} \qquad b_{1j}{}^1 = \frac{b_{1j}}{a_{11}}$$

$$R_i{}^1 = R_i - a_{i1}R_1{}^1 \qquad a_{ij}{}^1 = a_{ij} - a_{i1}a_{1j}{}^1 \qquad b_{ij}{}^1 = b_{ij} - a_{i1}b_{1j}{}^1 \qquad (i \neq 1)$$

† If $a_{kk}^{k-1} = 0$, a row interchange is necessary to avoid division by zero.

Stage 2. Eliminate column 2 of A except element 2, 2.

$$A^{(2)} = \begin{bmatrix} 1 & 0 & a_{13}{}^2 & a_{14}{}^2 \\ 0 & 1 & a_{23}{}^2 & a_{24}{}^2 \\ 0 & 0 & a_{33}{}^2 & a_{34}{}^2 \\ 0 & 0 & a_{43}{}^2 & a_{44}{}^2 \end{bmatrix} \qquad B^{(2)} = \begin{bmatrix} b_{11}{}^2 & b_{12}{}^2 & b_{13}{}^2 & b_{14}{}^2 \\ b_{21}{}^2 & b_{22}{}^2 & b_{23}{}^2 & b_{24}{}^2 \\ b_{31}{}^2 & b_{32}{}^2 & b_{33}{}^2 & b_{34}{}^2 \\ b_{41}{}^2 & b_{42}{}^2 & b_{43}{}^2 & b_{44}{}^2 \end{bmatrix}$$

$$R_2{}^2 = \frac{R_2{}^1}{a_{22}{}^1} \qquad a_{2j}{}^2 = \frac{a_{2j}{}^1}{a_{22}{}^1} \qquad b_{2j}{}^2 = \frac{b_{2j}{}^1}{a_{22}{}^1}$$

$$R_i{}^2 = R_i{}^1 - a_{i2}{}^1 R_2{}^2 \qquad a_{ij}{}^2 = a_{ij}{}^1 - a_{i2}{}^1 a_{2j}{}^2 \qquad b_{ij}{}^2 = b_{ij}{}^1 - a_{i2}{}^1 b_{2j}{}^2 \qquad (i \neq 2)$$

Stage 3. Eliminate column 3 of A except element 3, 3.

$$A^{(3)} = \begin{bmatrix} 1 & 0 & 0 & a_{14}{}^3 \\ 0 & 1 & 0 & a_{24}{}^3 \\ 0 & 0 & 1 & a_{34}{}^3 \\ 0 & 0 & 0 & a_{44}{}^3 \end{bmatrix} \qquad B^{(3)} = \begin{bmatrix} b_{11}{}^3 & b_{12}{}^3 & b_{13}{}^3 & b_{14}{}^3 \\ b_{21}{}^3 & b_{22}{}^3 & b_{23}{}^3 & b_{24}{}^3 \\ b_{31}{}^3 & b_{32}{}^3 & b_{33}{}^3 & b_{34}{}^3 \\ b_{41}{}^3 & b_{42}{}^3 & b_{43}{}^3 & b_{44}{}^3 \end{bmatrix}$$

$$R_3{}^3 = \frac{R_3{}^2}{a_{33}{}^2} \qquad a_{3j}{}^3 = \frac{a_{3j}{}^2}{a_{33}{}^2} \qquad b_{3j}{}^3 = \frac{b_{3j}{}^2}{a_{33}{}^2}$$

$$R_i{}^3 = R_i{}^2 - a_{i3}{}^2 R_3{}^3 \qquad a_{ij}{}^3 = a_{ij}{}^2 - a_{i3}{}^2 a_{3j}{}^3 \qquad b_{ij}{}^3 = b_{ij}{}^2 - a_{i3}{}^2 b_{3j}{}^3, \qquad (i \neq 3)$$

Stage 4. Eliminate column 4 of A except element 4, 4.

$$A^{(4)} = \begin{bmatrix} 1 & 0 & 0 & 0 \\ 0 & 1 & 0 & 0 \\ 0 & 0 & 1 & 0 \\ 0 & 0 & 0 & 1 \end{bmatrix} = I, \qquad B^{(4)} = \begin{bmatrix} b_{11}{}^4 & b_{12}{}^4 & b_{13}{}^4 & b_{14}{}^4 \\ b_{21}{}^4 & b_{22}{}^4 & b_{23}{}^4 & b_{24}{}^4 \\ b_{31}{}^4 & b_{32}{}^4 & b_{33}{}^4 & b_{34}{}^4 \\ b_{41}{}^4 & b_{42}{}^4 & b_{43}{}^4 & b_{44}{}^4 \end{bmatrix} = A^{-1}$$

$$R_4{}^4 = \frac{R_4{}^3}{a_{44}{}^3} \qquad a_{4j}{}^4 = \frac{a_{4j}{}^3}{a_{44}{}^3} \qquad b_{4j}{}^4 = \frac{b_{4j}{}^3}{a_{44}{}^3}$$

$$R_i{}^4 = R_i{}^3 - a_{i4}{}^3 R_4{}^4 \qquad a_{ij}{}^4 = a_{ij}{}^3 - a_{i4}{}^3 a_{4j}{}^4 \qquad b_{ij}{}^4 = b_{ij}{}^3 - a_{i4}{}^3 b_{4j}{}^4 \qquad (i \neq 4)$$

Numerical Example of Inversion by Elimination for $n = 4$

$$A = \begin{bmatrix} 4 & 8 & 2 & 1 \\ 1 & 5 & 3 & 8 \\ 2 & 7 & 1 & 4 \\ 3 & 8 & 2 & 1 \end{bmatrix} \qquad \begin{bmatrix} 1 & 0 & 0 & 0 \\ 0 & 1 & 0 & 0 \\ 0 & 0 & 1 & 0 \\ 0 & 0 & 0 & 1 \end{bmatrix} = B$$

Stage 1

$$A^{(1)} = \begin{bmatrix} 1 & 2 & 1/2 & 1/4 \\ 0 & 3 & 5/2 & 31/4 \\ 0 & 3 & 0 & 7/2 \\ 0 & 2 & 1/2 & 1/4 \end{bmatrix} \qquad \begin{bmatrix} 1/4 & 0 & 0 & 0 \\ -1/4 & 1 & 0 & 0 \\ -1/2 & 0 & 1 & 0 \\ -3/4 & 0 & 0 & 1 \end{bmatrix} = B^{(1)}$$

Stage 2

$$A^{(2)} = \begin{bmatrix} 1 & 0 & -7/6 & -59/12 \\ 0 & 1 & 5/6 & 31/12 \\ 0 & 0 & -15/6 & -17/4 \\ 0 & 0 & -7/6 & -59/12 \end{bmatrix} \quad \begin{bmatrix} 5/12 & -2/3 & 0 & 0 \\ -1/12 & 1/3 & 0 & 0 \\ -3/12 & -1 & 1 & 0 \\ -7/12 & -2/3 & 0 & 1 \end{bmatrix} = B^{(2)}$$

Stage 3

$$A^{(3)} = \begin{bmatrix} 1 & 0 & 0 & -44/15 \\ 0 & 1 & 0 & 7/6 \\ 0 & 0 & 1 & 17/10 \\ 0 & 0 & 0 & -44/15 \end{bmatrix} \quad \begin{bmatrix} 8/15 & -1/5 & -7/15 & 0 \\ -1/6 & 0 & 1/3 & 0 \\ 1/10 & 6/15 & -6/15 & 0 \\ -7/15 & -1/5 & -7/15 & 1 \end{bmatrix} = B^{(3)}$$

Stage 4

$$A^{(4)} = \begin{bmatrix} 1 & 0 & 0 & 0 \\ 0 & 1 & 0 & 0 \\ 0 & 0 & 1 & 0 \\ 0 & 0 & 0 & 1 \end{bmatrix} = I \quad \begin{bmatrix} 1 & 0 & 0 & -1 \\ -31/88 & -7/88 & 13/88 & 35/88 \\ -15/88 & 25/88 & -59/88 & 51/88 \\ 7/44 & 3/44 & 7/44 & -15/44 \end{bmatrix} = B^{(4)} = A^{-1}$$

Check Solution.

$$AA^{-1} = \begin{bmatrix} 4 & 8 & 2 & 1 \\ 1 & 5 & 3 & 8 \\ 2 & 7 & 1 & 4 \\ 3 & 8 & 2 & 1 \end{bmatrix} \begin{bmatrix} 1 & 0 & 0 & -1 \\ -31/88 & -7/88 & 13/88 & 35/88 \\ -15/88 & 25/88 & -59/88 & 51/88 \\ 7/44 & 3/44 & 7/44 & -15/44 \end{bmatrix}$$

$$= \begin{bmatrix} 1 & 0 & 0 & 0 \\ 0 & 1 & 0 & 0 \\ 0 & 0 & 1 & 0 \\ 0 & 0 & 0 & 1 \end{bmatrix}$$

Concurrent Computation of $|A|$

Initially set $\Delta = 1$ (Δ will be the value of the cumulative product at each stage of the elimination process).

For $k = 1$, $\Delta = 1a_{11}$.
For $k = 2$, $\Delta = 1a_{11}a_{22}^{1}$.
For $k = 3$, $\Delta = 1a_{11}a_{22}^{1}a_{33}^{2}$
For $k = 4$, $\Delta = 1a_{11}a_{22}^{1}a_{33}^{2}a_{44}^{3} = |A|$.

In the foregoing example the 4×4 matrix A was inverted by elimination, with the computations in each stage being performed without roundoff error (the numbers were carried in exact fractional form).

However, in FORTRAN, floating-point computations are performed with rounding, and this roundoff error is propagated from one computation to the next in much the same way that interest is compounded. Naturally we are interested in minimizing the roundoff error that occurs in each stage of the elimination process, and which in turn is propagated to the next stage. An analysis of the error expressions corresponding to the recursion formulas for the elimination process will reveal how this roundoff error can be minimized.

The recursion formula (5.51.1) for generating the $a_{ij}{}^k$ in the kth stage of the elimination process can be written in the form (where $a_{ij}{}^k$ is denoted by a_{ij}' and the a_{ij}^{k-1} are denoted by a_{ij})

$$(5.51.3) \qquad a_{ij}' = a_{ij} - \frac{a_{ik}}{a_{kk}} \cdot a_{kj}.$$

To simplify the development of the corresponding relative-error expression, we first rewrite the above recursion formula as

$$a_{ij}' = -\frac{a_{ik}}{a_{kk}} \cdot a_{kj} + a_{ij}.$$

Let r_{ij}' and r_{ij} denote the relative errors in a_{ij}' and the a_{ij}, respectively, and let δ, μ, σ denote the relative roundoff errors for division, multiplication, and subtraction in the kth stage of the elimination. Then the relative-error expression for the a_{ij}' recursion formula can be written directly as

$$r_{ij}' = [(r_{ik}\cdot 1 + r_{kk}\cdot(-1) + \delta)\cdot 1 + r_{kj}\cdot 1 + \mu]\left(\frac{(-a_{ik}/a_{kk})a_{kj}}{a_{ij}'}\right) + r_{ij}\frac{a_{ij}}{a_{ij}'} + \sigma$$

$$\underbrace{\quad N \;\; O \quad N \quad\;\; O \qquad R}\qquad\qquad$$
$$\underbrace{\qquad\qquad N \qquad\qquad O \quad N \; O \quad R}\qquad$$
$$\underbrace{\qquad\qquad\qquad\quad N \qquad\qquad\qquad O \qquad\quad N \; O \quad R}.$$
$$N$$

Using the relation (5.51.3), and multiplying the relative-error expression through by a_{ij}', we find

$$\varepsilon_{ij}' = [r_{ik} - r_{kk} + \delta + r_{kj} + \mu]\frac{-a_{ik}}{a_{kk}}a_{kj} + r_{ij}a_{ij} + \sigma\left[a_{ij} - \frac{a_{ik}}{a_{kk}}a_{kj}\right]$$

$$= [r_{ik} - r_{kk} + \delta + r_{kj} + \mu + \sigma]\frac{-a_{ik}}{a_{kk}}a_{kj} + (r_{ij} + \sigma)a_{ij}.$$

If the $|\delta|$, $|\mu|$, $|\sigma|$ are bounded by $5\cdot 10^{-d}$ and the $|r_{ij}|$ by $K\cdot 10^{-d}$, where $K \geq 5$, then the error bound for $|a_{ij}'|$ is

$$|\varepsilon_{ij}'| \leq \left[3(K + 5)|a_{kj}|\left|\frac{-a_{ik}}{a_{kk}}\right| + (K + 5)|a_{ij}|\right]\cdot 10^{-d}.$$

Examination of the error-bound expression reveals that the bound for $|\varepsilon_{ij}'|$ can be made smaller by proper choice of the ratio a_{ik}/a_{kk}, which is the ratio of column k elements in row i and row k. Therefore, if the element of maximum magnitude in column k on or below the main diagonal† say $a_{imax,k}$, is located, then rows *imax* and k can be interchanged so that the new element $|a_{kk}| \geq |a_{ik}|$, $i > k$. It follows then that the multipliers a_{ik}/a_{kk}, $i > k$, will each be less than or equal to one in magnitude, and consequently the error bound for $|\varepsilon_{ij}'|$ will be reduced. This process of interchanging rows *imax* and k, to preserve accuracy in the elimination process, is known as *pivotal elimination*, sometimes referred to as pivotal condensation. As a general rule, it is quite effective in preserving the accuracy in inverting a matrix by elimination or in solving a system of linear algebraic equations by elimination (see the Gauss-Jordan method, Sec. 5.7).

---

† Elements above the main diagonal cannot be considered because adding a multiple of row i, where $i < k$, to any other row could cause elements already zeroed to become non-zero. Therefore, since $|a_{ik}|$ may be larger than $|a_{kk}|$ for $i < k$, the magnitude of the multiplier $(-a_{ik}/a_{kk})$ may be larger than 1 for $i < k$.

Computational Summary for Inversion by Pivotal Elimination

To conserve storage, we will store all stages of A in the same array. (The same applies to B.)

Step 0. Input and initialization. Read ε, n [n = number of rows (columns) of A]. Read a_{ij} ($i, j = 1, n$). Set

$$b_{ij} = \delta_{ij} = \begin{pmatrix} 1, i = j \\ 0, i \neq j \end{pmatrix}, i, j = 1, n.$$

Set stage counter $k = 1$.

Set $\Delta = 1.0$.†

Step 1. Find pivot element (maximum magnitude element in column k on or below main diagonal). Compare $|a_{kk}|$, $|a_{k+1,k}|$, ..., $|a_{nk}|$ to find largest, say $|a_{imax,k}|$. Interchange row *imax* of A with row k of A (ditto for B).

if imax $\neq k$, $\Delta = -\Delta$.

Element $|a_{kk}|$ is now largest of set $\{|a_{kk}|, a_{k+1,k}|, ..., |a_{nk}|\}$.

Step 2. Test for near-zero maximum element. If $|a_{kk}| \leq \varepsilon$, go to error exit (singular or near-singular matrix). If $|a_{kk}| > \varepsilon$, continue to Step 3.

Step 3. Perform stage k of reduction process.

(a) Row $k = \dfrac{\text{Row } k}{a_{kk}}$

$$\Delta = a_{kk} \cdot \Delta.$$

Let $div = a_{kk}$.

$$a_{kj} = \frac{a_{kj}}{div}$$

$$b_{kj} = \frac{b_{kj}}{div} \quad (j = 1, n).$$

(b) Row i = Row $i - a_{ik}$ Row k. Let *mult* = a_{ik}.

$$a_{ij} = a_{ij} - mult \ a_{kj} \quad (i \neq k, \ j = 1, n)$$

$$b_{ij} = b_{ij} - mult \ b_{kj}.$$

Step 4. Test stage counter k. If $k < n$, set $k = k + 1$ and return to Step 1. If $k \geq n$, continue to Step 5.

Step 5. Write Output.

$$A^{-1} = B$$

$$|A| = \Delta.$$

† Statements in bold italics are those to be added for computation of determinant, $|A|$.

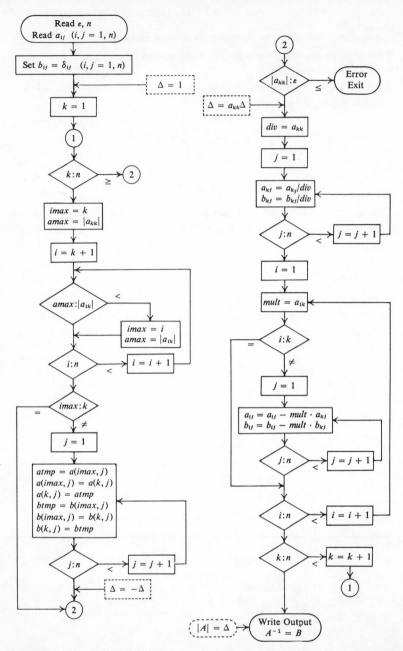

Figure 5-4. Flow chart for inversion by pivotal elimination. For concurrent computation of the determinant of A, add the detached blocks (in dashed lines) at points indicated by arrows.

```
      PROGRAM INVERSE
C     MATRIX INVERSION BY ELIMINATION WITH PARTIAL PIVOTING
C     ORIGINAL MATRIX = A INVERSE MATRIX = B
      DIMENSION A(20, 20), B(20, 20)
      READ (5,100) EPS, N
      READ (5,101) ((A(I,J), I=1,N), J=1,N)
C     CONSTRUCT IDENTITY MATRIX B(I,J) = I
      DO 6 I =1, N
      DO 5 J=1, N
      IF (I-J) 4,3,4
    3 B(I,J) = 1.0
      GO TO 5
    4 B(I, J) = 0.0
    5 CONTINUE
    6 CONTINUE
C     LOCATE MAXIMUM MAGNITUDE A(I, K) ON OR BELOW MAIN DIAGONAL
      DEL = 1.0
      DO 45 K = 1, N
      IF (K-N) 12, 30, 30
   12 IMAX = K
      AMAX = ABS (A(K, K))
      KP1 = K + 1
      DO 20 I = KP1, N
      IF (AMAX - ABS (A(I, K))) 15, 20, 20
   15 IMAX = I
      AMAX = ABS (A(I,K))
   20 CONTINUE
C     INTERCHANGE ROWS IMAX AND K IF IMAX NOT EQUAL TO K
      IF (IMAX - K) 25, 30, 25
   25 DO 29 J = 1, N
      ATMP = A (IMAX, J)
      A (IMAX, J) = A(K,J)
      A(K,J) = ATMP
      BTMP = B(IMAX, J)
      B(IMAX,J) = B(K,J)
   29 B(K,J)= BTMP
      DEL = -DEL
   30 CONTINUE
C     TEST FOR SINGULAR MATRIX
      IF (ABS(A(K,K)) - EPS) 93, 93, 35
   35 DEL = A(K,K) * DEL
C     DIVIDE PIVOT ROW BY ITS MAIN DIAGONAL ELEMENT
      DIV = A(K,K)
      DO 38 J = 1, N
      A(K,J) = A(K,J)/DIV
   38 B(K,J)= B(K,J)/DIV
C     REPLACE EACH ROW BY LINEAR COMBINATION WITH PIVOT ROW
      DO 43 I = 1, N
      AMULT = A(I,K)
      IF (I-K) 39, 43, 39
   39 DO 42 J = 1, N
      A(I,J) = A(I,J) - AMULT * A(K,J)
   42 B(I,J) = B(I,J) - AMULT * B(K,J)
   43 CONTINUE
   45 CONTINUE
      WRITE (6,120)
      WRITE (6,110) ((B(I,J), I=1,N), J=1,N)
      WRITE (6,121)
      WRITE (6,110) DEL
   99 STOP
   93 WRITE (6,113) K
      GO TO 99
  100 FORMAT (F10.8, I2)
  101 FORMAT (F15.8)
  110 FORMAT (2X, E15.8)
  113 FORMAT (25H SINGULAR MATRIX FOR K =, I2)
  120 FORMAT (20H ELEMENTS OF INVERSE)
  121 FORMAT (21H VALUE OF DETERMINANT)
      END
```

5.52 Hotelling's Method of Improving an Approximate Inverse

The use of pivotal condensation to minimize the effects of roundoff error in inverting a matrix by elimination was discussed in the preceding section. Probably in the majority of cases the inverse computed by pivotal condensation will be a fairly accurate approximation of the true inverse. However, if the elements of the inverse are sensitive to small changes in the elements of the original matrix, the matrix is said to be *ill-conditioned*. For example, the 2×2 matrix

$$A = \begin{bmatrix} 100 & 10 \\ 9.5 & 1 \end{bmatrix}$$

has as its inverse

$$\begin{bmatrix} 0.2 & -2.0 \\ -1.9 & 20.0 \end{bmatrix}.$$

If an element is changed by a small amount, say a_{21} becomes 9.9 instead of 9.5, the new matrix

$$\begin{bmatrix} 100 & 10 \\ 9.9 & 1 \end{bmatrix}$$

has for its inverse

$$\begin{bmatrix} 1 & -10 \\ -9.9 & 100 \end{bmatrix}.$$

The sensitivity of the elements of the inverse to small changes in elements of the original matrix is an indication of the ill-conditioning of the matrix.

In the case of an ill-conditioned matrix, the computed inverse may bear little resemblance to the true inverse, especially for large-order matrices wherein the errors are propagated through a number of computational stages. In general, the larger the order of the matrix, the greater the effect of the error propagation. And in ill-conditioned matrices, the effects of this can be devastating.

The effects of the propagation of roundoff error in the inversion of a matrix are illustrated in the following examples, wherein the computed inverse is actually only an approximation of the true inverse. An iterative method of improving such an approximate inverse, known as Hotelling's method, will be presented in this section.

Before developing the recursion formulas for Hotelling's iterative method of improving an approximate inverse, let us examine the analogous problem of improving an approximate reciprocal (inverse) of a real number.

Problem. Given an approximate reciprocal d_1 of the real number a, where $|1 - ad_1| < 1$, compute a sequence of successive approximations $d_2, d_3, \ldots, d_k$ which converge to the reciprocal of a, *without using division*.

Solution. If d_k is the kth approximation of the reciprocal of a, the corresponding error is

(5.52.1) $$e_k = 1 - ad_k.$$

Now, if $|e_1| = |1 - ad_1| < 1$, it follows that $\lim_{m \to \infty} e_1{}^m \to 0$.

If we can determine a recursion formula for the reciprocal approximations d_k such that the corresponding errors e_k satisfy a relation of the form

$$e_k = e_1{}^m$$

then the sequence $\{e_k\} = \{1 - ad_k\} \to 0$, provided that $|e_1| < 1$, and the sequence $\{d_k\} \to 1/a$.

Using the definition of the error e_k we see that

$$\begin{aligned} e_k{}^2 &= (1 - ad_k)(1 - ad_k) \\ &= 1 - 2ad_k + (ad_k)^2 \\ &= 1 - ad_k[1 + (1 - ad_k)] \\ &= 1 - ad_k[1 + e_k] \\ &= 1 - a[d_k(1 + e_k)]. \end{aligned}$$

It follows that if the d_k satisfy the recursion relation

(5.52.2) $$d_{k+1} = d_k(1 + e_k)$$

then

$$e_k{}^2 = 1 - ad_{k+1} = e_{k+1}.$$

That is, the $(k + 1)$st error equals the square of the kth error, so that

$$e_{k+1} = e_k{}^2 = e_{k-1}^4 = e_{k-2}^8 = \cdots = e_1{}^{2^k} \to 0, \qquad \text{when } |e_1| < 1.$$

Therefore, the sequence $\{d_{k+1}\}$ satisfying (5.52.2) will converge to the reciprocal of a, provided that $|e_1| < 1$, since in this case the sequence $\{e_{k+1}\} = \{1 - ad_{k+1}\}$ converges to zero.

NUMERICAL EXAMPLE

Compute an accurate reciprocal of $a = 13$, given that $d_1 = 0.05$. The initial error $e_1 = 1 - 13(0.05) = 0.35$ is less than 1 in magnitude.

$$\begin{aligned} d_2 &= 0.05(1 + 0.35) = 0.0675 \\ e_2 &= 1 - 13(0.0675) = 0.1225 \\ d_3 &= 0.0675(1 + 0.1225) = 0.07576875 \\ e_3 &= 1 - 13(0.07576875) = 0.01500625 \\ d_4 &= 0.07576875(1 + 0.01500625) = 0.07690575 \\ e_4 &= 1 - 13(0.07690575) = 0.00022525 \\ d_5 &= 0.07690575(1 + 0.00022525) = 0.076923073 \\ e_5 &= 1 - 13(0.076923073) = 0.000000051. \end{aligned}$$

Suppose that an approximate inverse D_1 of a square matrix A has been calculated by a direct inversion method, such as the elimination method. A simple iterative procedure, Hotelling's method, uses the approximate inverse D_1 as a starter matrix and calculates a sequence of matrices $\{D_k\}$, which converge to the actual inverse A^{-1} provided certain conditions are met. These conditions will be explained in the development of the method.

The derivation of Hotelling's method is based on the following theorem.

Theorem. If the norm† of E, denoted $N(E)$, is such that $N(E) < 1$, then

$$\lim_{m \to \infty} N(E^m) \to 0.$$

Proof of theorem. Using the fact that

$$N(E^m) = N(E^{m-1}E) \le N(E^{m-1})N(E)‡$$

we write

$$N(E^m) \le N(E^{m-1})N(E) \le N(E^{m-2})[N(E)]^2 \le \cdots \le [N(E)]^m$$

so that $N(E) < 1$ implies that $\lim [N(E)]^m \to 0$. Consequently, $N(E^m) \to 0$.

Using this theorem, we can derive Hotelling's method as follows: Assume that we have computed matrix D_1 such that

(5.52.3) $D_1 \doteq A^{-1}.$

Starting with this approximate inverse D_1, we want to develop an iterative procedure that will generate a sequence of matrices $D_2, D_3, \ldots, D_{k+1}$ that converges to A^{-1}.

For each iteration of the procedure we obtain an error matrix E_k defined by the relation

(5.52.4) $E_k = I - AD_k$

where $k = 1, 2, \ldots$ is the iteration index. Note that (5.52.4) can also be written in the form

(5.52.5) $AD_k = I - E_k.$

Now, if sequence $\{E_k\} \to 0$, it is evident from (5.52.5) that the sequence $\{D_k\} \to A^{-1}$.

Now, if $N(E_k) < 1$, it follows that

(5.52.6) $N(E_k^2) \le N(E_k)N(E_k) < 1N(E_k).$

Therefore, if the algorithm that generates $\{D_k\}$ produces error matrices $E_{k+1} = E_k^2$, then by (5.52.6), it follows that

(5.52.7) $N(E_{k+1}) < N(E_k)$

---

† We will use the following definition for the norm $N(E)$ of a matrix E:

$$N(E) = \max_i \sum_j |e_{ij}|.$$

‡ Fadeeva [4, p. 61].

provided $N(E_k) < 1$, so that successive error matrices will have smaller norms.

Using definition (5.52.4), we can write

$$(5.52.8) \qquad \begin{aligned} E_k{}^2 &= (I - AD_k)(I - AD_k) \\ &= I^2 - 2IAD_k + (AD_k)^2 \\ &= I - AD_k[2I - AD_k] \\ &= I - AD_k[I + (I - AD_k)] \\ &= I - AD_k[I + E_k] \\ &= I - A[D_k(I + E_k)]. \end{aligned}$$

Now, if we define

$$(5.52.9) \qquad\qquad D_{k+1} = D_k(I + E_k)$$

we find from (5.52.8) that

$$(5.52.10) \qquad\qquad E_k{}^2 = I - AD_{k+1}$$

but by definition of the error matrices

$$(5.52.11) \qquad\qquad E_{k+1} = I - AD_{k+1}.$$

Equating these last two expressions, we see that

$$(5.52.12) \qquad\qquad E_{k+1} = E_k{}^2.$$

We now have a recursion formula for generating a sequence of successive approximations $D_2, D_3, \ldots, D_k$, such that $E_k = E_{k-1}^2$. The algorithm for Hotelling's method can be stated as follows:

Given an approximate inverse D_1, compute error matrix $E_1 = I - AD_1$. If $N(E_1) < 1$, then generate the sequences of successive inverse approximations and error matrices defined by the recursion formulas

$$(5.52.13) \qquad \begin{aligned} D_k &= D_{k-1}(I + E_{k-1}) \\ E_k &= I - AD_k \end{aligned} \qquad (k = 2, 3, \ldots).$$

To show that the sequence $\{D_k\} \to A^{-1}$, we use the above algorithm to compute the following sequences of matrices, using D_1 as a starter for the recursion:

Approx. Inverses	Error Matrices
D_1	$E_1 = I - AD_1$
$D_2 = D_1(I + E_1)$	$E_2 = I - AD_2 = E_1{}^2$
$D_3 = D_2(I + E_2)$	$E_3 = I - AD_3 = E_2{}^2 = E_1{}^4$
$D_4 = D_3(I + E_3)$	$E_4 = I - AD_4 = E_3{}^2 = E_2{}^4 = E_1{}^8$
$\cdot \quad \cdot \quad \cdot \quad \cdot \quad \cdot$	$\cdot \quad \cdot \quad \cdot \quad \cdot \quad \cdot \quad \cdot \quad \cdot \quad \cdot$
$D_k = D_{k-1}(I + E_{k-1})$	$E_k = I - AD_k = E_{k-1}^2 = \cdots = E_1^{2^{(k-1)}}$

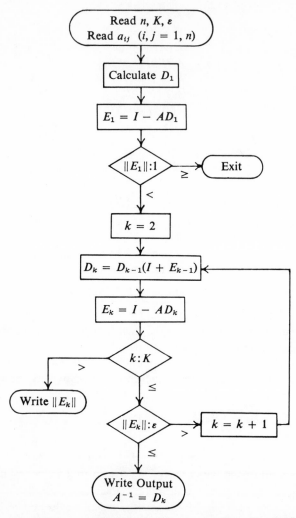

Figure 5-5. Flow chart for Hotelling's method: Here ε = convergence term; K = maximum number of iterations; n = order of matrix A. $\|E_k\|$ denotes $N(E_k)$.

Therefore, if $N(E_1) < 1$, it follows from theorem (p. 164) that $N(E_1^{2^{(k-1)}}) \to 0$. Hence, $N(E_k) \to 0$, which implies that $e_{ij} \to 0$, $(i,j = 1,n)$, so that $\{D_k\} \to A^{-1}$.

A representative flow chart (Figure 5-5) illustrates the Hotelling algorithm.

EXAMPLE.   Compute an approximate inverse D_1 of the 3×3 matrix

$$A = \begin{bmatrix} 9 & 10 & 10 \\ 8 & 9 & 10 \\ 7 & 8 & 9 \end{bmatrix} \qquad \textbf{Note:} \quad \text{Exact } A^{-1} = \begin{bmatrix} -1 & 10 & -10 \\ 2 & -11 & 10 \\ -1 & 2 & -1 \end{bmatrix}$$

by the elimination method, carrying two decimal places to the right of the point. Then compute improved approximations $D_2, D_3, \ldots$ by Hotelling's method.

Solution. Part 1. Computing approximate inverse D_1 by elimination:

$$A = \begin{bmatrix} 9 & 10 & 10 \\ 8 & 9 & 10 \\ 7 & 8 & 9 \end{bmatrix} \qquad \begin{bmatrix} 1 & 0 & 0 \\ 0 & 1 & 0 \\ 0 & 0 & 1 \end{bmatrix} = I$$

Stage 1 of inversion:

$$\begin{bmatrix} 1 & 1.11 & 1.11 \\ 0 & 0.12 & 0.12 \\ 0 & 0.23 & 1.23 \end{bmatrix} \qquad \begin{bmatrix} 0.11 & 0 & 0 \\ -0.88 & 1 & 0 \\ -0.77 & 0 & 1 \end{bmatrix}$$

Stage 2 of inversion:

$$\begin{bmatrix} 1 & 0 & -9.25 \\ 0 & 1 & 9.33 \\ 0 & 0 & -0.92 \end{bmatrix} \qquad \begin{bmatrix} 8.25 & -9.25 & 0 \\ -7.33 & 8.33 & 0 \\ 0.92 & -1.92 & 1 \end{bmatrix}$$

Stage 3 of inversion:

$$\begin{bmatrix} 1 & 0 & 0 \\ 0 & 1 & 0 \\ 0 & 0 & 1 \end{bmatrix} \qquad \begin{bmatrix} -1.00 & 10.08 & -10.08 \\ 2.00 & -11.17 & 10.16 \\ -1.00 & 2.09 & -1.09 \end{bmatrix} = D_1$$

Part 2. Compute improved approximate inverses by Hotelling:

$$AD_1 = \begin{bmatrix} 9 & 10 & 10 \\ 8 & 9 & 10 \\ 7 & 8 & 9 \end{bmatrix} \begin{bmatrix} -1.00 & 10.08 & -10.08 \\ 2.00 & -11.17 & 10.16 \\ -1.00 & 2.09 & -1.09 \end{bmatrix} = \begin{bmatrix} 1.00 & -0.08 & -0.02 \\ 0.00 & 1.01 & -0.10 \\ 0.00 & 0.01 & 0.91 \end{bmatrix}$$

$$E_1 = I - AD_1 = \begin{bmatrix} 0.00 & 0.08 & 0.02 \\ 0.00 & -0.01 & 0.10 \\ 0.00 & -0.01 & 0.09 \end{bmatrix}$$

$$D_2 = D_1(I + E_1) = \begin{bmatrix} -1.00 & 10.00 & -10.00 \\ 2.00 & -11.00 & 9.99 \\ -1.00 & 2.00 & -1.00 \end{bmatrix}$$

$$E_2 = I - AD_2 = \begin{bmatrix} 0.00 & 0.00 & 0.10 \\ 0.00 & 0.00 & 0.09 \\ 0.00 & 0.00 & 0.08 \end{bmatrix}$$

$$D_3 = D_2(I + E_2) = \begin{bmatrix} -1.00 & 10.00 & -10.00 \\ 2.00 & -11.00 & 10.00 \\ -1.00 & 2.00 & -1.00 \end{bmatrix}$$

Summary

The elimination method for computing an inverse and a determinant concurrently offers excellent accuracy, efficiency, and minimal storage requirements if properly programmed.

Inversion by elimination requires approximately n^3 multiplications and divisions for an arbitrary $n \times n$ matrix. The storage requirement of $2n^2$ locations, used in the program in this book, can be almost halved by using in-place inversion (i.e., A^{-1} is stored in the same array as A), such as used in the Gauss-Jordan-Rutishauser total pivotal elimination method.

The elimination method for computing a matrix inverse is quite accurate provided that pivotal elimination is used to minimize the effects of roundoff error. Wilkinson† indicates that total pivotal elimination may in some cases have to replace partial (column) pivotal elimination to obtain the required accuracy. If the matrix is ill-conditioned, Hotelling's method can be used effectively to compute an improved inverse if the norm of the first error matrix is less than 1. The use of double-precision arithmetic with pivotal elimination can also be used to minimize the cumulative effects of roundoff error.

An excellent discussion of "Matrix Inversion and Related Topics by Direct Methods" is given in Ralston and Wilf [24, p. 39] together with a variation of the Gauss-Jordan elimination technique wherein the product form of the inverse is computed.

A number of other methods for inverting matrices are available. Two representative methods are the method of rank annihilation (see Ralston-Wilf [24, p. 73]) and the Monte Carlo method (Ralston and Wilf, p. 78). The first method requires approximately $\frac{7}{3} n^3$ multiplications and divisions for inverting an arbitrary matrix, and storage of $2n^2 + 4n$ locations for matrix and vector arrays. The latter can be used to compute an approximate inverse, which can then be improved by the Hotelling method. The Monte Carlo method requires approximately $2n^2 + 3n$ locations for required arrays.

† J. H. Wilkinson, "Error Analysis of Direct Methods of Matrix Inversion," *J. ACM*, Vol. 8, 1961, p. 281.

Froberg [37, p. 94] gives an excellent illustration of the technique of in-place inversion for the minimization of storage requirements for the Gauss-Jordan elimination method.

SYSTEMS OF LINEAR ALGEBRAIC EQUATIONS

Outline

In this the second half of Chapter 5, methods are developed for directly solving systems of linear, non-homogeneous algebraic equations *instead* of inverting the matrix of coefficients. The classical solution of the system of algebraic equations $AX = B$ by Cramer's rule is first presented. However, because of the impracticality of this method for large systems, representative numerical methods of the following types are covered.

1. Direct methods—those methods that produce an exact solution (*provided no roundoff error occurs*) in a predetermined number of computational steps. The Gauss-Jordan elimination method, representative of this type, is efficient (requiring approximately $n^3/2$ multiplications and divisions) and accurate (provided that pivotal elimination is used to minimize the effects of roundoff).

2. Repeated use of direct methods—If the system of equations is ill-conditioned, as evidenced by a sensitivity of the solution to small changes in the elements of the coefficient matrix, then the solution obtained by direct methods (such as Gauss-Jordan) may be a poor representation of the true solution because of the accumulative effects of roundoff error. In such cases, the repeated use of direct methods may be employed effectively to produce an accurate solution.

3. Iterative methods—those methods that produce a sequence of successive approximations which, under specified conditions, converge to the true solution vector X. The Jacobi and Gauss-Seidel methods, representative of this type, are developed in Sec. 5.9. Iterative methods are well suited for dominant-diagonal systems, such as those systems arising in the solution of partial-differential equations by difference equations.

5.6   Classical Solution of Systems of Linear Algebraic Equations by Cramer's Rule

The system of n non-homogeneous linear algebraic equations in the n variables $x_1, x_2, \ldots, x_n$

(5.6.1)
$$
\begin{aligned}
a_{11}x_1 + a_{12}x_2 + \cdots + a_{1n}x_n &= b_1 \\
a_{21}x_1 + a_{22}x_2 + \cdots + a_{2n}x_n &= b_2 \\
\cdot \quad \cdot \quad \cdot \quad \cdot \quad \cdot \quad \cdot \quad \cdot \quad \cdot \quad \cdot \quad \cdot \\
a_{n1}x_1 + a_{n2}x_2 + \cdots + a_{nn}x_n &= b_n
\end{aligned}
$$

can be written as the single matrix equation

(5.6.2)
$$
\begin{bmatrix}
a_{11} & a_{12} & \cdots & a_{1n} \\
a_{21} & a_{22} & \cdots & a_{2n} \\
\cdot & \cdot & \cdot & \cdot \\
a_{n1} & a_{n2} & & a_{nn}
\end{bmatrix}
\begin{bmatrix}
x_1 \\
x_2 \\
\vdots \\
x_n
\end{bmatrix}
=
\begin{bmatrix}
b_1 \\
b_2 \\
\vdots \\
b_n
\end{bmatrix}
$$

or symbolically as the matrix equation

(5.6.3) $AX = B$

where $A = (a_{ij})$ is the matrix of coefficients, $X = (x_i)$ is the matrix of unknowns, and $B = (b_i)$ is the matrix of constants.

If the determinant of the matrix of coefficients is non-zero, then there exists a unique solution X of this system of equations. That is, since $|A| \neq 0$, there exists a unique inverse matrix A^{-1}, so that we can obtain the solution X by simply premultiplying both sides of (5.6.3) by A^{-1}, i.e.,

(5.6.4) $A^{-1}AX = A^{-1}B$

and since $A^{-1}A = I$, we obtain the solution

(5.6.5) $X = A^{-1}B.$

We see then that *the solution of a system of linear algebraic equations* of the form of (5.6.1) *can be obtained by computing the inverse* (by some appropriate method as described in Sec. 5.5) *and forming the product of the matrices A^{-1} and B. In the remaining sections of this chapter, we will describe other practical methods for computing the solution of such a system of linear equations.*

Cramer's Rule

Using the definition

$$
A^{-1} = \frac{1}{|A|} \, \text{Adj}(A)
$$

we can express the solution given in (5.6.5) in the form

(5.6.6) $X = \dfrac{1}{|A|} \, \text{Adj}(A)B$

This equation can then be written as

(5.6.7)
$$
X = \frac{1}{|A|}
\begin{bmatrix}
A_{11} & A_{21} & \cdots & A_{n1} \\
A_{12} & A_{22} & \cdots & A_{n2} \\
\cdot & \cdot & \cdot & \cdot \\
A_{1n} & A_{2n} & & A_{nn}
\end{bmatrix}
\begin{bmatrix}
b_1 \\
b_2 \\
\vdots \\
b_n
\end{bmatrix}
$$

where the A_{ij} are the cofactors of elements a_{ij}. Carrying out the indicated matrix multiplication, we obtain

$$(5.6.8) \qquad X = \frac{1}{|A|} \begin{bmatrix} b_1 A_{11} + b_2 A_{21} + \cdots + b_n A_{n1} \\ b_1 A_{12} + b_2 A_{22} + \cdots + b_n A_{n2} \\ \cdot \quad \cdot \quad \cdot \quad \cdot \quad \cdot \quad \cdot \quad \cdot \quad \cdot \\ b_1 A_{1n} + b_2 A_{2n} + \cdots + b_n A_{nn} \end{bmatrix}.$$

The ith row of the expression above is

$$(5.6.9) \qquad x_i = \frac{1}{|A|} (b_1 A_{1i} + b_2 A_{2i} + \cdots + b_n A_{ni}).$$

In Sec. 5.4 we stated that the determinant of the matrix obtained by replacing column i of matrix A by column $(b_1 \quad b_2 \quad \ldots \quad b_n)$ can be expressed as

$$(5.6.10) \qquad |A_i| = b_1 A_{1i} + b_2 A_{2i} + \cdots + b_n A_{ni}.$$

Therefore, we can substitute (5.6.10) into (5.6.9) to obtain Cramer's rule for computing *the solution of system (5.6.1)*:

$$x_i = \frac{|A_i|}{|A|} \qquad (i = 1, 2, \ldots, n).$$

Theoretically, the solution defined in the form of Cramer's rule can be used to solve systems of n non-homogeneous linear equations in n unknowns. However, this method of solution requires the evaluation of $n + 1$ determinants of order n, and is impractical for n greater than 3 or 4. Therefore, for large n, Cramer's rule is only of academic interest. In practice, more efficient methods (such as the elimination methods discussed in the following section) are used to solve large systems.

5.7   A Direct Numerical Method: The Gauss-Jordan Method

The Gauss-Jordan method for solving simultaneous non-homogeneous linear algebraic equations is a variation of the basic Gauss elimination method described in numerous texts. We will first develop this method by explaining the theory in terms of premultiplication by elementary matrices, then illustrate the elimination process in terms of row operations, and finally present the recursion formulas by which the solution is calculated in practice. It should be kept in mind that these are three equivalent methods of accomplishing the elimination process for reducing a system of linear equations with a square matrix of coefficients to an equivalent system of equations with a diagonal matrix of coefficients.

Premultiplication by Elementary Matrices

Analytically, the Gauss-Jordan method solves the matrix equation

$$(5.7.1) \qquad\qquad AX = B$$

by premultiplying both sides of the equation by a sequence† of elementary matrices $E_K E_{K-1} \ldots E_2 E_1$ which reduce matrix A to the identity I; i.e., such that

$$(5.7.2) \qquad\qquad E_K E_{K-1} \ldots E_2 E_1 A = I.$$

Hence, premultiplying both sides of (5.7.1) by the indicated sequence of elementary matrices, we obtain

$$(5.7.3) \qquad E_K E_{K-1} \ldots E_2 E_1 A X = E_K E_{K-1} \ldots E_2 E_1 B.$$

Then, by (5.7.2), this equation reduces to the form

$$(5.7.4) \qquad\qquad IX = E_K E_{K-1} \ldots E_2 E_1 B$$

and we see that the solution of (5.7.1) is

$$(5.7.5) \qquad\qquad X = E_K E_{K-1} \ldots E_2 E_1 B.$$

Row Operations on the Augmented Matrix

Corresponding to the sequence $E_K E_{K-1} \ldots E_2 E_1$ of elementary matrices is a sequence of row operations that reduces the augmented matrix A, B to the form I, B^+. To simplify the recursion formulas developed later, we define $b_i \equiv a_{i,n+1}$, so that we can write the system of equations corresponding to (5.7.1) in the form

$$(5.7.6) \qquad
\begin{aligned}
a_{11}x_1 + a_{12}x_2 + \cdots + a_{1n}x_n &= a_{1,n+1} \\
a_{21}x_1 + a_{22}x_2 + \cdots + a_{2n}x_n &= a_{2,n+1} \\
&\cdot \\
a_{n1}x_1 + a_{n2}x_2 + \cdots + a_{nn}x_n &= a_{n,n+1}
\end{aligned}$$

and we can write the augmented matrix of this system of equations as

$$(5.7.7) \qquad
\begin{bmatrix}
a_{11} & a_{12} & \cdots & a_{1n} & a_{1,n+1} \\
a_{21} & a_{22} & \cdots & a_{2n} & a_{2,n+1} \\
\cdot & \cdot & \cdot & \cdot & \cdot \\
a_{n1} & a_{n2} & \cdots & a_{nn} & a_{n,n+1}
\end{bmatrix}.$$

† Such a sequence exists, provided that matrix A is non-singular.

Now, corresponding to the sequence of elementary matrices $E_K E_{K-1} \ldots$ $E_2 E_1$ is a sequence of elementary row operations that will reduce augmented matrix (5.7.7) to the form (superscripts denote stage of reduction)

$$
\begin{bmatrix}
1 & 0 & \cdots & 0 & a^n_{1,n+1} \\
0 & 1 & \cdots & 0 & a^n_{2,n+1} \\
\cdot & \cdot & \cdot & \cdot & \cdot \\
0 & 0 & \cdots & 1 & a^n_{n,n+1}
\end{bmatrix}.
$$

This augmented matrix corresponds to equation (5.7.4). Then the solution is

(5.7.8) $\qquad\qquad x_i = a^n_{i,n+1} \qquad (i = 1, n).$

Recursion Formulas for Computing the Solution

As with the indicated sequence of row operations, we can write the recursion formulas for stage k $(k = 1, n)$ of this reduction-by-elimination process as follows:

(5.7.9) $\quad \left. \begin{aligned} a_{kj}{}^k &= a_{kj}^{k-1}/a_{kk}^{k-1} \\ a_{ij}{}^k &= a_{ij}^{k-1} - a_{ik}^{k-1} a_{kj}{}^k \quad (i \neq k) \end{aligned} \right\} (j = k, n+1) \qquad (k = 1, n)$

where $a_{ij}{}^0 \equiv a_{ij}$ are the elements of the original matrix.

Note: The range of the index j can be $(k, n+1)$ instead of range $(1, n+1)$ because $a_{kj}{}^k = 0$ when $j < k$. This change of index range saves computing time by not recomputing elements that are zeroed in a preceding stage.

Numerical Example

The Gauss-Jordan method is a *direct method* of solving a system of linear algebraic equations, i.e., a method by which the solution can be calculated by a finite predetermined number of computational steps. In the example below, we solve a system of equations with integer coefficients and constants so that the coefficients and constants in each stage of the reduction are rational numbers (represented in *exact* form as fractions), so that no roundoff error occurs in this example. This example illustrates that a direct method computes an *exact* solution provided no roundoff error occurs.

The system of equations to be solved is

$$
\begin{aligned}
4x_1 + x_2 + 2x_3 &= 16 \\
x_1 + 3x_2 + x_3 &= 10 \\
x_1 + 2x_2 + 5x_3 &= 12
\end{aligned}
$$

and the augmented matrix of this system of equations is

$$\begin{bmatrix} 4 & 1 & 2 & 16 \\ 1 & 3 & 1 & 10 \\ 1 & 2 & 5 & 12 \end{bmatrix}.$$

Stage 1 of the reduction eliminates off-diagonal elements in column 1:

$$\begin{bmatrix} 1 & 1/4 & 1/2 & 4 \\ 0 & 11/4 & 1/2 & 6 \\ 0 & 7/4 & 9/2 & 8 \end{bmatrix}.$$

Stage 2 of the reduction eliminates off-diagonal elements in column 2:

$$\begin{bmatrix} 1 & 0 & 5/11 & 38/11 \\ 0 & 1 & 2/11 & 24/11 \\ 0 & 0 & 46/11 & 46/11 \end{bmatrix}.$$

Stage 3 of the reduction eliminates off-diagonal elements in column 3:

$$\begin{bmatrix} 1 & 0 & 0 & 3 \\ 0 & 1 & 0 & 2 \\ 0 & 0 & 1 & 1 \end{bmatrix}.$$

The solution then is $(x_1, x_2, x_3) = (3, 2, 1)$. It should be noted again that the solution is exact because we have used a direct method and numbers such that no roundoff error occurs.

The step-by-step reduction of three equations in three unknowns is illustrated below, with the required row operations indicated on the right. The system of equations

$$a_{11}x_1 + a_{12}x_2 + a_{13}x_3 = a_{14}$$
$$a_{21}x_1 + a_{22}x_2 + a_{23}x_3 = a_{24}$$
$$a_{31}x_1 + a_{32}x_2 + a_{33}x_3 = a_{34}$$

has augmented matrix

$$\begin{bmatrix} a_{11} & a_{12} & a_{13} & a_{14} \\ a_{21} & a_{22} & a_{23} & a_{24} \\ a_{31} & a_{32} & a_{33} & a_{34} \end{bmatrix} \quad \begin{matrix} R_1 \\ R_2 \\ R_3. \end{matrix}$$

Stage 1 of the reduction eliminates off-diagonal elements in column 1:

$$\begin{bmatrix} 1 & a_{12}^1 & a_{13}^1 & a_{14}^1 \\ 0 & a_{22}^1 & a_{23}^1 & a_{24}^1 \\ 0 & a_{32}^1 & a_{33}^1 & a_{34}^1 \end{bmatrix} \quad \begin{matrix} R_1^1 = R_1/a_{11} & ①† \\ R_2^1 = R_2 - a_{21}R_1^1 & ② \\ R_3^1 = R_3 - a_{31}R_1^1 & ③ \end{matrix}$$

† Circled numbers indicate order of row operations within stage k ($k = 1, 2, 3$).

Stage 2 of the reduction eliminates off-diagonal elements in column 2:

$$\begin{bmatrix} 1 & 0 & a_{13}^2 & a_{14}^2 \\ 0 & 1 & a_{23}^2 & a_{24}^2 \\ 0 & 0 & a_{33}^2 & a_{34}^2 \end{bmatrix} \quad \begin{array}{ll} R_1^2 = R_1^1 - a_{12}^1 R_2^2 & ② \\ R_2^2 = R_2^1/a_{22}^1 & ① \\ R_3^2 = R_3^1 - a_{32}^1 R_2^2 & ③ \end{array}$$

Stage 3 of the reduction eliminates off-diagonal elements in column 3:

$$\begin{bmatrix} 1 & 0 & 0 & a_{14}^3 \\ 0 & 1 & 0 & a_{24}^3 \\ 0 & 0 & 1 & a_{34}^3 \end{bmatrix} \quad \begin{array}{ll} R_1^3 = R_1^2 - a_{13}^2 R_3^3 & ② \\ R_2^3 = R_2^2 - a_{23}^2 R_3^3 & ③ \\ R_3^3 = R_3^2/a_{33}^2 & ① \end{array}$$

The reduced system of equations can therefore be written in the form

$$x_1 + 0x_2 + 0x_3 = a_{14}^3$$
$$0x_1 + x_2 + 0x_3 = a_{24}^3$$
$$0x_1 + 0x_2 + x_3 = a_{34}^3$$

so that the solution of the system of equations is

$$x_1 = a_{14}^3$$
$$x_2 = a_{24}^3$$
$$x_3 = a_{34}^3$$

The actual computation of the elements in each stage of the Gauss-Jordan method illustrated above would be accomplished by the recursion formulas (5.7.9).

Numerical Example Illustrating Effects of Roundoff Error

Consider now the solution of the system of equations

$$15.0x_1 + 15.0x_2 + 14.0x_3 = 58.0$$
$$15.0x_1 + 14.0x_2 + 13.0x_3 = 55.0$$
$$14.0x_1 + 13.0x_2 + 12.0x_3 = 51.0.$$

To illustrate the effects of roundoff error in the Gauss-Jordan method, we will represent the coefficients and constants in decimal form, and the computations will be rounded in the 6th place to the right of the decimal point. The augmented matrix of the original system of equations is

$$\begin{bmatrix} 15.0 & 15.0 & 14.0 & 58.0 \\ 15.0 & 14.0 & 13.0 & 55.0 \\ 14.0 & 13.0 & 12.0 & 51.0 \end{bmatrix}.$$

The stages of the Gauss-Jordan elimination method are as follows:

Stage 1

$$\begin{bmatrix} 1.0 & 1.0 & 0.933333 & 3.866667 \\ 0.0 & -1.0 & -0.999995 & -3.000005 \\ 0.0 & -1.0 & -1.066662 & -3.133338 \end{bmatrix}$$

Stage 2

$$\begin{bmatrix} 1.0 & 0.0 & -0.066662 & 0.866662 \\ 0.0 & 1.0 & 0.999995 & 3.000005 \\ 0.0 & 0.0 & 0.066667 & 0.133333 \end{bmatrix}$$

Stage 3

$$\begin{bmatrix} 1.0 & 0.0 & 0.0 & 0.999985 \\ 0.0 & 1.0 & 0.0 & 1.000030 \\ 0.0 & 0.0 & 1.0 & 1.999985 \end{bmatrix}$$

so that the solution by the Gauss-Jordan method carrying six decimal places is:

$$x_1 = 0.999985$$
$$x_2 = 1.000030$$
$$x_3 = 1.999985.$$

It can be shown by direct substitution that the exact solution is (x_1, x_2, x_3) = $(1, 1, 2)$. Hence, we see that the computed solution differs from the exact solution because of the accumulative effects of roundoff error, even though a direct method of solution is used.

5.8 Repeated Application of a Direct Method: A Method of Solving Ill-Conditioned Systems of Linear Equations

An ill-conditioned system of linear equations is characterized by a matrix of coefficients whose determinant is small in comparison with the size of the elements of the matrix. If we interpret the equations of the system as hyperplanes in an n-dimensional coordinate system, then the elements $(a_{i1}, a_{i2}, \ldots, a_{in})$ of row i represent the *direction* numbers of the "normal" to the ith hyperplane. In this context, we can interpret an ill-conditioned system of linear equations as a set of n hyperplanes whose "normals" are very nearly "parallel."

If the system of linear equations is ill-conditioned, then the solution obtained by a direct method (such as the Gauss or the Gauss-Jordan method) may be only a poor approximation of the true solution, primarily because of the accumulative effects of roundoff in the successive stages of the solution process. As noted in the preceding section, this roundoff error can be limited to some degree by using pivotal condensation. Additionally, the roundoff can be reduced by carrying more decimal places (guarding figures) in each computational step.

However, if the system of equations is terribly ill-conditioned, it may be necessary to obtain the solution by the following iterative procedure. Let

X_1 denote the solution (approximate) of $AX = B$ obtained by a direct method such as the Gauss-Jordan algorithm. Substituting X_1 into system

(5.8.1) $AX = B$

we can compute an error vector E_1 defined by the relation

(5.8.2) $AX_1 - B = E_1$.

Now, if we could determine a correction vector δX such that

(5.8.3) $A\, \delta X = -E_1$

then the result of adding the correction vector δX to vector X_1 would be

(5.8.4) $A(X_1 + \delta X) = B$.

The truth of this statement can be verified by adding (5.8.3) and (5.8.2). If this could be done, we could compute the solution by simply calculating X_1 and δX and adding them, since by (5.8.4) the sum $X_1 + \delta X$ is a solution of system (5.8.1).

However, since the system of equations is ill-conditioned, the solution of the correction equation (5.8.3) by a direct method will probably be some approximation δX_1 of true solution δX that satisfies the relation

(5.8.5) $A\, \delta X_1 - (-E_1) = E_2$.

Now, if we add (5.8.2) and (5.8.5), we obtain

(5.8.6) $A(X_1 + \delta X_1) - B = E_2$.

Therefore, if the magnitude of vector E_2 is smaller than the magnitude of vector E_1, then the approximation $X_1 + \delta X_1$ is better than the approximation X_1. It appears then that we can establish an iterative procedure of reapplying a direct method to the error vectors just as a direct method was applied to the constant vector B.

That is, we can continue this iterative procedure by solving the equation

(5.8.7) $A\, \delta X = -E_2$

obtaining an approximate solution δX_2 which is defined by the relation

(5.8.8) $A\, \delta X_2 - (-E_2) = E_3$.

Again, we add equations (5.8.6) and (5.8.8) to find

(5.8.9) $A(X_1 + \delta X_1 + \delta X_2) - B = E_3$.

The process is continued in this manner so that in the kth application of the method we solve the error equation $A\, \delta X = -E_k$ obtaining δX_k, such that

$$A\, \delta X_k - (-E_k) = E_{k+1}.$$

Then the kth approximation of the solution X of system $AX = B$ is the vector

$$X_1 + \delta X_1 + \delta X_2 + \cdots + \delta X_k$$

which satisfies the relation

$$A(X_1 + \delta X_1 + \delta X_2 + \cdots + \delta X_k) - B = E_{k+1}.$$

Therefore, if vector E_{k+1} has all near-zero elements, the approximate solution $X_1 + \delta X_1 + \delta X_2 + \cdots + \delta X_k$ converges to the true solution X.

In practice, there is a limitation on the number of iterations that continue to improve the solution. Normally, one or two iterations will suffice. If the number of leading zeros in the error components approaches the number of decimal places carried in the computations, then the amount of improvement in the next iteration is negligible.

5.9   Iterative Methods

Jacobi's Iterative Method

The recursion formulas for Jacobi's iterative method will be developed for solving three equations in three unknowns. The recursion formulas for solving n equations in n unknowns are then obtained by direct extension.

If the system of linear algebraic equations

(5.9.1)
$$\begin{aligned}
a_{11}x_1 + a_{12}x_2 + a_{13}x_3 &= b_1 \\
a_{21}x_1 + a_{22}x_2 + a_{23}x_3 &= b_2 \\
a_{31}x_1 + a_{32}x_2 + a_{33}x_3 &= b_3
\end{aligned}$$

has non-zero diagonal elements a_{ii} ($i = 1, 2, 3$), then it can be rewritten in the form

(5.9.2)
$$\begin{aligned}
x_1 &= (1/a_{11})[b_1 \qquad\quad - a_{12}x_2 - a_{13}x_3] \\
x_2 &= (1/a_{22})[b_2 - a_{21}x_1 \qquad\quad - a_{23}x_3] \\
x_3 &= (1/a_{33})[b_3 - a_{31}x_1 - a_{32}x_2 \qquad\quad]
\end{aligned}$$

i.e., such that in the ith equation variable x_i is expressed in terms of the remaining variables and b_i.

An iterative procedure for solving these equations can be established as follows: *First*, choose arbitrary values x_1^0, x_2^0, x_3^0 and substitute these values for x_1, x_2, x_3 on the right side of (5.9.2). Compute the values of the expressions on the right sides to obtain values for x_1, x_2, x_3 on the left, which we denote x_1^1, x_2^1, x_3^1. *Next*, these x_i^1 values can be substituted into the right sides of (5.9.2) to produce values x_i^2 on the left.

$$
\begin{array}{ll}
& x_1^{k+1} = (1/a_{11})[b_1 \qquad\quad - a_{12}x_2{}^k - a_{13}x_3{}^k] \\
(5.9.3) & x_2^{k+1} = (1/a_{22})[b_2 - a_{21}x_1{}^k \qquad\quad - a_{23}x_3{}^k] \qquad (k = 0, 1, \ldots) \\
& x_3^{k+1} = (1/a_{33})[b_3 - a_{31}x_1{}^k - a_{32}x_2{}^k \qquad\quad]
\end{array}
$$

where vector $(x_1{}^0, x_2{}^0, x_3{}^0)$ is used to start the iteration (for $k = 0$), and vector $(x_1{}^k, x_2{}^k, x_3{}^k)$ denotes the kth successive approximation of the true solution vector which we denote by $(\alpha_1, \alpha_2, \alpha_3)$.

Example of Jacobi's Method

Consider the following system of linear equations

$$
\begin{array}{rcrcrcl}
4x_1 &+& x_2 &+& 2x_3 &=& 16 \\
x_1 &+& 3x_2 &+& x_3 &=& 10 \\
x_1 &+& 2x_2 &+& 5x_3 &=& 12.
\end{array}
$$

Note that the order of the equations has been selected such that the main diagonal coefficients predominate over the off-diagonal coefficients. Since the diagonal elements are non-zero, we can express variable x_i in the ith equation in terms of the remaining variables and the constant. The result is written in the iterative form of (5.9.3) as follows:

$$
\begin{array}{ll}
x_1^{k+1} = (1/4)[16 \qquad\quad - x_2{}^k - 2x_3{}^k] \\
x_2^{k+1} = (1/3)[10 - x_1{}^k \qquad\quad - x_3{}^k] \\
x_3^{k+1} = (1/5)[12 - x_1{}^k - 2x_2{}^k \qquad\quad].
\end{array}
$$

This system of equations can also be written as the matrix equation

$$
\begin{bmatrix} x_1^{k+1} \\ x_2^{k+1} \\ x_3^{k+1} \end{bmatrix} =
\begin{bmatrix} 0 & -1/4 & -1/2 \\ -1/3 & 0 & -1/3 \\ -1/5 & -2/5 & 0 \end{bmatrix}
\begin{bmatrix} x_1{}^k \\ x_2{}^k \\ x_3{}^k \end{bmatrix} +
\begin{bmatrix} 4 \\ 10/3 \\ 12/5 \end{bmatrix}.
$$

The first few iterations of the solution are given below, starting with $(x_1{}^0, x_2{}^0, x_3{}^0) = (0, 0, 0)$ as initial solution approximation:

$$
k = 0 \quad
\begin{bmatrix} x_1^1 \\ x_2^1 \\ x_3^1 \end{bmatrix} =
\begin{bmatrix} 0 & -1/4 & -1/2 \\ -1/3 & 0 & -1/3 \\ -1/5 & -2/5 & 0 \end{bmatrix}
\begin{bmatrix} 0 \\ 0 \\ 0 \end{bmatrix} +
\begin{bmatrix} 4 \\ 10/3 \\ 12/5 \end{bmatrix} =
\begin{bmatrix} 4 \\ 10/3 \\ 12/5 \end{bmatrix}
$$

$$
k = 1 \quad
\begin{bmatrix} x_1^2 \\ x_2 \\ x_3 \end{bmatrix} =
\begin{bmatrix} 0 & -1/4 & -1/2 \\ -1/3 & 0 & -1/3 \\ -1/5 & -2/5 & 0 \end{bmatrix}
\begin{bmatrix} 4 \\ 10/3 \\ 12/5 \end{bmatrix} +
\begin{bmatrix} 4 \\ 10/3 \\ 12/5 \end{bmatrix} =
\begin{bmatrix} 59/30 \\ 18/15 \\ 4/15 \end{bmatrix}
$$

$$
k = 2 \quad
\begin{bmatrix} x_1^3 \\ x_2^3 \\ x_3^3 \end{bmatrix} =
\begin{bmatrix} 0 & -1/4 & -1/2 \\ -1/3 & 0 & -1/3 \\ -1/5 & -2/5 & 0 \end{bmatrix}
\begin{bmatrix} 59/30 \\ 18/15 \\ 4/15 \end{bmatrix} +
\begin{bmatrix} 4 \\ 10/3 \\ 12/5 \end{bmatrix} =
\begin{bmatrix} 107/30 \\ 233/90 \\ 229/150 \end{bmatrix}
$$

$$
k = 3 \quad
\begin{bmatrix} x_1^4 \\ x_2^4 \\ x_3^4 \end{bmatrix} =
\begin{bmatrix} 0 & -1/4 & -1/2 \\ -1/3 & 0 & -1/3 \\ -1/5 & -2/5 & 0 \end{bmatrix}
\begin{bmatrix} 107/30 \\ 233/90 \\ 229/150 \end{bmatrix} +
\begin{bmatrix} 4 \\ 10/3 \\ 12/5 \end{bmatrix} =
\begin{bmatrix} 4661/1800 \\ 736/450 \\ 313/450 \end{bmatrix}
$$

If the iterative solution is continued in this manner, the approximation $(x_1^{k+1}, x_2^{k+1}, x_3^{k+1})$ converges to exact solution (3, 2, 1).

Jacobi's method for n equations. The system of n linear algebraic equations in n variables

$$
\begin{aligned}
a_{11}x_1 + a_{12}x_2 + a_{13}x_3 + \cdots + a_{1n}x_n &= b_1 \\
a_{21}x_1 + a_{22}x_2 + a_{23}x_3 + \cdots + a_{2n}x_n &= b_2 \\
a_{31}x_1 + a_{32}x_2 + a_{33}x_3 + \cdots + a_{3n}x_n &= b_3 \\
\cdot \quad \cdot \quad \cdot \quad \cdot \quad \cdot \quad \cdot \quad \cdot \quad \cdot \quad \cdot \quad \cdot \quad \cdot \quad \cdot \quad \cdot \\
a_{n1}x_1 + a_{n2}x_2 + a_{n3}x_3 + \cdots + a_{nn}x_n &= b_n
\end{aligned}
$$

(5.9.4)

can be solved by Jacobi's iterative method provided that certain conditions are satisfied. One such *sufficient* condition for the convergence of Jacobi's method for n equations in n variables is the following:

(5.9.5) $$ \max_i \left(\frac{1}{|a_{ii}|} \sum_{j \neq i} |a_{ij}| \right) < 1 \qquad (i = 1, 2, \ldots, n). $$

The recursion formulas for Jacobi's method for n equations in n variables are obtained as a direct extension of the formulas for three equations in three variables. That is,

$$
\begin{aligned}
x_1^{k+1} &= (1/a_{11})[b_1 \qquad\qquad\;\; - a_{12}x_2^{k} - a_{13}x_3^{k} - \cdots - a_{1n}x_n^{k}] \\
x_2^{k+1} &= (1/a_{22})[b_2 - a_{21}x_1^{k} \qquad\quad - a_{23}x_3^{k} - \cdots - a_{2n}x_n^{k}] \\
x_3^{k+1} &= (1/a_{33})[b_3 - a_{31}x_1^{k} - a_{32}x_2^{k} \qquad\quad - \cdots - a_{3n}x_n^{k}] \\
\cdot \quad \cdot \quad \cdot \quad \cdot \quad \cdot \quad \cdot \quad \cdot \quad \cdot \quad \cdot \quad \cdot \quad \cdot \quad \cdot \\
x_n^{k+1} &= (1/a_{nn})[b_n - a_{n1}x_1^{k} - a_{n2}x_2^{k} - a_{n3}x_3^{k} - \cdots - a_{n,n-1}x_{n-1}^{k}].
\end{aligned}
$$
(5.9.6)

If (5.9.5) is satisfied, then Jacobi's recursion formulas (5.9.6) will generate a sequence of successive approximations that converge to the exact solution of system (5.9.4), starting with an arbitrary approximation $(x_1^0, x_2^0, x_3^0, \ldots x_n^0)$. Since (5.9.5) is a sufficient, not a necessary, condition, Jacobi's method may converge when (5.9.5) is not satisfied.

The Gauss-Seidel Iterative Method

This iterative method for solving systems of linear algebraic equations is a simple modification of Jacobi's method. If Jacobi's recursion formulas are changed so that *as* each improved value x_i^{k+1} is computed it is used in the calculation of $x_{i+1}^{k+1}, x_{i+2}^{k+1}, \ldots, x_n^{k+1}$, then we obtain the following Gauss-Seidel recursion formulas.

$$x_1^{k+1} = \frac{1}{a_{11}} [b_1 - a_{12}x_2^{k} - a_{13}x_3^{k} \qquad\qquad - \cdots - a_{1n}x_n^{k}]$$

$$x_2^{k+1} = \frac{1}{a_{22}} [b_2 - a_{21}x_1^{k+1} - a_{23}x_3^{k} \qquad\qquad - \cdots - a_{2n}x_n^{k}]$$

$$x_3^{k+1} = \frac{1}{a_{33}} [b_3 - a_{31}x_1^{k+1} - a_{32}x_2^{k+1} - a_{34}x_4^{k} - \cdots - a_{3n}x_n^{k}]$$

$$x_n^{k+1} = \frac{1}{a_{nn}} [b_n - a_{n1}x_1^{k+1} - a_{n2}x_2^{k+1} \qquad\qquad - \cdots - a_{n,n-1}x_{n-1}^{k+1}].$$
(5.9.7)

A *sufficient* condition for the convergence of the Gauss-Seidel method is that

$$(5.9.8) \qquad \max_i \left(\frac{1}{|a_{ii}|} \sum_{j \neq i} |a_{ij}| \right) < 1 \qquad (i = 1, 2, \ldots, n).$$

Note that condition (5.9.8) is the same as condition (5.9.5). However, for a given initial solution approximation, the Gauss-Seidel method may converge to the true solution whereas the Jacobi method will not, and conversely.

Todd [28, p. 404] shows that the rate of convergence for the Gauss-Seidel method is twice that of Jacobi's method. For this reason, the Gauss-Seidel method is preferred over Jacobi's method, provided that the initial approximation leads to convergence. For this reason, we present the computational summary, flow chart (Figure 5-6), and program for the Gauss-Seidel method. The reader may wish to summarize, flow chart, and program Jacobi's method as a worthwhile exercise.

Computational Summary of Gauss-Seidel Method

Step 0. Input parameters and data.

n = number of equations;
kc = maximum number of iterations;
ε = convergence criteria.

Augmented matrix A, B read into $n \times n + 1$ array A so that $a_{i,n+1} = b_i$ ($i = 1, n$).

Step 1. Divide ith equation by diagonal element a_{ii} ($a_{ii} \neq 0$).

$$a_{ij} = \frac{a_{ij}}{a_{ii}}, \qquad (i = 1, n; j = 1, n + 1).$$

Step 2. Set iteration counter $k = 1$. Set initial values for components $x_i^1 = 0.0$ ($i = 1, n$).

Step 3. Calculate successive iterates x_i^{k+1} using the computational formula

$$x_i^{k+1} = a_{i,n+1} - \sum_{j=1}^{i-1} a_{ij}x_j^{k+1} - \sum_{j=i+1}^{n} a_{ij}x_j^{k} \qquad (i = 1, n).$$

Step 4. Test convergence.

(a) If any $|x_i^{k+1} - x_i^k| > \varepsilon$,   go to Step 5.

(b) If all $|x_i^{k+1} - x_i^k| \leq \varepsilon$,   go to Step 6.

Step 5. Test iteration counter.

(a) If $k < kc$, increment k by 1, and return to Step 3.
(b) If $k \geq kc$, go to Step 7.

Step 6. Convergence exit. Write out solutions

$$x_i = x_i^{k+1} \qquad (i = 1, n).$$

Exit.

Step 7. Exit. Write "Program fails to converge in kc iterations." Exit.

```
      PROGRAM GAUSS
C     SOLVE AX = B BY GAUSS-SEIDEL ITERATIVE METHOD
C     AUGMENTED MATRIX A, B IS INPUT INTO N X N+1 ARRAY A
      DIMENSION A(20,21), XK(20), XKP1(20)
      READ (5,100) N, NP1, KC, EPS
      READ (5,101) ((A(I,J), I=1,N), J=1,NP1)
C     DIVIDE I TH EQUATION BY DIAGONAL TERM A(I, I)
      DO 15 I = 1, N
      DIV = A (I, I)
      DO 10 J = 1,NP1
   10 A(I,J) = A(I,J)/DIV
   15 CONTINUE
C     SET INITIAL VALUES
      K = 1
      DO 20 I = 1, N
   20 XK (I) = 0.0
C     CALCULATE (K+1)ST ITERATES XKP1(I) OF VARIABLE X(I)
   21 DO 30 I = 1, N
      XKP1(I) = A(I,NP1)
      DO 25 J = 1, N
      IF(J-I) 22, 25, 24
   22 XKP1(I) = XKP1(I) - A(I,J) * XKP1(J)
      GO TO 25
   24 XKP1(I) = XKP1(I) - A(I,J) * XK(J)
   25 CONTINUE
   30 CONTINUE
C     TEST CONVERGENCE OF ITERATION
      DO 40 I = 1, N
      IF (ABS(XKP1(I) - XK(I)) - EPS) 40,40,50
   40 CONTINUE
C     CONVERGENCE EXIT. WRITE OUT SOLUTIONS X (I)
      WRITE (6,120)
      WRITE (6,110) (XKP1(I), I=1,N)
   99 STOP
C     REPLACE K TH ITERATES BY K + 1 ST ITERATES
   50 IF(K-KC) 51, 55, 55
   51 K = K+1
      DO 52 I = 1, N
   52 XK(I) = XKP1(I)
      GO TO 21
   55 WRITE (6,113) KC
      GO TO 99
  100 FORMAT (3I2, F10.8)
  101 FORMAT (F19.9)
  110 FORMAT (2X, E19.9)
  113 FORMAT (28H0 FAILURE TO CONVERGE AFTER I2, 11H ITERATIONS)
  120 FORMAT (20H0 SOLUTIONS X(I) ARE)
      END
```

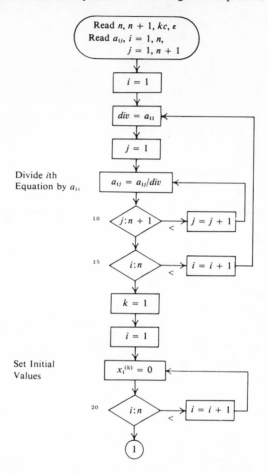

Figure 5-6. Flow chart for Gauss-Seidel method.

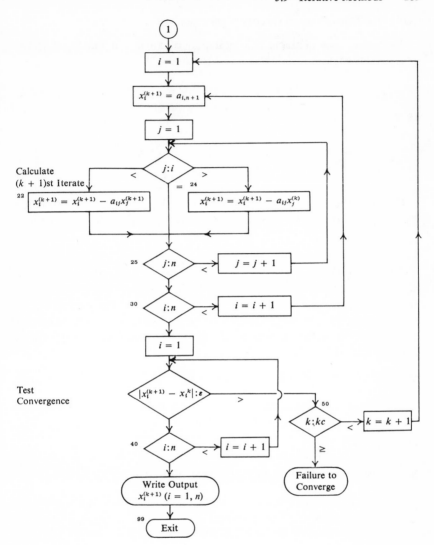

5.10   The Eigen Problem (Matrix Eigen Methods)

Definition. The characteristic equation of an $n \times n$ matrix A is the polynomial equation of degree n in the variable λ:

$$(5.10.1) \quad |A - \lambda I| = \begin{vmatrix} a_{11} - \lambda & a_{12} & \cdots & a_{1n} \\ a_{21} & a_{22} - \lambda & \cdots & a_{2n} \\ \cdot & \cdot & \cdots & \cdot \\ a_{n1} & a_{n2} & \cdots & a_{nn} - \lambda \end{vmatrix}$$

$$= (-1)^n [\lambda^n + c_1 \lambda^{n-1} + \cdots + c_{n-1} \lambda + c_n] = 0.$$

Denoting the polynomial in brackets by $P_n(\lambda)$ and dividing the equation by $(-1)^n$, we obtain an equivalent characteristic equation in the form

$$(5.10.2) \qquad P_n(\lambda) = \lambda^n + b_1 \lambda^{n-1} + \cdots + b_{n-1} \lambda + b_n = 0.$$

The *factor theorem* (see Sec. 4.0) states that an algebraic (polynomial) equation of degree n of the form of (5.10.2) has exactly n roots λ_i ($i = 1, 2, \ldots, n$). The roots λ_i are assumed distinct.† These roots $\lambda_1, \lambda_2, \ldots, \lambda_n$ of the characteristic equation are called eigenvalues or characteristic numbers of the square matrix A. By the factor theorem, we can express $P_n(\lambda)$ as a product of n linear factors, in the form

$$(5.10.3) \qquad P_n(\lambda) = (\lambda - \lambda_1)(\lambda - \lambda_2) \ldots (\lambda - \lambda_n).$$

Associated with each eigenvalue λ_i is an n-dimensional vector $X_i = (x_{1i}, x_{2i}, \ldots, x_{ni})$ which satisfies the equation

$$(5.10.4) \qquad A X_i = \lambda_i X_i \qquad (i = 1, 2, \ldots, n).$$

Such a vector X_i is called an eigenvector of matrix A.‡

(Note. The first subscript denotes the component number and the second subscript denotes the vector number.)

The characteristic equation of A^t, the transpose of A, is

$$(5.10.5) \quad |A^t - \lambda I| = \begin{vmatrix} a_{11} - \lambda & a_{21} & \cdots & a_{n1} \\ a_{12} & a_{22} - \lambda & \cdots & a_{n2} \\ \cdot & \cdot & \cdots & \cdot \\ a_{1n} & a_{2n} & \cdots & a_{nn} - \lambda \end{vmatrix} = 0.$$

† We shall consider herein only matrices with distinct eigenvalues.
‡ If the eigenvalues λ_i of A are distinct, then the eigenvectors X_i of A are linearly independent.

If the determinant of (5.10.1) is expanded in terms of cofactors of the first row, and if the determinant of (5.10.5) is expanded in terms of cofactors of the first column, it is seen that the characteristic equation of A and the characteristic equation of A^t are the same. Hence, the eigenvalues of A^t coincide with the eigenvalues of A.

The characteristic vectors $\underline{X}_i$ of A^t are n-dimensional vectors that satisfy the equation

$$A^t \underline{X}_i = \lambda_i \underline{X}_i \qquad (i = 1, 2, \ldots, n).$$

In practice, the matrix eigen problem is usually not solved by reducing the characteristic equation to *explicit* polynomial form. Instead, most numerical methods for calculating the eigenvalues and eigenvectors of the characteristic matrix work with the problem in matrix form. In this section, two such numerical methods are presented. The iterative method is included because of its simplicity. Householder's method is included because it is one of the most efficient eigen methods for computation using a digital computer.

Iterative Method for Computing Eigenvalue of Maximum Modulus

Let A be an arbitrary square matrix that has n real distinct eigenvalues $\gamma_1, \gamma_2, \ldots, \gamma_n$ such that $|\gamma_1| > |\gamma_2| > \cdots > |\gamma_n|$. Now, since the γ_i are distinct, there exist n linearly independent eigenvectors $V_1, V_2, \ldots, V_n$ satisfying the relation

$$AV_i = \gamma_i V_i \qquad (i = 1, n).$$

By the vector span theorem any arbitrary n-dimensional vector X_0 can be expressed as a linear combination of the V_i, in the form

$$X_0 = c_1 V_1 + c_2 V_2 + \cdots + c_n V_n.$$

Multiplying this successively by $A, A^2, \ldots, A^k$, we obtain the following

$$
\begin{aligned}
AX_0 &= c_1 A V_1 &&+ c_2 A V_2 &&+ \cdots + c_n A V_n \\
&= c_1 \gamma_1 V_1 &&+ c_2 \gamma_2 V_2 &&+ \cdots + c_n \gamma_n V_n \\
A^2 X_0 &= c_1 \gamma_1 A V_1 &&+ c_2 \gamma_2 A V_2 &&+ \cdots + c_n \gamma_n A V_n \\
&= c_1 \gamma_1^2 V_1 &&+ c_2 \gamma_2^2 V_2 &&+ \cdots + c_n \gamma_n^2 V_n \\
&\quad \cdot \quad \cdot \quad \cdot \\
A^k X_0 &= c_1 \gamma_1^{k-1} A V_1 &&+ c_2 \gamma_2^{k-1} A V_2 &&+ \cdots + c_n \gamma_n^{k-1} A V_n \\
&= c_1 \gamma_1^k V_1 &&+ c_2 \gamma_2^k V_2 &&+ \cdots + c_n \gamma_n^k V_n \\
&= \gamma_1^k \left[c_1 V_1 \right. &&+ c_2 \left(\frac{\gamma_2}{\gamma_1} \right)^k V_2 &&+ \cdots + c_n \left(\frac{\gamma_n}{\gamma_1} \right)^k V_n \left. \right].
\end{aligned}
$$

As k becomes large, the quantity in brackets approaches $c_1 V_1$, where V_1 is the eigenvector corresponding to γ_1, so that

(5.10.6)
$$
\begin{aligned}
A^k X_0 &\doteq \gamma_1^k c_1 V_1 \\
A^{k+1} X_0 &\doteq \gamma_1^{k+1} c_1 V_1.
\end{aligned}
$$

It follows then that for large k

$$\gamma_1 \doteq \frac{(A^{k+1}X_0)_i}{(A^k X_0)_i} \qquad (i = 1, n)$$

where $(\ )_i$ denotes the ith component of the enclosed vector.

The sequence $AX_0, A^2 X_0, \ldots, A^k X_0$ can be generated without actually forming powers of A. This is accomplished by starting with an arbitrary normalized vector $X_0^t = [1, x_{20}, x_{30}, \ldots, x_{n0}]$ and constructing

$$AX_0 = AX_0 = Y_1 = m_1 X_1$$

$$\frac{1}{m_1} A^2 X_0 = AX_1 = Y_2 = m_2 X_2$$

$$\frac{1}{m_2 m_1} A^3 X_0 = AX_2 = Y_3 = m_3 X_3$$

$$\cdot \ \ \cdot \ \ \cdot \ \ \cdot \ \ \cdot \ \ \cdot \ \ \cdot \ \ \cdot \ \ \cdot \ \ \cdot \ \ \cdot$$

$$\frac{1}{m_{k-1} \ldots m_1} A^k X_0 = AX_{k-1} = Y_k = m_k X_k$$

$$\frac{1}{m_k \ldots m_1} A^{k+1} X_0 = AX_k = Y_{k+1} = m_{k+1} X_{k+1}$$

where $Y_k^t = [y_{1k}, y_{2k}, \ldots, y_{nk}]$ and $X_k^t = [1, x_{2k}, \ldots, x_{nk}]$.

It follows that (by the use of the above and 5.10.6)

$$Y_{k+1} = \frac{1}{m_k \ldots m_1} A^{k+1} X_0 \doteq \frac{1}{m_k \ldots m_1} \gamma_1^{k+1} c_1 V_1$$

$$X_k = \frac{1}{m_k \ldots m_1} A^k X_0 \doteq \frac{1}{m_k \ldots m_1} \gamma_1^{k} c_1 V_1.$$

The eigenvalue γ_1 is obtained by the ratio of the components

$$\frac{y_{1,k+1}}{x_{1k}} \doteq \frac{(1/m_k \ldots m_1) \gamma_1^{k+1} (c_1 V_1)_1}{(1/m_k \ldots m_1) \gamma_1^{k} (c_1 V_1)_1} = \gamma_1$$

and, since $x_{1k} \equiv 1$, it follows that

(5.10.7) $y_{1,k+1} \doteq \gamma_1.$

The corresponding eigenvector for γ_1 is computed, using the relation

(5.10.8) $X_{k+1} = \frac{1}{m_{k+1} m_k \ldots m_1} A^{k+1} X_0 \doteq \frac{1}{m_{k+1} m_k \ldots m_1} \gamma_1^{k+1} c_1 V_1$

so that X_{k+1} is the normalized eigenvector corresponding to γ_1.

The computation of (5.10.7) and (5.10.8) can be accomplished in an efficient manner by the recursion formulas (note: $m_k = y_{1k}$)

$$AX_{k-1} = Y_k$$

(5.10.9)

$$X_k = \frac{1}{y_{1k}} Y_k \qquad (k = 1, 2, \ldots).$$

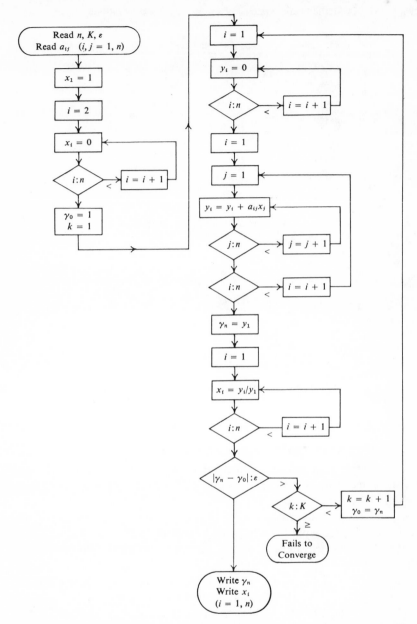

Figure 5-7. Iterative method for computing eigenvalue of maximum modulus and corresponding eigenvector. Here n = order of matrix A; k = iteration counter; K = max iterations; ε = convergence term; y_i = components of AX; x_i = components of successive-approximation vector; γ_0 = old estimate of eigenvalue; γ_n = new estimate of eigenvalue.

```
       PROGRAM EIGEN
C      ITERATIVE METHOD FOR CALCULATING EIGENVALUES/EIGENVECTORS
       DIMENSION A(20, 20), X(20), Y(20)
       READ (5,100) N, KC, EPS
       READ (5, 101) ((A(I,J), I=1, N), J=1, N)
C      SET INITIAL EIGENVECTOR (1,0,0,...,0)
       X (1) = 1.0
       DO 5 I = 2, N
     5 X (I) = 0.0
       GAMOLD = 1.0
       K = 1
     7 DO 10 I = 1, N
    10 Y (I) = 0.0
       DO 15 I = 1, N
       DO 15 J = 1, N
    15 Y(I) = Y(I) + A(I,J) * X(J)
       GAMNEW = Y(1)
       DO 20 I = 1, N
    20 X (I) = Y (I) / Y (1)
C      TEST CONVERGENCE OF EIGENVALUE
       IF (ABS(GAMNEW - GAMOLD) - EPS) 30, 30, 25
C      TEST ITERATION COUNTER K AGAINST MAX ITERATIONS KC
    25 IF (K-KC) 26, 81, 81
    26 K = K +1
       GAMOLD = GAMNEW
       GO TO 7
    30 WRITE (6, 110) GAMNEW
       WRITE (6, 111)
       WRITE (6, 112) (X(I), I=1, N)
    32 STOP
    81 WRITE (6, 113)
       GO TO 32
   100 FORMAT (I2, 3X, I2, 3X, F10.8)
   101 FORMAT (F15.7)
   110 FORMAT (14H EIGENVALUE =, E15.7)
   111 FORMAT (23H EIGENVECTOR COMPONENTS)
   112 FORMAT (1X, E15.7)
   113 FORMAT (26H PROGRAM FAILS TO CONVERGE)
       END
```

See Figure 5-7 for a representative flow chart of this method.

Householder's Method for Calculating Eigenvalues and Eigenvectors

One of the most efficient numerical methods for solving the characteristic-value problem (for real symmetric matrices†) is that of Householder. This method can be organized into four distinct parts.

Part I. The original matrix $A_1 = (a_{ij})$ is reduced to a tridiagonal matrix A_{n-1} by a sequence of orthogonal transformations $A_k = B_k^{-1} A_{k-1} B_k$.

Part II. The eigenvalues $\gamma_1, \gamma_2, \ldots, \gamma_n$ of A_{n-1} are computed by forming the Sturm sequence $1, f_1(\gamma), f_2(\gamma), \ldots, f_n(\gamma)$, where $f_i(\gamma)$ is the ith principal minor of $|\gamma I - A_{n-1}|$, and then calculating the roots of the characteristic equation $f_n(\gamma) = 0$ by the method of successive bisection.

Part III. Computation of the eigenvectors X_j of A_{n-1} that satisfy the relations

$$A_{n-1} X_j = \gamma_j X_j \qquad (j = 1, n).$$

† The method can be modified for unsymmetric matrices. See J. H. Wilkinson, "Householder's Method for the Solution of the Algebraic Eigenproblem," *Computer Journal 3*, No. 1, Apr. 1960, pp. 23–27.

Part IV. The eigenvectors V_j of the original matrix A_1, satisfying the relation

$$A_1 V_j = \gamma_j V_j \qquad (j = 1, n)$$

are computed by performing the sequence of transformations on the vectors X_j, where the transformations are performed in reverse order; i.e.,

$$V_j = B_2 B_3 B_4 \ldots B_{n-1} X_j \qquad (j = 1, n).$$

In the following pages of this subsection, the formulas for Householder's method will be derived, using a real symmetric 5×5 matrix $A_1 = (a_{ij})$; these formulas can then be directly extended for the general case of a real symmetric $n \times n$ matrix.

Part I. The original matrix

$$(5.10.10) \qquad A_1 = \begin{bmatrix} a_{11} & a_{12} & a_{13} & a_{14} & a_{15} \\ a_{21} & a_{22} & a_{23} & a_{24} & a_{25} \\ a_{31} & a_{32} & a_{33} & a_{34} & a_{35} \\ a_{41} & a_{42} & a_{43} & a_{44} & a_{45} \\ a_{51} & a_{52} & a_{53} & a_{54} & a_{55} \end{bmatrix}$$

is reduced to a tridiagonal matrix of the form

$$(5.10.11) \qquad A_4 = \begin{bmatrix} \alpha_1 & \beta_1 & 0 & 0 & 0 \\ \beta_1 & \alpha_2 & \beta_2 & 0 & 0 \\ 0 & \beta_2 & \alpha_3 & \beta_3 & 0 \\ 0 & 0 & \beta_3 & \alpha_4 & \beta_4 \\ 0 & 0 & 0 & \beta_4 & \alpha_5 \end{bmatrix}$$

by a sequence of orthogonal transformations

$$(5.10.12) \qquad A_k = B_k^{-1} A_{k-1} B_k \qquad (k = 2, 3, 4)$$

where

$$B_k = I - 2 W_k W_k^{\,t}$$
$$W_k = [0, \ldots, 0, w_k, \ldots, w_5]$$
$$W_k^{\,t} W_k = w_k^2 + \cdots + w_5^2 = 1.$$

Since B_k is symmetric ($B_k^{\,t} = B_k$) and orthogonal ($B_k^{\,t} = B_k^{-1}$), (5.10.12) can be written $A_k = B_k A_{k-1} B_k$.

In the first stage of the reduction, the off-tridiagonal elements in the first row (and column) are zeroed as follows: First, note that premultiplication of any matrix A by B_2 does *not* alter the elements of the first row of A; therefore, the off-tridiagonal elements of the first row of $B_2A_1B_2$ can be zero if and only if the off-tridiagonal elements of A_1B_2 are zero.

$$(5.10.13) \qquad \begin{aligned} A_1B_2 &= A_1[I - 2W_2W_2{}^t] \\ &= A_1 - 2A_1W_2W_2{}^t \\ &= A_1 - 2PW_2{}^t \qquad \text{where } P = A_1W_2. \end{aligned}$$

In expanded form

$$(5.10.14) \quad A_1B_2 = \begin{bmatrix} a_{11} & a_{12} & a_{13} & a_{14} & a_{15} \\ a_{21} & a_{22} & a_{23} & a_{24} & a_{25} \\ a_{31} & a_{32} & a_{33} & a_{34} & a_{35} \\ a_{41} & a_{42} & a_{43} & a_{44} & a_{45} \\ a_{51} & a_{52} & a_{53} & a_{54} & a_{55} \end{bmatrix} - 2 \cdot \begin{bmatrix} p_1 \\ p_2 \\ p_3 \\ p_4 \\ p_5 \end{bmatrix} [0 \ w_2 \ w_3 \ w_4 \ w_5]$$

Carrying out the indicated multiplication and subtraction, we obtain

$$A_1B_2 = \begin{bmatrix} a_{11} & a_{12} - 2p_1w_2 & a_{13} - 2p_1w_3 & a_{14} - 2p_1w_4 & a_{15} - 2p_1w_5 \\ a_{21} & a_{22} - 2p_2w_2 & a_{23} - 2p_2w_3 & a_{24} - 2p_2w_4 & a_{25} - 2p_2w_5 \\ a_{31} & a_{32} - 2p_3w_2 & a_{33} - 2p_3w_3 & a_{34} - 2p_3w_4 & a_{35} - 2p_3w_5 \\ a_{41} & a_{42} - 2p_4w_2 & a_{43} - 2p_4w_3 & a_{44} - 2p_4w_4 & a_{45} - 2p_4w_5 \\ a_{51} & a_{52} - 2p_5w_2 & a_{53} - 2p_5w_3 & a_{54} - 2p_5w_4 & a_{55} - 2p_5w_5 \end{bmatrix}.$$

Vector W_2 is to be determined such that

$$(5.10.15) \qquad \begin{aligned} a_{13} - 2p_1w_3 &= 0 \\ a_{14} - 2p_1w_4 &= 0 \\ a_{15} - 2p_1w_5 &= 0. \end{aligned}$$

The sum of the squares of the elements of a row of a matrix is invariant under an orthogonal transformation BAB, and since the elements of the first row $B_2A_1B_2$ are the same as those of A_1B_2, it follows that

$$a_{11}{}^2 + a_{12}{}^2 + a_{13}{}^2 + a_{14}{}^2 + a_{15}{}^2 = a_{11}{}^2 + (a_{12} - 2p_1w_2)^2 + 0^2 + 0^2 + 0^2.$$

When we define $S = a_{12}{}^2 + a_{13}{}^2 + a_{14}{}^2 + a_{15}{}^2$, it is seen that

$$(5.10.16) \qquad \pm S^{\frac{1}{2}} = a_{12} - 2p_1w_2$$

and it follows from (5.10.15) and (5.10.16) that

$$\begin{aligned} \pm w_2S^{\frac{1}{2}} &= w_2(a_{12} - 2p_1w_2) + w_3(a_{13} - 2p_1w_3) + w_4(a_{14} - 2p_1w_4) + w_5(a_{15} - 2p_1w_5) \\ &= a_{12}w_2 + a_{13}w_3 + a_{14}w_4 + a_{15}w_4 - 2p_1(w_2{}^2 + w_3{}^2 + w_4{}^2 + w_5{}^2) \\ &= p_1 \qquad\qquad\qquad\qquad\qquad - 2p_1(1). \end{aligned}$$

Then

$$(5.10.17) \qquad p_1 = \mp w_2S^{\frac{1}{2}}.$$

Substituting (5.10.17) into (5.10.16), we obtain

$$\pm S^{1/2} = a_{12} \pm 2w_2{}^2 S^{1/2}$$

so that

$$w_2{}^2 = \frac{1}{2}\left[1 \mp \frac{a_{12}}{S^{1/2}}\right]$$

and substituting (5.10.17) into (5.10.15), we obtain

$$w_3 = \mp \frac{a_{13}}{2w_2 S^{1/2}}$$

$$w_4 = \mp \frac{a_{14}}{2w_2 S^{1/2}}$$

$$w_5 = \mp \frac{a_{15}}{2w_2 S^{1/2}}.$$

Since the computation of w_i ($i = 3, 4, 5$) involves division by w_2, it is best to make w_2 as large as possible; this is accomplished as follows:

$$w_2{}^2 = \frac{1}{2}\left[1 + \frac{a_{12}\ \text{sgn}\ a_{12}}{S^{1/2}}\right]$$

$$w_i = \frac{a_{1i}\ \text{sgn}\ a_{12}}{2w_2 S^{1/2}} \qquad (i = 3, 4, 5).$$

A_2 is obtained from A_1 by the orthogonal transformation

$$\begin{aligned}
A_2 &= B_2 A_1 B_2 \\
&= [I - 2W_2 W_2{}^t]A_1[I - 2W_2 W_2{}^t] \\
&= [A_1 - 2W_2 W_2{}^t A_1][I - 2W_2 W_2{}^t] \\
&= A_1 - 2W_2 W_2{}^t A_1 - 2A_1 W_2 W_2{}^t + 4W_2(W_2{}^t A_1 W_2)W_2{}^t \\
&= A_1 - 2W_2[W_2{}^t A_1 - (W_2{}^t A_1 W_2)W_2{}^t] \\
&\qquad\qquad\qquad - 2[A_1 W_2 - W_2(W_2{}^t A_1 W_2)]W_2{}^t.
\end{aligned}$$

Define

$$\begin{aligned}
Q &= A_1 W_2 - W_2(W_2{}^t A_1 W_2) \\
&= P - W_2 C
\end{aligned}$$

where C is a scalar:

$$C = W_2{}^t A_1 W_2$$

The components of P and Q are computed by the formulas

$$p_j = \sum_i a_{ji} w_i \qquad (j = 2, 5)$$

$$C = \sum_j w_j p_j$$

$$q_j = p_j - C w_j \qquad (j = 2, 5).$$

Finally, A_2 can be expressed as

$$A_2 = A_1 - 2W_2Q^t - 2QW_2{}^t.$$

In expanded form, the elements of A_2 are (where superscripts denote stage of reduction)

$$
\begin{bmatrix}
\alpha_1 & \beta_1 & 0 & 0 & 0 \\
\beta_1 & a_{22}{}^2 & a_{23}{}^2 & a_{24}{}^2 & a_{25}{}^2 \\
0 & a_{32}{}^2 & a_{33}{}^2 & a_{34}{}^2 & a_{35}{}^2 \\
0 & a_{42}{}^2 & a_{43}{}^2 & a_{44}{}^2 & a_{45}{}^2 \\
0 & a_{52}{}^2 & a_{53}{}^2 & a_{54}{}^2 & a_{55}{}^2
\end{bmatrix}
$$

$$
= \begin{bmatrix}
a_{11} & a_{12} & a_{13} & a_{14} & a_{15} \\
a_{21} & a_{22} & a_{23} & a_{24} & a_{25} \\
a_{31} & a_{32} & a_{33} & a_{34} & a_{35} \\
a_{41} & a_{42} & a_{43} & a_{44} & a_{45} \\
a_{51} & a_{52} & a_{53} & a_{54} & a_{55}
\end{bmatrix}
- 2 \begin{bmatrix} 0 \\ w_2 \\ w_3 \\ w_4 \\ w_5 \end{bmatrix}
[q_1 \quad q_2 \quad q_3 \quad q_4 \quad q_5]
$$

$$
- 2 \begin{bmatrix} q_1 \\ q_2 \\ q_3 \\ q_4 \\ q_5 \end{bmatrix}
[0 \quad w_2 \quad w_3 \quad w_4 \quad w_5].
$$

Equating corresponding terms on left and right, we obtain formulas for computing α_1, β_1, and the elements $a_{ij}{}^2$ of A_2:

$$
\begin{aligned}
\alpha_1 &= a_{11} \\
\beta_1 &= a_{12} - 2q_1w_2 = \begin{cases} +S^{\frac{1}{2}} & \text{if } a_{12} < 0 \\ -S^{\frac{1}{2}} & \text{if } a_{12} \geq 0 \end{cases} \\
a_{ij}{}^2 &= a_{ij} - 2w_iq_j - 2q_iw_j \qquad (i, j = 2, 5).
\end{aligned}
$$

Now that A_2 has been computed by the transformation $A_2 = B_2A_1B_2$, A_3 can be computed in like manner by transformation $A_3 = B_3A_2B_3$, and A_4 can be computed by $A_4 = B_4A_3B_4$. The matrix A_4 is the required tridiagonal matrix.

Part II (*Computation of the eigenvalues of the tridiagonal matrix*). The eigenvalues $\gamma_1, \gamma_2, \gamma_3, \gamma_4, \gamma_5$ of tridiagonal matrix A_4 are the roots of the characteristic equation

$$|\gamma I - A_4| = 0$$

which can be written in expanded form as

$$
\begin{vmatrix}
\gamma - \alpha_1 & -\beta_1 & 0 & 0 & 0 \\
-\beta_1 & \gamma - \alpha_2 & -\beta_2 & 0 & 0 \\
0 & -\beta_2 & \gamma - \alpha_3 & -\beta_3 & 0 \\
0 & 0 & -\beta_3 & \gamma - \alpha_4 & -\beta_4 \\
0 & 0 & 0 & -\beta_4 & \gamma - \alpha_5
\end{vmatrix} = 0.
$$

If this determinant were expanded and the terms collected, then a quintic polynomial equation would be obtained in the form

$$\gamma^5 + c_2\gamma^4 + c_3\gamma^3 + c_4\gamma^2 + c_5\gamma + c_6 = 0.$$

The roots of this polynomial equation are the eigenvalues γ_j of A_4.

However, instead of computing the roots of the *explicit* polynomial equation, the eigenvalues γ_j are computed by the following steps:

1. Form the Sturm sequence (see Froberg [37, pp. 116–117]):

$$
\begin{aligned}
f_0(\gamma) &= 1 \\
f_1(\gamma) &= \gamma - \alpha_1 \\
f_i(\gamma) &= (\gamma - \alpha_i)f_{i-1}(\gamma) - \beta_{i-1}^2 f_{i-2}(\gamma) \qquad (i = 2, 5)
\end{aligned}
$$

where $f_i(\gamma)$ is the ith principal minor of $|\gamma I - A_4|$, and $f_5(\gamma) = 0$ is the characteristic equation.

2. Calculate the eigenvalues, i.e., the roots of $f_5(\gamma) = 0$, by the method of successive bisection. This is accomplished by evaluating the recursion formulas $f_i(\gamma)$ for $\gamma = a$ and $\gamma = b$ to obtain values of $f_5(a)$ and $f_5(b)$. The eigenvalues in (a, b) are then computed by successive bisection.

The bounds for these eigenvalues are determined, using the following theorem (Fadeeva [4, p. 61]): "No eigenvalue of a matrix exceeds any of its norms in modulus;" we define the norm N_0 of a general matrix $A = (a_{ij})$ as

$$N_0 = \max_i \sum_j |a_{ij}|.$$

If $V(\gamma)$ denotes the number of sign changes in the sequence $1, f_1(\gamma), \ldots, f_n(\gamma)$, the number of eigenvalues between a and b is equal to

$$V(a) - V(b),$$

provided $f_n(a) \neq 0$ and $f_n(b) \neq 0$.

Part III (Calculating the eigenvectors of the tridiagonal matrix). The eigenvectors X_j of A_4 satisfy the relation

$$A_4 X_j = \gamma_j X_j \qquad (j = 1, 5).$$

If this relation is written in expanded form

$$
\begin{bmatrix}
\alpha_1 & \beta_1 & 0 & 0 & 0 \\
\beta_1 & \alpha_2 & \beta_2 & 0 & 0 \\
0 & \beta_2 & \alpha_3 & \beta_3 & 0 \\
0 & 0 & \beta_3 & \alpha_4 & \beta_4 \\
0 & 0 & 0 & \beta_4 & \alpha_5
\end{bmatrix}
\begin{bmatrix}
x_{1j} \\ x_{2j} \\ x_{3j} \\ x_{4j} \\ x_{5j}
\end{bmatrix}
= \gamma_j
\begin{bmatrix}
x_{1j} \\ x_{2j} \\ x_{3j} \\ x_{4j} \\ x_{5j}
\end{bmatrix}
$$

then it is seen that the components of the normalized eigenvector X_j satisfy the following recursion formulas:

$$x_{1j} = 1$$

$$x_{2j} = \frac{1}{\beta_1}[\gamma_j - \alpha_1]x_{1j}$$

$$x_{i+1,j} = \frac{1}{\beta_i}[(\gamma_j - \alpha_i)x_{ij} - \beta_{i-1}x_{i-1,j}] \qquad (i = 2, 3, 4).$$

Wilkinson ["The Calculation of Eigenvectors of Codiagonal Matrices," *Comp. Jour.* Vol. I, p. 90] points out that the eigenvector components computed by these recursion formulas are *highly sensitive* to roundoff error. For more accurate computation of these components the reader is referred to Wilkinson's paper.

Part IV (Calculating the eigenvectors of the original matrix). The eigenvectors V_j of the original matrix A_1 satisfy the relations

$$A_1 V_j = \gamma_j V_j \qquad (j = 1, 5).$$

Assuming that the eigenvectors X_j of tridiagonal matrix A_4 have been computed (Part III), we have

$$A_4 X_j = \gamma_j X_j \qquad (j = 1, 5),$$

and since $A_4 = B_4[B_3(B_2 A_1 B_2)B_3]B_4$, it follows that

$$B_4 B_3 B_2 A_1 B_2 B_3 B_4 X_j = \gamma_j X_j \qquad (j = 1, 5).$$

If this equation is premultiplied successively by B_4, B_3, B_2 we obtain

$$B_2[B_3(B_4 B_4)B_3]B_2 A_1 B_2 B_3 B_4 X_j = \gamma_j B_2[B_3(B_4 X_j)]$$

which, by the orthogonality relation $B_k{}^t B_k = I$, reduces (since $B^t = B$) to

$$A_1(B_2 B_3 B_4 X_j) = \gamma_j(B_2 B_3 B_4 X_j).$$

If we define $V_j = B_2 B_3 B_4 X_j$, we find that

$$A_1 V_j = \gamma_j V_j$$

so that V_j is the eigenvector of A_1 corresponding to eigenvalue γ_j.

A representative flow chart for Householder's method is shown in Figure 5-8.

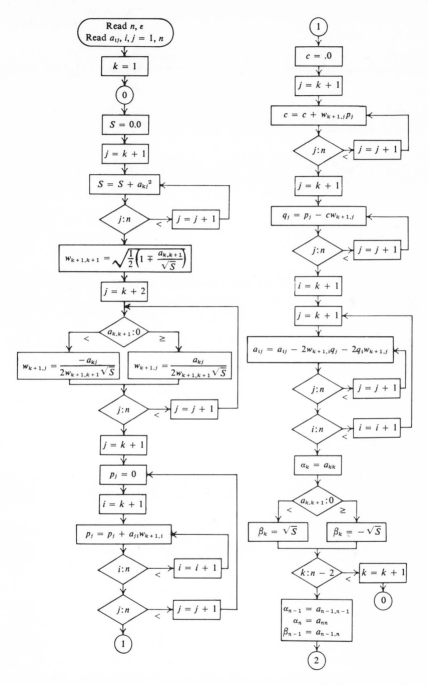

Figure 5-8. Flow chart for Householder's method.

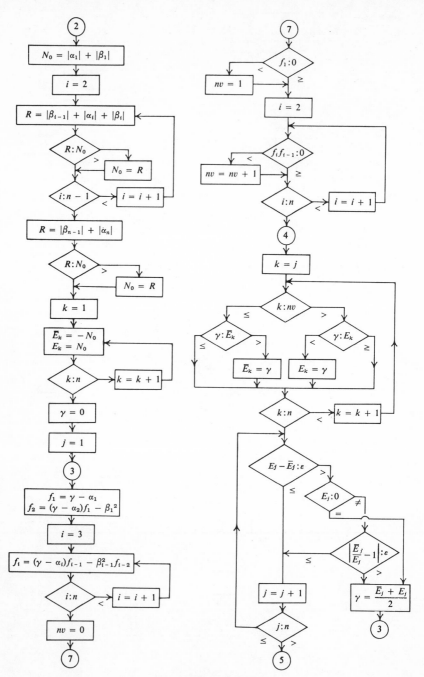

Figure 5-8 (*Continued*)

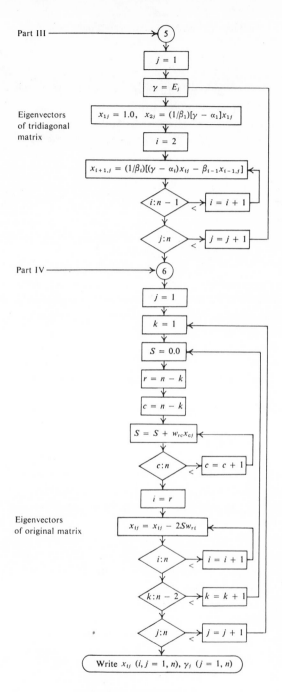

Figure 5-8 (*Continued*)

Summary

The elimination methods are widely used because procedures for inverting A, for calculating $|A|$, and for solving $AX = B$ can be incorporated into a single computer program that is accurate, efficient, and requires minimal storage. Such a program is the Gauss-Jordan-Rutishauser method which uses in-place inversion to minimize storage and total pivotal elimination to minimize the effects of roundoff error. (See also SHARE routine SDA 3316 for in-place inversion by pivotal elimination.)

The Gauss-Jordan method requires approximately $n^3/2$ multiplications and divisions for solving $AX = B$, as opposed to $n^3/3$ for the basic Gauss elimination method and $2n^3$ for the conjugate gradient method of Hestenes and Stiefel (for arbitrary real matrices).

Again, the Gauss-Jordan method is quite accurate if pivotal elimination is used, and is especially suited for the solution of dense systems, i.e., systems in which the matrix of coefficients has few zero elements. Naturally, the accuracy of the method can be improved by the use of double-precision arithmetic.

If the system of linear equations is ill-conditioned, two or three repetitions of the Gauss-Jordan method, using augmented matrix $[A, -E_k]$, will usually produce an accurate solution. It should be noted that the residual vector E_k may have to be computed in double precision. A typical program that utilizes repeated Gauss elimination is SHARE routine SDA 3194.

Hartree [6, p. 168] gives an excellent discussion of ill-conditioned equations, and shows that the effects of roundoff in the elimination process are minimized if the multiplier a_{ik}/a_{kk} is of modulus less than one. McCracken and Dorn [16] give an excellent analysis, using the concept of process graphs, of the stage-by-stage accumulation of roundoff errors.

The Jacobi and Gauss-Seidel methods, representative of the iterative methods, are well suited for the solution of sparse systems, i.e., systems in which the matrix of coefficients has many zero elements. The Gauss-Seidel method is a special case of the Liebmann iterative method, which has accelerated-convergence properties.

COMPUTATIONAL EXERCISES

1. Calculate the point of intersection of the lines whose equations are

$$x + 2y = 3$$
$$2x - 3y = 7.$$

2. Calculate the point common to the planes whose equations are

$$x + 2y + 3z = 5$$
$$2x - y + z = 6$$
$$x + 3y - 5z = 2.$$

3. Given the matrices

$$A = \begin{bmatrix} 1 & 2 & 3 \\ 4 & 5 & -6 \\ 2 & -1 & 3 \end{bmatrix} \quad B = \begin{bmatrix} 2 & 5 \\ 3 & 1 \\ 1 & 7 \end{bmatrix} \quad C = \begin{bmatrix} 2 & 8 & 5 \\ 1 & -2 & 4 \\ 6 & 3 & 1 \end{bmatrix}$$

calculate (a) the matrix sum $A + C$; (b) the matrix product AB; (c) the matrix product AC.

4. Use the Gauss-Jordan elimination method to solve the system of equations

$$4x + y + 2z = 16$$
$$x + 3y + z = 10$$
$$x + 2y + 5z = 12.$$

5. Use the Gauss-Jordan elimination method to calculate the inverse of the matrix

$$A = \begin{bmatrix} 4 & 1 & 2 \\ 1 & 3 & 1 \\ 1 & 2 & 5 \end{bmatrix}.$$

6. Use repeated elimination to improve the solution of the system of equations

$$7x + 7y + 6z = 34$$
$$7x + 6y + 5z = 31$$
$$6x + 5y + 4z = 26$$

carrying (a) three digits to the right of the decimal point; and (b) six digits to the right of the decimal point.

7. Use the Gauss-Jordan elimination method, carrying three digits to the right of the decimal point, to calculate an approximate inverse of the matrix

$$A = \begin{bmatrix} 7 & 7 & 6 \\ 7 & 6 & 5 \\ 6 & 5 & 4 \end{bmatrix}.$$

Improve this approximate inverse by the Hotelling method.

8. Use the iterative method to calculate to three-decimal-place accuracy the maximum-modulus eigenvalue and the corresponding eigenvector of the matrix

$$A = \begin{bmatrix} 1 & 3 \\ -2 & -4 \end{bmatrix}.$$

PROGRAMMING EXERCISES

1. Modify the matrix-addition program to conserve storage, by using array B instead of array C to store the elements of the sum of matrices A and B.

2. Modify the matrix-multiplication program to conserve storage (*hint:* use $1 \times J$ array to store row i of product matrix; then store this array in row i of A array before computing row $i + 1$ of product).

3. Write recursion formulas, flow chart, and FORTRAN program for algorithm which calculates the determinant of matrix A by reducing A to upper-triangular form, and then calculates product of main diagonal elements to obtain the value of the determinant.

4. Modify inversion-by-elimination with partial pivoting in order to use total pivoting.

5. Prepare computational summary, flow chart, and FORTRAN program for Jacobi's iterative method to solve a system of linear algebraic equations.

6. Program the matrix addition as a subroutine, with array names A, B, C and parameters I and J as arguments in the calling sequence.

7. Program the matrix multiplication as a subroutine with array names A, B, C and parameters I, J, K as arguments in the calling sequence.

8. Program the matrix inversion as a subroutine with array names A, $AINV$, and parameters N, EPS, $DETERM$ as arguments in the calling sequence.

6

Interpolation

6.0 Introduction

The reader in all probability is familiar with the method of *linear interpolation* used in trigonometry to calculate the value of y corresponding to a given x, using tabulated values (x_k, y_k) and (x_{k+1}, y_{k+1}). If x lies inside interval (x_k, x_{k+1}), then we *interpolate* for $y(x)$. On the other hand, if x lies outside the interval (x_k, x_{k+1}), then we *extrapolate* for $y(x)$, *or* use a pair of tabulated values that bracket x and interpolate.

The formula for linear interpolation is obtained by simply constructing the straight line through points (x_k, y_k) and (x_{k+1}, y_{k+1}) and, using the relations in the similar triangles in Figure 6-1, writing the two-point formula for this straight line:

(6.0.1)
$$\frac{y - y_k}{x - x_k} = \frac{y_{k+1} - y_k}{x_{k+1} - x_k}.$$

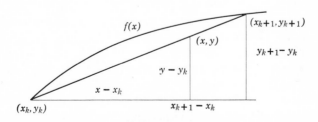

Figure 6-1.

Equation (6.0.1) can be solved for y, and the result

(6.0.2) $$y = y_k + \left(\frac{y_{k+1} - y_k}{x_{k+1} - x_k}\right)(x - x_k)$$

can be simplified and written in the form

(6.0.3) $$y = \frac{y_k[x - x_{k+1}] - y_{k+1}[x - x_k]}{x_k - x_{k+1}}.$$

Equation (6.0.3) is one standard form of the linear interpolation formula for approximating y for a given x.

Traditionally, interpolation was used to obtain values of elementary functions (trigonometric, hyperbolic, logarithmic, etc.) which were tabulated at discrete values of the independent variable, or argument. Today, in computing centers, values of elementary functions are usually generated by standard subroutines that evaluate some type of convergent series in the argument, e.g., Taylor series, Chebyshev economized series, continued fractions, and Hasting's approximations. The function values are thus generated *as required* so that the computer memory is not filled with unnecessary tables of values of the elementary functions.

This does not mean that there no longer is a need for interpolation in this day of high-speed computing. On the contrary, if the explicit form of a function is not known and cannot be obtained by analytic means, then we will have to work with samples (discrete values) of the function that are known or can be computed. Interpolation then provides us a means of obtaining a simple approximating function that can be easily differentiated, integrated, or evaluated, as required to obtain information about the original function whose explicit form is not known.

Additionally, a number of numerical integration (quadrature) methods and methods of solving ordinary differential equations are derived by replacing the integrand or derivative, respectively, by an interpolating polynomial. For example, the Newton-Coates quadrature formulas are based on Newton's interpolating polynomials; Romberg's method of numerical integration is based, in part, on the principle of repeated linear interpolation, as is the Euler-Romberg method of solving ordinary differential equations. The Adams-Bashforth-Moulton methods are based on Newton's backward-form interpolating polynomial.

In this chapter, we will develop representative interpolation algorithms of several distinct types, such as Lagrange's ordinate formulas, Newton's finite-difference formulas, and the Aitken-Neville repeated linear interpolation formulas. All of these basic types of interpolation use given values of (x_k, y_k) to generate distinct forms of a unique interpolating polynomial.

In this chapter interpolation methods are developed which use function points (x_i, y_i) $(i = 0, n)$ to determine a unique polynomial $P_n(x)$ which satisfies the constraints

(6.0.4) $P_n(x_i) = y_i$ $(i = 0, n)$.

For practical computational methods, this unique interpolating polynomial is usually represented in one of the following forms.

1. *Ordinate Form*—The Lagrange interpolating polynomial is the best-known example of this type, and is derived by assuming that $P_n(x)$ can be expressed in the form

$$P_n(x) = y_0 b_0(x) + y_1 b_1(x) + \cdots + y_n b_n(x)$$

where each $b_k(x)$ $(k = 0, n)$ is a polynomial of degree n. The $b_k(x)$ are determined by solving the constraint equations (6.0.4).

2. *Difference Form*—The Newton interpolating polynomials are the simplest of the finite-difference forms. Newton's forward-form (NFF) polynomial will be derived by assuming that $P_n(x)$ can be expressed in the form

$$P_n(x) = c_0 + c_1(x - x_0) + c_2(x - x_0)(x - x_1) + \cdots$$
$$+ c_n(x - x_0)(x - x_1)\dots(x - x_{n-1})$$

where the c_k $(k = 0, n)$ are real constants determined by solving the constraint equations (6.0.4).

3. *Iterated Form*—The Aitken-Neville techniques use repeated linear interpolation to compute the value of $P_n(X)$, for X in (x_0, x_n), subject to the constraints that $P_n(x_i) = y_i$ $(i = 0, n)$.

It should be noted that each of these three forms is simply an algebraic rearrangement of the unique interpolating polynomial that satisfies constraints (6.0.4). A myriad of variations within each of these three sub-categories exist and are presented in a number of numerical-analysis texts.

Other categories of interpolation methods, such as the function-derivative formulas of Hermite and various forms of trigonometric interpolation, appear throughout the literature but will not be included here.

6.1   The Unique Interpolating Polynomial

Suppose that we are given $n + 1$ data pairs $(x_0, y_0), (x_1, y_1), \ldots, (x_n, y_n)$ representing $n + 1$ points of the graph of a function $y = f(x)$ where *the explicit form of $f(x)$ is not known*. The x_i $(i = 0, n)$ are assumed to be distinct values of the independent variable x.

We would like to approximate $f(x)$ by some simple function $P(x)$ that can be easily manipulated mathematically, and can be evaluated at any

$x = X$ in an interval I that contains $x_0, x_1, \ldots, x_n$. The value of $P(X)$ is then used to approximate $f(X)$. Polynomials are well-suited for this purpose because they are easily differentiated, integrated, and evaluated.

Since we are given $n + 1$ function values y_i ($i = 0, n$), we can impose $n + 1$ conditions to determine the coefficients in the approximating polynomial. That is, we can determine a polynomial $P_n(x)$ of max-degree† n with the coefficients being determined by the $n + 1$ conditions

$$(6.1.1) \qquad P_n(x_i) = y_i \qquad (i = 0, n).$$

A polynomial $P_n(x)$ that approximates $f(x)$ over an interval (x_0, x_n) and satisfies (6.1.1) is usually referred to as an interpolating polynomial.

Assume then that there exists a polynomial $P_n(x)$ of the form

$$(6.1.2) \qquad P_n(x) = a_0 + a_1 x + a_2 x^2 + \cdots + a_n x^n$$

which satisfies the conditions (constraints) imposed in (6.1.1).

Equations (6.1.1), called the constraint equations, can be written in the expanded form

$$(6.1.3) \qquad \begin{aligned} a_0 + a_1 x_0 + a_2 x_0{}^2 + \cdots + a_n x_0{}^n &= y_0 \\ a_0 + a_1 x_1 + a_2 x_1{}^2 + \cdots + a_n x_1{}^n &= y_1 \\ a_0 + a_1 x_2 + a_2 x_2{}^2 + \cdots + a_n x_2{}^n &= y_2 \\ \cdot \quad \cdot \quad \cdot \quad \cdot \quad \cdot \quad \cdot \quad \cdot \quad \cdot \quad \cdot \quad \cdot \quad \cdot \quad \cdot \\ a_0 + a_1 x_n + a_2 x_n{}^2 + \cdots + a_n x_n{}^n &= y_n. \end{aligned}$$

These constraint equations, which can also be written in matrix form as

$$(6.1.4) \qquad \begin{bmatrix} 1 & x_0 & x_0{}^2 & \ldots & x_0{}^n \\ 1 & x_1 & x_1{}^2 & \ldots & x_1{}^n \\ 1 & x_2 & x_2{}^2 & \ldots & x_2{}^n \\ \cdot & \cdot & \cdot & \cdot & \cdot \\ 1 & x_n & x_n{}^2 & \ldots & x_n{}^n \end{bmatrix} \begin{bmatrix} a_0 \\ a_1 \\ a_2 \\ \vdots \\ a_n \end{bmatrix} = \begin{bmatrix} y_0 \\ y_1 \\ y_2 \\ \vdots \\ y_n \end{bmatrix}$$

have a unique solution $(a_0, a_1, a_2, \ldots, a_n)$ since the determinant of the matrix of coefficients (the well-known Vandermonde determinant) is non-zero for distinct x_i. Hence, there exists a unique polynomial of the form of (6.1.2) that satisfies (6.1.1).

However, it is not convenient to approximate $f(x)$ by a polynomial of the form

$$P_n(x) = a_0 + a_1 x + a_2 x^2 + \cdots + a_n x^n$$

because it requires solution of the system of simultaneous linear algebraic equations (6.1.3).

---

† A polynomial of max-degree n is a polynomial of degree not exceeding n.

In practice, therefore, *more convenient forms of the interpolating polynomial are used.* Commonly used interpolating polynomials are expressed in terms of either the *ordinates* y_i, in terms of *finite differences of the ordinates*, or in repeated (iterated) linear-interpolation form.

The Lagrangian polynomial is the classical example of an interpolating polynomial expressed in terms of the ordinates y_i. In this case, $P_n(x)$ has the form

$$(6.1.5) \qquad P_n(x) = y_0 b_0(x) + y_1 b_1(x) + \cdots + y_n b_n(x).$$

Newton's forward-form interpolating polynomial is the simplest example of the interpolating polynomials that are expressed in terms of finite differences of the ordinates. In this case, $P_n(x)$ has the form

$$(6.1.6) \quad P_n(x) = c_0 + c_1(x - x_0) + c_2(x - x_0)(x - x_1)$$
$$+ \cdots + c_n(x - x_0)(x - x_1) \ldots (x - x_{n-1}).$$

In either case, the coefficients b_k and c_k are uniquely determined by the constraint equations $P_n(x_i) = y_i$ $(i = 0, n)$.

Error Estimation in Polynomial Interpolation

The $n + 1$ constraints $P_n(x_i) = y_i$ $(i = 0, n)$ require that the interpolating polynomial coincide with the function $y = f(x)$ at the $n + 1$ distinct points $x_0, x_1, \ldots, x_n$. But how well does the polynomial approximate the function $f(x)$ at other points of an interval I which contains $x_0, x_1, \ldots, x_n$? Before we blindly approximate $f(x)$ by an nth-degree interpolating polynomial, we should have some idea of the accuracy of such an approximation. The reader will recall that the accuracy of approximation by a Taylor polynomial (truncated Taylor series) can be estimated by defining an error function $E_n(x)$ as the difference between $f(x)$ and the Taylor approximation. It seems reasonable, then, that we can use a similar approach to determine the accuracy of approximation by polynomial interpolation.

Let the interpolation error function, denoted by $E_n(x)$, be defined by the relation

$$(6.1.7) \qquad E_n(x) = f(x) - P_n(x).$$

Now, we know from the constraint equations that $P_n(x_i) = f(x_i)$ $(i = 0, n)$. It follows then that $E_n(x_i) = 0$ $(i = 0, n)$. That is, the error function $E_n(x)$ has $n + 1$ distinct zeros $x_0, x_1, \ldots, x_n$. Note that the function

$$(6.1.8) \qquad \pi(x) = (x - x_0)(x - x_1) \ldots (x - x_n)$$

has zeros at $x_0, x_1, \ldots, x_n$, which are coincident with the zeros of $E_n(x)$.

For an arbitrary x^+ in I, distinct from $x_0, x_1, \ldots, x_n$, a value of K can be determined such that an auxiliary function $F(x)$, defined by

(6.1.9) $F(x) = f(x) - P_n(x) - K(x - x_0)(x - x_1)\ldots(x - x_n)$

has a zero at $x = x^+$, i.e., such that

(6.1.10) $F(x^+) = 0.$

Note also that

$$F(x_i) = 0 \qquad (i = 0, n)$$

because

$$E_n(x_i) = \pi(x_i) = 0 \qquad (i = 0, n).$$

Therefore, auxiliary function $F(x)$ has $n + 2$ distinct zeros at $x^+, x_0, x_1, \ldots, x_n$. By repeated application of Rolle's theorem, we obtain the following:

$F'(x)$ has at least $n + 1$ zeros in I, if $F(x)$ and $F'(x)$ continuous over I

$F''(x)$ has at least n zeros in I, if $F'(x)$ and $F''(x)$ continuous over I

⋮

$F^{(n+1)}(x)$ has at least 1 zero in I, if $F^{(n)}(x)$ and $F^{(n+1)}(x)$ continuous over I

i.e., $F^{(n+1)}(x^*) = 0$, for at least one x^* in I. All of these conditions will be met if $f(x)$ and its first $n + 1$ derivatives are continuous over I, because $P_n(x)$ and $\pi(x)$ and their derivatives are continuous over I.

If these assumptions hold, then we can successively compute the first $n + 1$ derivatives of (6.1.9), obtaining, at $x = x^*$,

(6.1.11) $F^{(n+1)}(x^*) = 0 = f^{(n+1)}(x^*) - P_n^{(n+1)}(x^*) - K\pi^{(n+1)}(x^*)$
$$= f^{(n+1)}(x^*) - 0 - K(n + 1)!$$

because $P_n^{(n+1)}(x) \equiv 0$, and $\pi^{(n+1)}(x) = (n + 1)!$.

We can then solve (6.1.11) for K, obtaining

(6.1.12) $K = \dfrac{f^{(n+1)}(x^*)}{(n + 1)!}$

Substituting this expression for K into (6.1.9) and evaluating $F(x)$ at $x = x^+$, we find

(6.1.13) $F(x^+) = 0 = f(x^+) - P_n(x^+)$
$$- \frac{f^{(n+1)}(x^*)}{(n + 1)!}(x^+ - x_0)(x^+ - x_1)\ldots(x^+ - x_n)$$

Now, since $E_n(x) = f(x) - P_n(x)$, we find from (6.1.13) that

(6.1.14) $E_n(x^+) = \dfrac{f^{(n+1)}(x^*)}{(n + 1)!}(x^+ - x_0)(x^+ - x_1)\ldots(x^+ - x_n)$

Noting that $\pi(x_i) = 0$, for $i = 0, n$, we can also write

(6.1.15) $E_n(x_i) = 0 = \dfrac{f^{(n+1)}(x^*)}{(n+1)!} (x_i - x_0)(x_i - x_1)\ldots(x_i - x_n)$

$$(i = 0, n)$$

Combining (6.1.14) and (6.1.15), we can write for any x in I the error function

(6.1.16) $E_n(x) = \dfrac{f^{(n+1)}(x^*)}{(n+1)!} (x - x_0)(x - x_1)\ldots(x - x_n).$

Formula (6.1.16) is the error inherent in the approximation of $f(x)$ by interpolating polynomial $P_n(x)$ that satisfies the $n + 1$ constraints $P_n(x_i) = f(x_i)$ ($i = 0, n$). This is true regardless of the explicit form of the interpolating polynomial $P_n(x)$. Therefore, this formula can be used to estimate the error in the Lagrangian, Newton, and Aitken-Neville forms of $P_n(x)$ presented in the following sections.

6.2   The Lagrange Interpolating Polynomial (Ordinate Form)

Given the $n + 1$ number pairs $(x_0, y_0), (x_1, y_1), \ldots, (x_n, y_n)$ representing points on the graph of the function $y = f(x)$, find an interpolating polynomial $P_n(x)$ of max-degree n which approximates $y = f(x)$ over an interval I containing $x_0, x_1, \ldots, x_n$. Further, we will require that the interpolating polynomial be expressed in terms of the ordinates $y_0, y_1, \ldots, y_n$ of the function. By doing so, we will obtain a different *explicit* form of the unique interpolating polynomial

$$P_n(x) = a_0 + a_1 x + a_2 x^2 + \cdots + a_n x^n.$$

Suppose then that we let the interpolating polynomial be of the form

(6.2.1) $P_n(x) = y_0 b_0(x) + y_1 b_1(x) + \cdots + y_n b_n(x)$

where each $b_k(x)$ is a polynomial of degree n. Then $P_n(x)$ is also a polynomial of degree not exceeding n since it is a linear combination of the $b_k(x)$ ($k = 0, n$).

The polynomials $b_k(x)$ can be uniquely determined by imposing $n + 1$ constraints on $P_n(x)$ of the form

(6.2.2) $P_n(x_i) = y_i$ $(i = 0, n)$

Substituting (6.2.1) into (6.2.2), we obtain the constraint equations in the expanded form

(6.2.3)
$$y_0 b_0(x_0) + y_1 b_1(x_0) + \cdots + y_n b_n(x_0) = y_0$$
$$y_0 b_0(x_1) + y_1 b_1(x_1) + \cdots + y_n b_n(x_1) = y_1$$
$$\cdot \quad \cdot \quad \cdot \quad \cdot \quad \cdot \quad \cdot \quad \cdot \quad \cdot \quad \cdot \quad \cdot \quad \cdot \quad \cdot$$
$$y_0 b_0(x_n) + y_1 b_1(x_n) + \cdots + y_n b_n(x_n) = y_n$$

Examining equations (6.2.3), we see that if the $b_k(x)$ are chosen such that

$$(6.2.4) \qquad b_k(x_j) = \delta_{kj} = \begin{cases} 1, & k = j \\ 0, & k \neq j \end{cases}$$

then the constraint equations (6.2.3) are satisfied.

Since each $b_k(x)$ $(k = 0, n)$ is a polynomial of degree n that has distinct zeros $x_0, x_1, \ldots, x_{k-1}, x_{k+1}, \ldots, x_n$ by (6.2.4), it can be expressed in the form

$$(6.2.5) \quad b_k(x) = K_k(x - x_0)(x - x_1)\ldots(x - x_{k-1})(x - x_{k+1})\ldots(x - x_n).$$

The constants K_k can then be determined by evaluating $b_k(x)$ at $x = x_k$. That is,

$$b_k(x_k) = K_k(x_k - x_0)(x_k - x_1)\ldots(x_k - x_{k-1})(x_k - x_{k+1})\ldots(x_k - x_n).$$
(6.2.6)

Also we see that $b_k(x_k) = 1$, from equation (6.2.4). It follows then that

$$(6.2.7) \quad K_k = \frac{1}{(x_k - x_0)(x_k - x_1)\ldots(x_k - x_{k-1})(x_k - x_{k+1})\ldots(x_k - x_n)}.$$

Substituting (6.2.7) into (6.2.5), we obtain

$$(6.2.8) \quad b_k(x) = \frac{(x - x_0)(x - x_1)\ldots(x - x_{k-1})(x - x_{k+1})\ldots(x - x_n)}{(x_k - x_0)(x_k - x_1)\ldots(x_k - x_{k-1})(x_k - x_{k+1})\ldots(x_k - x_n)}$$

for $k = 0, 1, 2, \ldots, n$.

Finally, substituting the $b_k(x)$ of (6.2.8) into (6.2.1), we obtain

$$P_n(x) = y_0 \frac{(x - x_1)(x - x_2)\ldots(x - x_n)}{(x_0 - x_1)(x_0 - x_2)\ldots(x_0 - x_n)}$$
$$+ y_1 \frac{(x - x_0)(x - x_2)\ldots(x - x_n)}{(x_1 - x_0)(x_1 - x_2)\ldots(x_1 - x_n)}$$
$$+ \cdots$$
$$+ y_k \frac{(x - x_0)(x - x_1)\ldots(x - x_{k-1})(x - x_{k+1})\ldots(x - x_n)}{(x_k - x_0)(x_k - x_1)\ldots(x_k - x_{k-1})(x_k - x_{k+1})\ldots(x_k - x_n)}$$
$$+ \cdots$$
$$+ y_n \frac{(x - x_0)(x - x_1)\ldots(x - x_{n-1})}{(x_n - x_0)(x_n - x_1)\ldots(x_n - x_{n-1})}.$$
(6.2.9)

Equation (6.2.9) is the classical form of Lagrange's interpolating polynomial.

Using summation notation, we can write Lagrange's interpolating polynomial of (6.2.9) in the form

$$P_n(x) = \sum_{k=0}^{n} y_k \frac{(x - x_0)(x - x_1)\ldots(x - x_{k-1})(x - x_{k+1})\ldots(x - x_n)}{(x_k - x_0)(x_k - x_1)\ldots(x_k - x_{k-1})(x_k - x_{k+1})\ldots(x_k - x_n)}.$$
(6.2.10)

To simplify the form still further, we define

$$L_k(x) = (x - x_0)(x - x_1)\ldots(x - x_{k-1})(x - x_{k+1})\ldots(x - x_n)$$

which, evaluated at $x = x_k$, is

$$L_k(x_k) = (x_k - x_0)(x_k - x_1)\ldots(x_k - x_{k-1})(x_k - x_{k+1})\ldots(x_k - x_n).$$

If we use the above definition, then Lagrange's interpolating polynomial can be written in the following simplified form:

$$(6.2.11) \qquad P_n(x) = \sum_{k=0}^{n} y_k \frac{L_k(x)}{L_k(x_k)}.$$

NUMERICAL EXAMPLE. Calculate the Lagrange interpolating polynomial for the following set of (x_i, y_i): $(0.0, 1.0)$, $(0.33, 1.391)$, $(0.66, 1.935)$, $(1.0, 2.718)$.

Using formula (6.2.11) for the given set of data, we obtain

$$P_3(x) = y_0 \frac{L_0(x)}{L_0(x_0)} + y_1 \frac{L_1(x)}{L_1(x_1)} + y_2 \frac{L_2(x)}{L_2(x_2)} + y_3 \frac{L_3(x)}{L_3(x_3)}$$

$$= 1.0 \frac{(x - 0.33)(x - 0.66)(x - 1.0)}{(0.0 - 0.33)(0.0 - 0.66)(0.0 - 1.0)}$$

$$+ 1.391 \frac{(x - 0.0)(x - 0.66)(x - 1.0)}{(0.33 - 0.0)(0.33 - 0.66)(0.33 - 1.0)}$$

$$+ 1.935 \frac{(x - 0.0)(x - 0.33)(x - 1.0)}{(0.66 - 0.0)(0.66 - 0.33)(0.66 - 1.0)}$$

$$+ 2.718 \frac{(x - 0.0)(x - 0.33)(x - 0.66)}{(1.0 - 0.0)(1.0 - 0.33)(1.0 - 0.66)}.$$

Note that the abscissas x_i in this example have been truncated after two significant figures. In practice, they should be *rounded* to the desired number of significant figures.

A representative flow chart for Lagrange's interpolating polynomial is shown in Figure 6-2.

6.3   Newton's Interpolating Polynomials (Difference Form)

In Sec. 6.1 it was proved that there exists a unique interpolating polynomial

$$(6.3.1) \qquad P_n(x) = a_0 + a_1 x + a_2 x^2 + \cdots + a_n x^n$$

which approximates function $y = f(x)$ over the interval (x_0, x_n) and satisfies the constraints $P_n(x_i) = y_i$ $(i = 0, n)$.

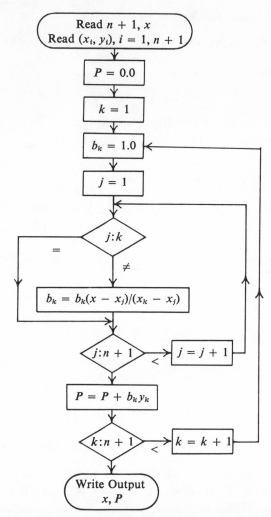

Figure 6-2. Flow chart for Lagrange interpolating polynomial. *Preliminary:* To eliminate the zero value of the subscript, adjust the indexes i and k from $(0, n)$ to $(1, n + 1)$. The Lagrange interpolating polynomial for points (x_1, y_1), $(x_2, y_2), \ldots, (x_{n+1}, y_{n+1})$ can then be written as

$$P_n(x) = y_1 b_1(x) + y_2 b_2(x) + \cdots + y_{n+1} b_{n+1}(x)$$

where for $k = 1, 2, \ldots, n + 1$,

$$b_k(x) = \frac{(x - x_1)(x - x_2)\ldots(x - x_{k-1})(x - x_{k+1})\ldots(x - x_{n+1})}{(x_k - x_1)(x_k - x_2)\ldots(x_k - x_{k-1})(x_k - x_{k+1})\ldots(x_k - x_{n+1})}$$

$$= \prod_{\substack{j=1 \\ j \neq k}}^{n+1} \frac{(x - x_j)}{(x_k - x_j)}$$

```
          PROGRAM LAGRANGE
  C       LAGRANGE INTERPOLATING POLYNOMIAL OF DEGREE N
  C       POLYNOMIAL IS EVALUATED AT X = XBAR
  C       NP1 = N + 1
          DIMENSION X (20), Y(20), B(20)
          READ (5,101) NP1, XBAR
          READ (5,102) (X(I), Y(I), I = 1,NP1)
          P = 0.0
          DO 25 K = 1,NP1
          B (K) = 1.0
          DO 20 J = 1,NP1
          IF (J - K) 15,20,15
   15     B(K) = B(K) * ((XBAR - X(J))/(X(K) - X(J)))
   20     CONTINUE
   25     P = P + B(K) * Y(K)
          WRITE (6, 110) XBAR, P
          STOP
  101     FORMAT (I2, F10.5)
  102     FORMAT (2F10.5)
  110     FORMAT (38H0 VALUE OF LAGRANGE POLYNOMIAL AT X = F10.5,
          X8H IS P = F10.5)
          END
```

It was shown in Sec. 6.2 that the unique interpolating polynomial $P_n(x)$ can be expressed in a *different explicit form* as the Lagrange interpolating polynomial

$$P_n(x) = \sum_{k=0}^{n} y_k \frac{(x - x_0)(x - x_1)\ldots(x - x_{k-1})(x - x_{k+1})\ldots(x - x_n)}{(x_k - x_0)(x_k - x_1)\ldots(x_k - x_{k-1})(x_k - x_{k+1})\ldots(x_k - x_n)}$$
(6.3.2)

in terms of the ordinates $y_0, y_1, \ldots, y_n$ of the function.

In this section we will show that the unique interpolating polynomial $P_n(x)$ can be represented in *other explicit forms* by expressing $P_n(x)$ in terms of differences (called finite differences). A number of different explicit forms of $P_n(x)$ can be found in terms of finite differences, depending on whether we use forward, backward, or central differences.

The forward form of *Newton's interpolating polynomial* will be derived in detail because of its simplicity. Other forms of finite-difference interpolating polynomials will be given in less detail.

Newton's forward-form interpolating polynomial is derived by assuming that $P_n(x)$ can be represented in the form

$$(6.3.3) \quad P_n(x) = c_0 + c_1(x - x_0) + c_2(x - x_0)(x - x_1) + \cdots$$
$$+ c_n(x - x_0)(x - x_1)\ldots(x - x_{n-1}).$$

The coefficients c_k $(k = 0, n)$ are uniquely determined by the constraints $P_n(x_i) = y_i$ $(i = 0, n)$. It will be shown that the c_k can be calculated in terms of forward finite differences. These finite differences are of the form

$$(6.3.4) \quad \begin{aligned} \Delta y_i &= y_{i+1} - y_i & (i = 0, n - 1) \\ \Delta^k y_i &= \Delta^{k-1} y_{i+1} - \Delta^{k-1} y_i & (k = 2, n)(i = 0, n - k). \end{aligned}$$

Details of finite differences are discussed in the following pages.

Finite Differences

Forward differences. Consider the set of values y_i $(i = 0, 1, \ldots, n)$ of the function $y = f(x)$, which correspond to the set x_i of the independent variable x. We define the differences of succeeding function values as†

(6.3.5) $\Delta y_i = y_{i+1} - y_i$ $(i = 0, 1, \ldots, n - 1)$.

These differences Δy_i are called *first-order differences* of the function $f(x)$ over the interval (x_0, x_n).

The differences of succeeding first-order differences are in turn defined as

(6.3.6) $\Delta^2 y_i = \Delta(\Delta y_i) = \Delta y_{i+1} - \Delta y_i$ $(i = 0, 1, \ldots, n - 2)$

and are called *second-order differences* of $f(x)$ over (x_0, x_n).

In general, higher-order differences of the function $y = f(x)$ are defined over the interval (x_0, x_n) as

(6.3.7) $\Delta^k y_i = \Delta^{k-1} y_{i+1} - \Delta^{k-1} y_i$ $(i = 0, 1, \ldots, n - k)$

and are called kth *order differences* of the function $y = f(x)$.

Diagonal-Difference Table (forward differences). If the finite differences $\Delta^k y_i$ are tabulated in the form shown in Table F, then that tabulation is

Table F

x_i	y_i	Δy_i $=$ $y_{i+1} - y_i$	$\Delta^2 y_i$ $=$ $\Delta y_{i+1} - \Delta y_i$	$\Delta^3 y_i$ $=$ $\Delta^2 y_{i+1} - \Delta^2 y_i$	$\Delta^4 y_i$ $=$ $\Delta^3 y_{i+1} - \Delta^3 y_i$
x_0	y_0				
		Δy_0			
x_1	y_1		$\Delta^2 y_0$		
		Δy_1		$\Delta^3 y_0$	
x_2	y_2		$\Delta^2 y_1$		$\Delta^4 y_0$
		Δy_2		$\Delta^3 y_1$	
x_3	y_3		$\Delta^2 y_2$		$\vdots$
		Δy_3		$\vdots$	
x_4	y_4		$\vdots$		
$\vdots$	$\vdots$	$\vdots$			

referred to as a *diagonal-difference table*. Note that the differences of any order are obtained from the differences of the next-lower order.

The $\Delta^k y_i$ are called *forward differences* because they are defined from x_0 forward, i.e., to the right with respect to the independent variable x.

---

† First-order differences Δy_i can also be written $\Delta^1 y_i$.

Ordinate-Form Difference Table. If the differences of the ordinates $y_0, y_1, y_2, \ldots, y_n$ are written in ordinate form, the first column of differences appear as $y_{i+1} - y_i$ $(i = 0, n - 1)$. Elements in the second column are formed as differences of successive elements in the first column, $(y_{i+2} - y_{i+1}) - (y_{i+1} - y_i)$, $(i = 0, n - 2)$. And elements in each successive column are computed by taking the difference of adjacent elements in the preceding column, expressed in *ordinate form*, as in Table O.

Table O

x_0	y_0			
		$y_1 - y_0$		
x_1	y_1		$y_2 - 2y_1 + y_0$	
		$y_2 - y_1$		$y_3 - 3y_2 + 3y_1 - y_0$
x_2	y_2		$y_3 - 2y_2 + y_1$	
		$y_3 - y_2$		$\cdot$
x_3	y_3		$\cdot$	
$\vdots$	$\vdots$	$\vdots$	$\cdot$	$\cdot$
x_{n-3}	y_{n-3}		$\cdot$	
		$y_{n-2} - y_{n-3}$		$\cdot$
x_{n-2}	y_{n-2}		$y_{n-1} - 2y_{n-2} + y_{n-3}$	
		$y_{n-1} - y_{n-2}$		$y_n - 3y_{n-1} + 3y_{n-2} - y_{n-3}$
x_{n-1}	y_{n-1}		$y_n - 2y_{n-1} + y_{n-2}$	
		$y_n - y_{n-1}$		
x_n	y_n			

A relation between the forward differences and the ordinates can be derived by equating corresponding elements of Table F and Table O. Of particular interest are the elements of the uppermost diagonal:

$$\Delta y_0 = y_1 - y_0$$
$$\Delta^2 y_0 = y_2 - 2y_1 + y_0$$
$$\Delta^3 y_0 = y_3 - 3y_2 + 3y_1 - y_0.$$

It can be proved by the finite-induction principle that, in general,

$$(6.3.8) \qquad \Delta^k y_0 = y_k - \binom{k}{1}y_{k-1} + \binom{k}{2}y_{k-2} - \cdots + (-1)^k y_0$$

where

$$\binom{k}{i} = \frac{k!}{i!(k - i)!}$$

is the binomial coefficient formula.

Relation (6.3.8) is particularly useful in deriving Newton's forward-form interpolating polynomial.

It should be noted that the corresponding elements of Table F and Table O are simply two different representations of the same set of numbers. Another representation of this set of numbers can be expressed in terms of backward differences.

Backward differences. Whereas forward differences were defined by starting at the beginning of a table and going forward, backward differences are defined by starting at the end of the table and going backward. Differences of successive ordinates are defined by the formula†

$$\nabla y_{n-i} = y_{n-i} - y_{n-i-1} \quad (i = 0, n - 1)$$

and the terms ∇y_{n-i} are called *first-order backward differences* of function $y = f(x)$ over the interval (x_0, x_n).

The differences of successive first-order backward differences are in turn defined as

$$\nabla^2 y_{n-i} = \nabla y_{n-i} - \nabla y_{n-i-1} \quad (i = 0, n - 2).$$

In general, higher-order backward differences are defined as

$$\nabla^k y_{n-i} = \nabla^{k-1} y_{n-i} - \nabla^{k-1} y_{n-i-1} \quad (i = 0, n - k).$$

A diagonal-difference table of backward differences appears in Table B.

Table B

x_i	y_i	∇y	$\nabla^2 y$	$\nabla^3 y$	$\nabla^4 y$
$\vdots$	$\vdots$				
		$\vdots$			
x_{n-4}	y_{n-4}		$\vdots$		
		∇y_{n-3}		$\vdots$	
x_{n-3}	y_{n-3}		$\nabla^2 y_{n-2}$		$\vdots$
		∇y_{n-2}		$\nabla^3 y_{n-1}$	
x_{n-2}	y_{n-2}		$\nabla^2 y_{n-1}$		$\nabla^4 y_n$
		∇y_{n-1}		$\nabla^3 y_n$	
x_{n-1}	y_{n-1}		$\nabla^2 y_n$		
		∇y_n			
x_n	y_n				

A relation between the backward differences and the ordinates can be derived by equating corresponding elements of the bottom backward diagonals of Table B and Table O:

$$\nabla y_n = y_n - y_{n-1}$$
$$\nabla^2 y_n = y_n - 2y_{n-1} + y_{n-2}$$
$$\nabla^3 y_n = y_n - 3y_{n-1} + 3y_{n-2} - y_{n-3}$$
$$\cdot \quad \cdot \quad \cdot \quad \cdot \quad \cdot \quad \cdot \quad \cdot \quad \cdot \quad \cdot \quad \cdot$$

---

† First-order differences ∇y_i can also be written $\nabla^1 y_i$.

And it can be proved that in general

(6.3.9) $\nabla^k y_n = y_n - \binom{k}{1} y_{n-1} + \binom{k}{2} y_{n-2} - \cdots + (-1)^k y_{n-k}.$

This relation is useful in deriving Newton's backward-form interpolating polynomial.

Derivation of Third Degree Newton Forward-Form Interpolating Polynomial

Given the set of data pairs (x_0, y_0), (x_1, y_1), (x_2, y_2), (x_3, y_3) of the function $y = f(x)$, determine a polynomial of the form

(6.3.10) $P_3(x) = c_0 + c_1(x - x_0) + c_2(x - x_0)(x - x_1)$
$$+ c_3(x - x_0)(x - x_1)(x - x_2)$$

which approximates the function $y = f(x)$ over interval (x_0, x_3) and such that the polynomial $P_3(x)$ coincides with the function at the mesh points x_i $(i = 0, 1, 2, 3)$; i.e., such that

(6.3.11) $\qquad\qquad P_3(x_i) = y_i \qquad (i = 0, 1, 2, 3)$

where the adjacent x_i are separated by evenly spaced intervals of length h, i.e., $h = x_{i+1} - x_i$ $(i = 0, 1, 2)$.

Equations (6.3.11) are the *constraint equations*, which, when written in expanded form using (6.3.10), reduce to

$$
\begin{aligned}
P_3(x_0) &= c_0 & &= y_0 \\
P_3(x_1) &= c_0 + c_1(x_1 - x_0) & &= y_1 \\
P_3(x_2) &= c_0 + c_1(x_2 - x_0) + c_2(x_2 - x_0)(x_2 - x_1) & &= y_2 \\
P_3(x_3) &= c_0 + c_1(x_3 - x_0) + c_2(x_3 - x_0)(x_3 - x_1) \\
& \quad + c_3(x_3 - x_0)(x_3 - x_1)(x_3 - x_2) = y_3.
\end{aligned}
$$

Substituting the relation $(x_j - x_i) = (j - i)h$ into these constraint equations, we obtain

$$
\begin{aligned}
c_0 & & &= y_0 \\
c_0 + c_1 h & & &= y_1 \\
c_0 + c_1 \cdot 2h + c_2 \cdot 2hh & & &= y_2 \\
c_0 + c_1 \cdot 3h + c_2 \cdot 3h \cdot 2h + c_3 \cdot 3h \cdot 2hh &= y_3
\end{aligned}
$$

which is a system of non-homogeneous algebraic equations linear in c_0, c_1, c_2, c_3. This system has a triangular matrix of coefficients and hence can be solved directly by successive substitution.

Solution of the foregoing constraint equations is accomplished by successive substitution as follows.

$$c_0 = y_0$$

$$c_1 = \frac{y_1 - c_0}{h} = \frac{y_1 - y_0}{h}$$

$$c_2 = \frac{y_2 - c_1 2h - c_0}{2h^2} = \frac{y_2 - 2y_1 + y_0}{2h^2}$$

$$c_3 = \frac{y_3 - c_2 6h^2 - c_1 3h - c_0}{6h^3} = \frac{y_3 - 3y_2 + 3y_1 - y_0}{6h^3}.$$

Substitution of relations (6.3.8) into the equations above produces

$$c_0 = y_0$$
$$c_1 = \Delta y_0 / h$$
$$c_2 = \Delta^2 y_0 / 2h^2$$
$$c_3 = \Delta^3 y_0 / 6h^3.$$

These values of the c_k are then substituted into (6.3.10) to obtain the polynomial

$$(6.3.12) \quad P_3(x) = y_0 + \frac{\Delta y_0}{h}(x - x_0) + \frac{\Delta^2 y_0}{2h^2}(x - x_0)(x - x_1)$$

$$+ \frac{\Delta^3 y_0}{6h^3}(x - x_0)(x - x_1)(x - x_2).$$

Equation (6.3.12) is the third-degree Newton's interpolating polynomial (forward form) which approximates $y = f(x)$ over (x_0, x_3).

It should be noted that the coefficients c_k of the Newton's interpolating polynomial above can be calculated directly, simply by constructing the diagonal-difference table for the given function values, namely:

$$
\begin{array}{llll}
x_0 & y_0 & & \\
 & & \Delta y_0 & \\
x_1 & y_1 & & \Delta^2 y_0 \\
 & & \Delta y_1 & & \Delta^3 y_0 \\
x_2 & y_2 & & \Delta^2 y_1 \\
 & & \Delta y_2 & \\
x_3 & y_3 & &
\end{array}
$$

Having constructed the diagonal-difference table, we can calculate the coefficients c_k of the interpolating polynomial directly, using the formulas

$$c_0 = y_0$$

$$c_k = \frac{\Delta^k y_0}{k! h^k} \quad (k = 1, 2, 3).$$

The construction of the diagonal-difference table thus obviates the need for the explicit solution of the constraint equations.

NUMERICAL EXAMPLE. Given the set of (x_i, y_i) data pairs $(0.0, 1.0)$, $(0.33, 1.391)$, $(0.66, 1.935)$, $(1.0, 2.718)$, calculate the third-degree Newton forward form interpolating polynomial which approximates $y = f(x)$ over the interval $(0, 1)$.

Substituting the given (x_i, y_i) values into the constraint equations (6.3.11), we obtain

$$
\begin{aligned}
P_3(0.0) &= c_0 & = 1.000 \\
P_3(0.33) &= c_0 + c_1(0.33 - 0.0) & = 1.391 \\
P_3(0.66) &= c_0 + c_1(0.66 - 0.0) & \\
& \quad + c_2(0.66 - 0.0)(0.66 - 0.33) & = 1.935 \\
P_3(0.10) &= c_0 + c_1(1.0 - 0.0) & \\
& \quad + c_2(1.0 - 0.0)(1.0 - 0.33) & \\
& \quad + c_3(1.0 - 0.0)(1.0 - 0.33)(1.0 - 0.66) & = 2.718
\end{aligned}
$$

The solution of the above constraint equations is

$$ c_0 = 1.0, \quad c_1 = 1.185, \quad c_2 = 0.689, \quad c_3 = 0.387. $$

The required third-degree Newton interpolating polynomial is

$$
\begin{aligned}
P_3(x) = 1.0 &+ 1.185(x - 0.0) + 0.689(x - 0.0)(x - 0.33) \\
&+ 0.387(x - 0.0)(x - 0.33)(x - 0.66).
\end{aligned}
$$

As stated earlier, a simpler way of determining the coefficients is by constructing the diagonal-difference table, which for this example is

0.00	1.000		
		0.391	
0.33	1.391		0.153
		0.544		0.086
0.66	1.935		0.239
		0.783	
1.00	2.718		

The coefficients c_k are then calculated directly as

$$
\begin{aligned}
c_0 &= y_0 & = 1.000 \\
c_1 &= \Delta y_0/h & = 0.391/0.33 & = 1.185 \\
c_2 &= \Delta^2 y_0/2h^2 & = 0.153/0.222 & = 0.689 \\
c_3 &= \Delta^3 y_0/6h^3 & = 0.086/0.222 & = 0.387
\end{aligned}
$$

so that the solution of the constraint equations is replaced by the simpler equivalent process of constructing the diagonal-difference table.

Derivation of nth-degree Newton Interpolating Polynomial

The derivation of the nth-degree Newton forward-form (NFF) interpolating polynomial is a direct extension of the derivation for the third-degree case.

Given the set of data pairs (x_i, y_i) $i = 0, 1, \ldots, n$ of the function $y = f(x)$, the explicit form of which is unknown, determine an nth-degree polynomial of the form

$$(6.3.13) \quad P_n(x) = c_0 + c_1(x - x_0) + c_2(x - x_0)(x - x_1) + \cdots$$
$$+ c_n(x - x_0)(x - x_1)\ldots(x - x_{n-1})$$

which approximates $y = f(x)$ over the interval (x_0, x_n) and which coincides with the function at the $n + 1$ evenly spaced mesh points x_i; i.e., the polynomial $P_n(x)$ satisfies the constraint equations

$$(6.3.14) \qquad P_n(x_i) = y_i \qquad (i = 0, 1, \ldots, n).$$

Writing the constraint equations in expanded form by using (6.3.13), we obtain the following triangular system of linear, non-homogeneous algebraic equations

$$
\begin{aligned}
&P_n(x_0) = c_0 && = y_0 \\
&P_n(x_1) = c_0 + c_1(x_1 - x_0) && = y_1 \\
(6.3.15) \quad &P_n(x_2) = c_0 + c_1(x_2 - x_0) + c_2(x_2 - x_0)(x_2 - x_1) && = y_2 \\
&\quad\cdot\quad\cdot\quad\cdot\quad\cdot\quad\cdot\quad\cdot\quad\cdot\quad\cdot\quad\cdot\quad\cdot\quad\cdot\quad\cdot\quad\cdot \\
&P_n(x_n) = c_0 + c_1(x_n - x_0) + c_2(x_n - x_0)(x_n - x_1) \\
&\qquad\qquad + \cdots + c_n(x_n - x_0)(x_n - x_1)\ldots(x_n - x_{n-1}) = y_n
\end{aligned}
$$

Substituting $(x_j - x_i) = (j - i)h$ into (6.3.15) we obtain

$$
\begin{aligned}
&c_0 && = y_0 \\
&c_0 + c_1 h && = y_1 \\
(6.3.16) \quad &c_0 + c_1\cdot 2h + c_2\cdot 2hh && = y_2 \\
&\quad\cdot\quad\cdot\quad\cdot\quad\cdot\quad\cdot\quad\cdot\quad\cdot\quad\cdot\quad\cdot\quad\cdot\quad\cdot\quad\cdot \\
&c_0 + c_1 nh + c_2 n(n - 1)h^2 + \cdots + c_n n! h^n = y_n
\end{aligned}
$$

Solving the triangular system of constraint equations (6.3.16) by successive substitution, and using the relations (6.3.8)

$$\Delta^k y_0 = y_k - \binom{k}{1} y_{k-1} + \binom{k}{2} y_{k-2} - \cdots + (-1)^k y_0$$

we obtain the coefficients c_k of the nth-degree Newton forward-form interpolating polynomial, as follows:

$$c_0 = y_0$$

$$c_1 = \frac{\Delta y_0}{h}$$

$$c_2 = \frac{\Delta^2 y_0}{2h^2}$$

(6.3.17)

$$\cdot \quad \cdot \quad \cdot \quad \cdot$$

$$c_k = \frac{\Delta^k y_0}{k!h^k}$$

$$\cdot \quad \cdot \quad \cdot \quad \cdot$$

$$c_n = \frac{\Delta^n y_0}{n!h^n}$$

Substituting the c_k into (6.3.13) we find the required nth-degree Newton interpolating polynomial

(6.3.18) $P_n(x) = y_0 + \dfrac{\Delta y_0}{h}(x - x_0) + \dfrac{\Delta^2 y_0}{2h^2}(x - x_0)(x - x_1)$

$$+ \cdots + \frac{\Delta^n y_0}{n!h^n}(x - x_0)(x - x_1)\ldots(x - x_{n-1})$$

that approximates $y = f(x)$ over the interval (x_0, x_n).

Again, we note that the coefficients c_k of Newton's forward-form interpolating polynomial can be calculated directly from the upper diagonal of the diagonal-difference table:

x_0	y_0					
		Δy_0				
x_1	y_1		$\Delta^2 y_0$			
		Δy_1		$\Delta^3 y_0$		
x_2	y_2		$\Delta^2 y_1$		$\cdot$ $\cdot$	
		Δy_2		$\Delta^3 y_1$		
x_3	y_3		$\Delta^2 y_2$			$\vdots$ $\Delta^n y_0$
		Δy_3				
x_4	y_4					
				$\Delta^3 y_{n-3}$	$\cdot$	
$\vdots$	$\vdots$		$\Delta^2 y_{n-2}$		$\cdot$	
		Δy_{n-1}				
x_n	y_n					

The coefficients c_k can then be calculated by formula (6.3.17), without solving explicitly the constraint equations.

Formula Adjustment to Eliminate Zero Subscript

Since most **FORTRAN** processors do not permit use of a zero subscript (index value of zero) we must adjust the subscripts to eliminate the zero value, and then correct the formulas for the finite differences and interpolating polynomial, accordingly.

Elimination of the zero subscript is accomplished by simply replacing old subscript i, which ranges from 0 through n, with *new* subscript i, which ranges from 1 through $n + 1$. The set of $\{(x_i, y_i)\ i = 0, n\}$ is then stored as the set $\{(x_i, y_i), i = 1, n + 1\}$.

Accordingly, the diagonal-difference table appears as follows (using *new* index i):

$$
\begin{array}{llll}
x_1 & y_1 & & \\
& & \Delta y_1 & \\
x_2 & y_2 & & \Delta^2 y_1 \\
& & \Delta y_2 & & \cdot \\
x_3 & y_3 & & \Delta^2 y_2 & & \cdot \\
& & \Delta y_3 & & & \cdot \\
x_4 & y_4 & & & & & \Delta^n y_1 \\
\vdots & \vdots & & & \cdot \\
& & & & \cdot \\
x_n & y_n & & \Delta^2 y_{n-1} \\
& & \Delta y_n \\
x_{n+1} & y_{n+1}
\end{array}
$$

and the original Newton forward-form interpolating polynomial of equation (6.2.18) becomes (using *new* index i)

$$P_n(x) = y_1 + \frac{\Delta y_1}{h}(x - x_1) + \frac{\Delta^2 y_1}{2h^2}(x - x_1)(x - x_2)$$

$$+ \cdots + \frac{\Delta^n y_1}{n!\,h^n}(x - x_1)(x - x_2)\ldots(x - x_n).$$

Note that the interpolating polynomial can be constructed directly from the diagonal-difference table, simply by dividing term $\Delta^k y_1$ of the upper diagonal by $k!\,h^k$ to obtain coefficient c_k.

Computational Summary for Newton Forward-Form Interpolation

Step 0. Input parameters and data. Here, $n + 1$ = number of data pairs (x_i, y_i); $h = x_{i+1} - x_i$, interval length for given x_i; $\delta x = \bar{x}_{j+1} - \bar{x}_j$, interval between interpolated points $\bar{x}_j$; $m = h/\delta x$ = number of subintervals between x_i and x_{i+1}; and (x_i, y_i) $i = 1, n + 1$, given set of data.

Step 1. Clear diagonal-difference table. Set

$$\Delta^k y_i = 0 \qquad (i = 1, n) \quad (k = 1, n).$$

Step 2. Construct diagonal-difference table using formulas

$$\Delta y_i = y_{i+1} - y_i \qquad (i = 1, n)$$
$$\Delta^k y_i = \Delta^{k-1} y_{i+1} - \Delta^{k-1} y_i \qquad (k = 2, n; i = 1, n + 1 - k).$$

Note: $\Delta y_i \equiv \Delta^1 y_i$.

Step 3. Generate NFF polynomial, evaluated at $\bar{x}_j$ ($j = 1, m \cdot n + 1$).
(a) Set $\bar{x}_1 = x_1$; Set index $j = 1$.
(b) Calculate values of $P_n(\bar{x}_j)$:

$$P_n(\bar{x}_j) = y_1 + \frac{\Delta y_1}{h}(\bar{x}_j - x_1) + \frac{\Delta^2 y_1}{2h^2}(\bar{x}_j - x_1)(\bar{x}_j - x_2)$$

$$+ \cdots + \frac{\Delta^n y_1}{n!h^n}(\bar{x}_j - x_1)(\bar{x}_j - x_2)\ldots(\bar{x}_j - x_n)$$

$$= y_1 + T_1 + T_2 + \cdots + T_n$$

where

$$T_1 = \frac{\Delta y_1}{h}(\bar{x}_j - x_1)$$

$$T_k = \frac{\Delta^k y_1}{k!h^k}(\bar{x}_j - x_1)(\bar{x}_j - x_2)\ldots(\bar{x}_j - x_k) \qquad (k = 2, n).$$

Step 4. Compare index j to $m \cdot n + 1$. If $j < m \cdot n + 1$, set $\bar{x}_{j+1} = \bar{x}_j + \delta x$, set $j = j + 1$, return to Step 3b. If $j = m \cdot n + 1$, go to Step 5.

Step 5. Write output. Write out

$$(\bar{x}_j, P_n(\bar{x}_j)) \qquad (j = 1, \ldots, m \cdot n + 1).$$

A representative flow chart for the Newton forward-form interpolating polynomial is shown in Figure 6-3.

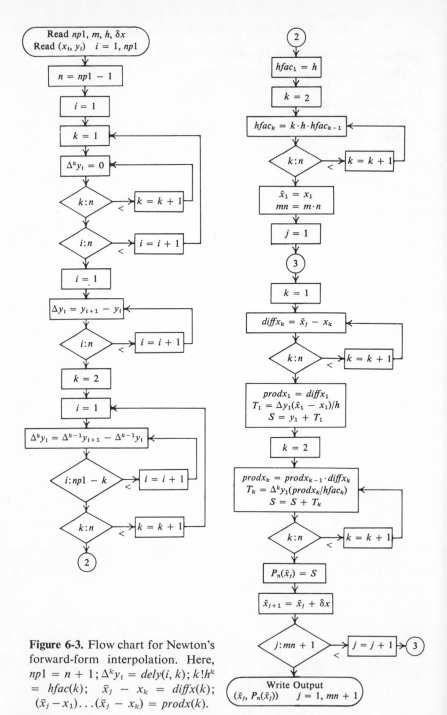

Figure 6-3. Flow chart for Newton's forward-form interpolation. Here, $np1 = n + 1$; $\Delta^k y_i = dely(i, k)$; $k!\,h^k = hfac(k)$; $\bar{x}_j - x_k = diffx(k)$; $(\bar{x}_j - x_1)\ldots(\bar{x}_j - x_k) = prodx(k)$.

224

```
      PROGRAM NFF
C     NEWTON FORWARD-FORM INTERPOLATING POLYNOMIAL
      DIMENSION X(20), Y(20), DELY (20,20), T(20), PN (400), XBAR (400),
     XDIFFX(20), PRODX(20), HFAC(20)
C     NP1 = N+1 = NUMBER OF GIVEN DATA POINTS
C     M = MULTIPLE OF DATA POINTS,  M * N + 1 = NUMBER INTERPOLATED POINTS
C     H = INTERVAL BETWEEN SUCCESSIVE GIVEN X (I)
C     DELX = H/M = INTERVAL BETWEEN INTERPOLATED POINTS
      READ (5,100) NP1, M, H, DELX
      READ (5,101) (X(I), Y(I), I = 1,NP1)
C     CLEAR DIAGONAL DIFFERENCE TABLE
      N = NP1 - 1
      DO 10 I = 1, N
      DO 5 K = 1, N
    5 DELY (I, K) = 0.0
   10 CONTINUE
C     CALCULATE FIRST DIFFERENCES DELY (I,1)
      DO 20 I = 1, N
   20 DELY (I,1) = Y(I+1) - Y(I)
C     CALCULATE HIGHER ORDER DIFFERENCES DELY (I,K)
      DO 30 K = 2, N
      NK = NP1 - K
      DO 25 I = 1, NK
   25 DELY (I,K) = DELY (I+1, K-1) - DELY (I,K-1)
   30 CONTINUE
C     GENERATE NEWTON FORWARD FORM POLYNOMIAL
C     EVALUATE POLYNOMIAL AT INTERVALS OF DELX FROM X (1)
      HFAC (1) = H
      DO 35 K = 2,N
   35 HFAC(K) = FLOAT(K) * H * HFAC(K-1)
      XBAR (1) = X(1)
      MN = M * N + 1
      DO 55 J = 1, MN
      DO 40 K = 1, N
   40 DIFFX (K) = XBAR (J) - X(K)
      PRODX (1) = DIFFX (1)
      T(1) = (DELY(1,1) * PRODX(1))/HFAC(1)
      SUM = Y (1) + T (1)
      DO 50 K = 2,N
      PRODX(K) = PRODX(K-1) * DIFFX(K)
      T(K) = (DELY(1,K) * PRODX(K))/HFAC(K)
   50 SUM = SUM + T (K)
      PN (J) = SUM
      XBAR (J+1) = XBAR (J) + DELX
   55 CONTINUE
      DO 60 K = 1, N
      NK = NP1 - K
      WRITE (6,120) K
   60 WRITE (6,104) (DELY(I,K), I=1,NK)
      WRITE (6,121)
      WRITE (6,110) (X(I), Y(I), I = 1,NP1)
      WRITE (6,111) (XBAR(I), PN(I), I = 1,MN)
      STOP
  100 FORMAT (2I2, 2F8.6)
  101 FORMAT (2F19.9)
  104 FORMAT (2X, F19.9)
  110 FORMAT (2X, 2F16.8)
  111 FORMAT (2X, F16.8, 18X, F16.8)
  120 FORMAT (2X, I2, 25H TH ORDER DIFFERENCES ARE)
  121 FORMAT(53H0   IND VAR X(I)      DEP VAR Y(I)      POLYNOMIAL PN(I))
      END
```

6.4 Aitken-Neville Repeated (Iterated) Linear Interpolation

Suppose that we are given the $n + 1$ data pairs $(x_0, y_0), (x_1, y_1), \ldots,$ (x_n, y_n) representing the points $p_0, p_1, \ldots, p_n$ of the graph of a function $y = f(x)$, where the explicit form of $f(x)$ is not known. As defined in Sec. 6.1, an interpolating polynomial of degree n for these $n + 1$ data pairs is that nth-degree polynomial $P_n(x)$ that satisfies the constraints

(6.4.1) $$P_n(x_i) = y_i \qquad (i = 0, 1, 2, \ldots, n).$$

From the preceding sections we know that Lagrange's form (ordinate type) of the unique interpolating polynomial of degree n is obtained by assuming a polynomial of the form

(6.4.2) $$P_n(x) = y_0 b_0(x) + y_1 b_1(x) + y_2 b_2(x) + \cdots + y_n b_n(x)$$

while Newton's forward form (finite-difference type) of the unique interpolating polynomial is obtained by assuming a polynomial of the form

$$P_n(x) = c_0 + c_1(x - x_0) + c_2(x - x_0)(x - x_1) + \cdots$$
$$+ c_n(x - x_0)(x - x_1)(x - x_2) \ldots (x - x_{n-1}).$$
(6.4.3)

In either case, the coefficients $[b_k(x)$ or $c_k]$ are uniquely *determined* by the linear constraint equations (6.4.1), though they are computed in practice by other means (see Secs. 6.2 and 6.3).

In this section we will show that *another distinct form of the unique interpolating polynomial $P_n(x)$* of degree n can be generated by repeated *(iterated) linear interpolation*. To do this we first need to derive a linear interpolation formula. Repeated application of this formula will then produce interpolating polynomials of degree $2, 3, \ldots$.

To interpolate linearly between two points $(x_k, y_k), (x_{k+m}, y_{k+m})$, we can write the two-point formula of the line through these points

(6.4.4) $$\frac{y(x) - y_k}{x - x_k} = \frac{y_{k+m} - y_k}{x_{k+m} - x_k}.$$

To calculate $y(x)$ for a given x, we simply solve (6.4.4) for $y(x)$, i.e.,

(6.4.5) $$y(x) = \frac{y_k(x - x_{k+m}) - y_{k+m}(x - x_k)}{x_k - x_{k+m}}.$$

Formula (6.4.5) is called the *basic linear interpolating formula* (BLIF).

The process of repeated linear interpolation can be nicely illustrated by Figure 6-4, which shows how a quadratic function is obtained by two repeated applications of linear interpolation. Let $(x_0, y_0), (x_1, y_1), (x_2, y_2)$ denote points on the graph of $y = f(x)$, where the x_i $(i = 0, 1, 2)$ are arbitrary distinct values of x.

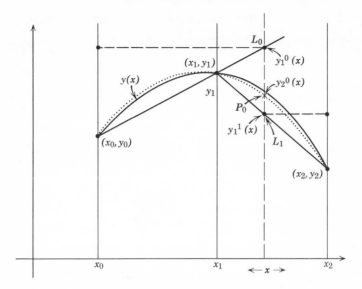

Figure 6-4. Repeated linear interpolation.

By linearly interpolating between (x_0, y_0) and (x_1, y_1), we obtain the *linear* function

$$y_1{}^0(x) = \frac{y_0[x - x_1] - y_1[x - x_0]}{x_0 - x_1}.$$

As the vertical dashed line (denoted by $\leftarrow x \rightarrow$) moves, the point L_0 takes on the values of the linear function $y_1{}^0(x)$. Similarly, by linearly interpolating between (x_1, y_1) and (x_2, y_2), we obtain the *linear* function

$$y_1{}^1(x) = \frac{y_1[x - x_2] - y_2[x - x_1]}{x_1 - x_2}$$

and the point L_1 takes on the values of linear function $y_1{}^1(x)$ as $\leftarrow x \rightarrow$ moves.

By repeating the process of linear interpolation, this time over the double interval (x_0, x_2), i.e., by linearly interpolating between the two linear functions $y_1{}^0(x)$ and $y_1{}^1(x)$, we obtain the *quadratic* function†

$$y_2{}^0(x) = \frac{y_1{}^0(x)[x - x_2] - y_1{}^1(x)[x - x_0]}{x_0 - x_2}$$

and the point P_0 takes on the value of the quadratic function $y_2{}^0(x)$ as the vertical line $\leftarrow x \rightarrow$ moves. Evaluating $y_2{}^0(x)$ successively at x_0, x_1, x_2, we

† $y_2{}^0(x)$ is a quadratic function because each term in the numerator is the product of a linear function $y_1{}^i(x)$ and a linear term $(x - x_j)$.

find that $y_2{}^0(x_0) = y_0$, $y_2{}^0(x_1) = y_1$, and $y_2{}^0(x_2) = y_2$, i.e., $y_2{}^0(x_i)$ satisfy the constraints of the quadratic interpolating polynomial determined by the points (x_0, y_0), (x_1, y_1), (x_2, y_2).

In the following pages, it will be shown that *three repetitions* of linear interpolation, starting with $p_0(x_0, y_0)$, $p_1(x_1, y_1)$, $p_2(x_2, y_2)$, $p_3(x_3, y_3)$, produce a *cubic* interpolating polynomial, and that n *repetitions* of linear interpolation, starting with the points (x_i, y_i) $(i = 0, n)$, produce an nth-*degree* interpolating polynomial.

We will first† show how a *cubic* interpolating polynomial $P_3(x)$ for points p_0, p_1, p_2, p_3 *can be generated by three repeated applications* of BLIF formula (6.4.5), where:

1. The first application of the BLIF generates three linear polynomials‡ $y_1{}^0(x)$, $y_1{}^1(x)$, $y_1{}^2(x)$ (see equations 6.4.6).

2. The second application of the BLIF generates two quadratic polynomials† $y_2{}^0(x)$, $y_2{}^1(x)$ (see equations 6.4.7).

3. The third application of the BLIF generates one cubic polynomial $y_3{}^0(x)$ (see equation 6.4.8), which is another distinct form of the unique interpolating polynomial $P_3(x)$ for p_0, p_1, p_2, p_3.

The *detailed generation by repeated linear interpolation* of this cubic interpolating polynomial is accomplished by the following steps:

Step 1. Generate three first-degree polynomials by applying the basic linear interpolating formula (6.4.5) successively to the three (x, y) pairs connected by $\frown$:

$$(x_0, \overgroup{y_0), (x_1, y_1), (x_2, y_2), (x_3}, y_3)$$

obtaining

(6.4.6a)
$$y_1{}^0(x) = \frac{y_0[x - x_1] - y_1[x - x_0]}{x_0 - x_1}$$

noting that $y_1{}^0(x_0) = y_0$, $y_1{}^0(x_1) = y_1$;

(6.4.6b)
$$y_1{}^1(x) = \frac{y_1[x - x_2] - y_2[x - x_1]}{x_1 - x_2}$$

noting that $y_1{}^1(x_1) = y_1$, $y_1{}^1(x_2) = y_2$;

(6.4.6c)
$$y_1{}^2(x) = \frac{y_2[x - x_3] - y_3[x - x_2]}{x_2 - x_3}$$

noting that $y_1{}^2(x_2) = y_2$, $y_1{}^2(x_3) = y_3$.

---

† These results can be directly extended to nth-degree case.

‡ The subscript of function $y_m{}^k$ denotes degree of function, while the superscript denotes the index of the left endpoint of interval (x_k, x_{k+m}) from which the function is defined. This notation was selected so that it could be consistent with the notation of the Romberg and Euler-Romberg methods (Secs. 8.4 and 9.3, respectively).

The functions $y_1{}^0(x)$, $y_1{}^1(x)$, $y_1{}^2(x)$ are linear interpolating polynomials for $y = f(x)$ over (x_0, x_1), (x_1, x_2), (x_2, x_3), respectively, because each is a first-degree polynomial satisfying the constraints of a linear interpolating polynomial.

Step 2. Next, generate two second-degree polynomials by applying BLIF formula (6.4.5) to the two (x, y) pairs connected by $\frown$:

$$(x_0, \overbrace{y_1{}^0(x)), (x_1, y_1{}^1(x)), (x_2, y_1{}^1(x)), (x_3, y_1{}^2(x))}$$

obtaining

(6.4.7a) $$y_2{}^0(x) = \frac{y_1{}^0(x)[x - x_2] - y_1{}^1(x)[x - x_0]}{x_0 - x_2}$$

noting $y_2{}^0(x_0) = y_0$, $y_2{}^0(x_1) = y_1$, $y_2{}^0(x_2) = y_2$;

(6.4.7b) $$y_2{}^1(x) = \frac{y_1{}^1(x)[x - x_3] - y_1{}^2(x)[x - x_1]}{x_1 - x_3}$$

noting $y_2{}^1(x_1) = y_1$, $y_2{}^1(x_2) = y_2$, $y_2{}^1(x_3) = y_3$.

Functions $y_2{}^0(x)$ and $y_2{}^1(x)$ are quadratic interpolating polynomials of $y = f(x)$ over the *double intervals* (x_0, x_2), (x_1, x_3), respectively, because each is a second-degree polynomial satisfying the constraints of a quadratic interpolating polynomial.

Step 3. Finally, generate one third-degree polynomial by applying the BLIF formula (6.4.5) to the one (x, y) pair $(x_0, \overbrace{y_2{}^0(x)), (x_3, y_2{}^1(x))}$, obtaining

(6.4.8) $$y_3{}^0(x) = \frac{y_2{}^0(x)[x - x_3] - y_2{}^1(x)[x - x_0]}{x_0 - x_3}$$

noting $y_3{}^0(x_0) = y_0$, $y_3{}^0(x_1) = y_1$, $y_3{}^0(x_2) = y_2$, $y_3{}^0(x_3) = y_3$. Function $y_3{}^0(x)$ is simply *another distinct form* of the cubic interpolating polynomial $P_3(x)$ for points p_0, p_1, p_2, p_3 because it is a polynomial of degree 3, which satisfies the constraints of the cubic interpolating polynomial.

We see then that repeated (iterated) application of the basic linear interpolating formula (6.4.5) to three pairs, two pairs, one pair of (x, y) generates a third-degree interpolating polynomial for $y = f(x)$ over (x_0, x_3). *The nth-degree interpolating polynomial $P_n(x)$ for $y = f(x)$ over interval (x_0, x_n) is obtained by a direct extension of the process illustrated above.*

Formulas (6.4.6) can be rewritten in the single formula

(6.4.9) $$y_1{}^k(x) = \frac{y_k[x - x_{k+1}] - y_{k+1}[x - x_k]}{x_k - x_{k+1}}   (k = 0, 1, 2).$$

Formulas (6.4.7) and (6.4.8) can be rewritten and *grouped* into the following single recursion formula

$$(6.4.10) \quad y_m{}^k(x) = \frac{y_{m-1}^k(x)[x - x_{k+m}] - y_{m-1}^{k+1}(x)[x - x_k]}{x_k - x_{k+m}}$$

$$(m = 2, 3; k = 0, 3 - m).$$

For $n + 1$ data pairs (x_k, y_k) $(k = 0, n)$, the *recursion formulas* for the method of repeated (iterated) linear interpolation are the same as formulas (6.4.9) and (6.4.10) with limits $k = 0, n - 1$ in (6.4.9) and limits $m = 2, n$ and $k = 0, n - m$ in (6.4.10).

Method of Repeated Linear Interpolation for Degree n

Given $n + 1$ data pairs (x_k, y_k) $(k = 0, n)$, let us first shift the index range from $(0, n)$ to $(1, n + 1)$ *to eliminate the zero-subscript* problem encountered with most FORTRAN processors. We then have the data pairs (x_k, y_k) $(k = 1, n + 1)$. The recursion formulas for generating the nth-degree interpolating polynomial are simply formulas (6.4.9) and (6.4.10) with indexes shifted to reflect range $(1, n + 1)$ for k. The formulas, with their new index ranges are

$$(6.4.9)^* \quad y_1{}^k(x) = \frac{y_k(x - x_{k+1}) - y_{k+1}(x - x_k)}{x_k - x_{k+1}} \quad (k = 1, n)$$

$$(6.4.10)^* \quad y_m{}^k(x) = \frac{y_{m-1}^k(x)(x - x_{k+m}) - y_{m-1}^{k+1}(x)(x - x_k)}{x_k - x_{k+m}}$$

$$(m = 2, n; k = 1, n + 1 - m).$$

We can now generate functions $y_1{}^k(x)$ and $y_m{}^k(x)$ $(m = 2, n)$ using formulas (6.4.9)* and (6.4.10)* by the method of repeated linear interpolation. The functions generated thereby are given in the following table.

x_1	y_1				
		y_1^1			
x_2	y_2		y_2^1		
		y_1^2		y_3^1	
x_3	y_3		y_2^2	$\cdot$	
		y_1^3		$\vdots$	
x_4	y_4		$\vdots$		$\cdot$
					$\cdot$
$\vdots$	$\vdots$	$\vdots$			y_n^1
				$\cdot$	
x_{n-2}	y_{n-2}		$\cdot$		
		y_1^{n-2}	y_2^{n-2}		
x_{n-1}	y_{n-1}			y_3^{n-2}	
		y_1^{n-1}	y_2^{n-1}		
x_n	y_n				
		y_1^n			
x_{n+1}	y_{n+1}				

The value of $P_n(x)$, for a specific value of $x = X$, is obtained by evaluating formulas (6.4.9)* and (6.4.10)* for $x = X$, as the recursion formulas are iterated over the given ranges of the indexes.

A representative flow chart (Figure 6-5) for the method of repeated (iterated) linear interpolation, for data pairs (x_k, y_k) $(k = 1, n + 1)$ is as follows [using formulas (6.4.9)* and (6.4.10)*]:

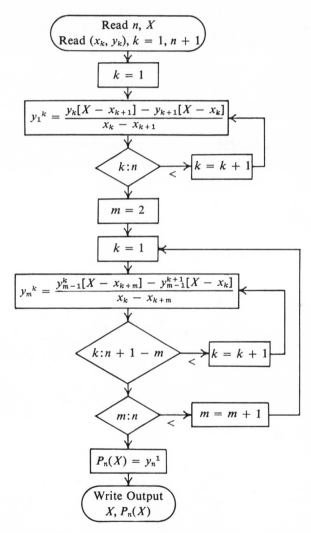

Figure 6-5. Note that $P_n(X)$ is generated, where X is a specific value of x. Values $y_m^k(X)$ are denoted in the flow chart simply by y_m^k.

```
      PROGRAM ITERINT
C         FORTRAN PROGRAM FOR AITKEN-NEVILLE ITERATED LINEAR INTERPOLATION
C         EVALUATE P (X) AT X = XC BY ITERATED INTERPOLATION
C         MAXIMUM NUMBER OF TABULATED DATA PAIRS = 10
          DIMENSION X(10), Y(10), VAL(10, 10)
          READ (5, 100) N, XC
          NP1 = N + 1
          READ (5, 101) (X(K), Y(K), K = 1, NP1)
          DO 5 K = 1, N
        5 VAL(1,K)=(Y(K) * (XC-X(K+1))-Y(K+1) * (XC-X(K)))/(X(K)-X(K+1))
          DO 15 M = 2, N
          NP1MM = N + 1 - M
          DO 10 K = 1, NP1MM
          KPM = K + M
       10 VAL(M,K) =(VAL(M-1,K) * (XC - X(KPM)) - VAL(M-1,K+1) * (XC - X(K))
          X)/(X(K) - X(KPM))
       15 CONTINUE
          POLY = VAL (N,1)
          WRITE (6, 110) XC, POLY
          STOP
      100 FORMAT (I2, F15.8)
      101 FORMAT (2F15.8)
      110 FORMAT (1X, 2E15.8)
          END
```

Numerical Example of Aitken-Neville Iterated Interpolation

Given the set of data pairs†

k	x_k	y_k
1	0.0	1.000000
2	0.1	1.105171
3	0.2	1.221403
4	0.3	1.349859
5	0.4	1.491825

calculate the value of the function at $X = 0.25$ by iterated interpolation, using the recursion formulas

$$y_1{}^k = \frac{y_k(X - x_{k+1}) - y_{k+1}(X - x_k)}{x_k - x_{k+1}}$$

$$y_m{}^k = \frac{y_{m-1}^k(X - x_{k+m}) - y_{m-1}^{k+1}(X - x_k)}{x_k - x_{k+m}} \qquad (m = 2, 3, \ldots; k = 1, 2, \ldots).$$

Solution. Construct the table of values (Table 6-1) for the iterated interpolation method, using the formulas above.

The required value of $y(0.25) = 1.284026$ which agrees to six significant digits with the value 1.284025 of $e^{0.25}$ taken from a table of the exponential function.

† The data pairs for this example were taken from a table of values of the function e^x.

Table 6-1

x_k	y_k	$y_1{}^k$	$y_2{}^k$	$y_3{}^k$	$y_4{}^k$
0.0	1.000000				
		1.262928			
0.1	1.105171		1.283667		
		1.279519		1.284030	
0.2	1.221403		1.284103		1.284026
		1.285631		1.284023	
0.3	1.349859		1.283942		
		1.278876			
0.4	1.491825				

Advantages and Disadvantages of Lagrangian Interpolation

Some of the more obvious advantages and disadvantages of Lagrangian (ordinate-form) interpolation can be readily seen by examining the structure of the Lagrangian polynomial

$$P_n(x) = \sum_{k=0}^{n} y_k \frac{(x - x_0)\ldots(x - x_{k-1})(x - x_{k+1})\ldots(x - x_n)}{(x_k - x_0)\ldots(x_k - x_{k-1})(x_k - x_{k+1})\ldots(x_k - x_n)}$$

$$\equiv \sum_{k=0}^{n} \frac{L_k(x)}{L_k(x_k)} y_k.$$

Advantages include:

A1. Lagrangian interpolation can be applied when the abscissas x_0, x_1, $\ldots$, x_n are arbitrary, provided that these values are distinct.

A2. Inverse interpolation can be accomplished by a simple interchange of the roles of x and y because of property A1 (provided y_k distinct).

A3. If more than one set of dependent variables $y_k, \bar{y}_k, \bar{\bar{y}}_k, \ldots$ ($k = 0, n$), are tabulated for the same set of independent x_k ($k = 0, n$), then Lagrangian polynomials $P_n(x), \bar{P}_n(x), \bar{\bar{P}}_n(x), \ldots$ can be generated by simply replacing y_k by $\bar{y}_k, \bar{\bar{y}}_k, \ldots$ in the formula above, without recomputing the coefficients $L_k(x)/L_k(x_k)$.

A4. For a given set of x_k and a given set of x at which $P_n(x)$ is to be evaluated, the Lagrangian coefficients $L_k(x)/L_k(x_k)$ can be precomputed and tabulated.

A5. The Lagrangian method of interpolation is easy to program for a stored-program digital computer because of the simplicity of its recursion formulas.

Disadvantages include:

D1. To add additional interpolating points (e.g., x_{n+1}, y_{n+1}), it is necessary to recompute the Lagrangian coefficients.

D2. To evaluate $P_n(x)$ at additional x values, it is necessary to recompute the Lagrangian coefficients.

D3. It is difficult in advance to determine the number of interpolating points required for a particular accuracy in a given table of (x_k, y_k) values.

Advantages and Disadvantages of Finite-Difference Interpolation

Some of the more obvious advantages and disadvantages of finite-difference-form interpolation can be readily seen by examining a particular formula of this type. To illustrate this, we will use the Newton forward-form interpolating polynomial

$$P_n(x) = y_0 + \frac{\Delta y_0}{h}(x - x_0) + \frac{\Delta^2 y_0}{2h^2}(x - x_0)(x - x_1)$$

$$+ \frac{\Delta^3 y_0}{3!h^3}(x - x_0)(x - x_1)(x - x_2)$$

$$+ \cdots + \frac{\Delta^n y_0}{n!h^n}(x - x_0)(x - x_1)\ldots(x - x_{n-1}).$$

Advantages include:

A1. Additional interpolating points, e.g., (x_{n+1}, y_{n+1}), can be added by simply adding another term to the above formula, in this case $(\Delta^{n+1} y_0/[(n + 1)!h^{n+1}])(x - x_0)(x - x_1)\ldots(x - x_n)$. In a similar manner, interpolating points can be deleted (on the end of the interval).

A2. The finite-difference table gives an indication of the highest-order interpolating polynomial required for a particular accuracy; i.e., if the given set of points can be represented exactly by an nth-degree polynomial, then the column of nth differences is constant.

A3. Random errors in function values are easily detected in the finite-difference table (see Nielsen [19, p. 41]). If differences are erratic, the user is warned that a polynomial fit may not be satisfactory.

Disadvantages include:

D1. The finite-difference-form interpolating polynomial requires that the abscissas $x_0, x_1, \ldots, x_n$ be evenly spaced. This requirement can be eliminated by using divided-difference interpolating polynomials, ordinate-form polynomials, or other suitable forms.

D2. The finite-difference-form interpolating polynomial is not adaptable to inverse interpolation because of property D1, except for the special case of linear interpolation.

D3. If more than one set of dependent variables $y_k, \bar{y}_k, \bar{\bar{y}}_k$ $(k = 0, n)$ are tabulated for the same set of x_k $(k = 0, n)$, then it is necessary to compute separate difference tables for each set.

Advantages and Disadvantages of Repeated Linear Interpolation

An examination of the recursion formulas for repeated (iterated) linear interpolation reveals some of the more obvious advantages and disadvantages of this form of interpolation. These formulas are

$$y_1{}^k(x) = \frac{y_k[x - x_{k+1}] - y_{k+1}[x - x_k]}{x_k - x_{k+1}} \quad (k = 0, n - 1)$$

$$y_m{}^k(x) = \frac{y_{m-1}^k(x)[x - x_{k+m}] - y_{m-1}^{k+1}(x)[x - x_k]}{x_k - x_{k+m}} \quad (m = 2, n; k = 0, n - m)$$

where the subscript m = degree, and the superscript k indicates the left end point of the interval of definition (x_k, x_{k+m}).

Advantages include:

A1. The method can be applied when the $x_0, x_1, \ldots, x_n$ are arbitrary, provided that these values are distinct.

A2. Inverse interpolation can be accomplished by simple interchange of the roles of x and y because of property A1 (provided y_k distinct).

A3. The number of repetitions of the recursion formula (i.e., the value of m = degree) required for a particular accuracy can be easily determined by comparing successive values $y_{m-1}^0(x)$ and $y_m{}^0(x)$ on the *upper diagonal* of the iterated interpolation array.

A4. The method is easy to program because of the simplicity of its recursion formulas. Additionally, a simple test can be incorporated in the program to determine the number of repetitions required for a particular accuracy.

Disadvantages include:

D1. For each value of x for which $P_n(x) \equiv y_n{}^0(x)$ is to be evaluated, it is necessary to cycle through the recursion formulas from the beginning. That is, we cannot precompute coefficients as can be done in ordinate-type (Lagrangian) interpolation.

D2. If more than one set of dependent variables $y_k, \bar{y}_k, \bar{\bar{y}}_k$ $(k = 0, n)$ are tabulated for the same set of x_k $(k = 0, n)$, then it is necessary to evaluate the recursion formulas for each set of dependent variables separately.

DERIVATION EXERCISES

1. Use backward differences, as defined on p. 216, to derive the third-degree Newton backward-form (NBF) interpolating polynomial

$$P_3(x) = y_n + \frac{\nabla^1 y_n}{h}(x - x_n) + \frac{\nabla^2 y_n}{2!h^2}(x - x_n)(x - x_{n-1})$$

$$+ \frac{\nabla^3 y_n}{3!h^3}(x - x_n)(x - x_{n-1})(x - x_{n-2})$$

Hint: Assume that

$$P_3(x) = c_0 + c_1(x - x_n) + c_2(x - x_n)(x - x_{n-1})$$
$$+ c_3(x - x_n)(x - x_{n-1})(x - x_{n-2})$$

and solve the constraint equations

$$P_3(x_{n-i}) = y_{n-i} \quad (i = 0, 1, 2, 3)$$

for the coefficients c_0, c_1, c_2, c_3.

2. Use backward differences, to derive the nth-degree Newton backward-form interpolating polynomial

$$P_n(x) = y_n + \frac{\nabla^1 y_n}{h}(x - x_n) + \frac{\nabla^2 y_n}{2!h^2}(x - x_n)(x - x_{n-1})$$
$$+ \cdots + \frac{\nabla^n y_n}{n!h^n}(x - x_n)(x - x_{n-1})\ldots(x - x_1).$$

3. Use the principle of finite induction, to derive the ordinate-difference relation†

$$\Delta^k y_0 = y_k - \binom{k}{1}y_{k-1} + \binom{k}{2}y_{k-2} - \cdots + (-1)^k y_0$$
$$= \sum_{i=0}^{k} (-1)^i \binom{k}{i} y_{k-i}.$$

4. Use successive substitution to derive the relation†

$$y_k = y_0 + \binom{k}{1}\Delta y_0 + \binom{k}{2}\Delta^2 y_0 + \cdots + \Delta^k y_0$$
$$= \sum_{i=0}^{k} \binom{k}{i}\Delta^i y_0.$$

5. Use the ordinate-difference relations

$$\Delta y_0 = y_1 - y_0$$
$$\Delta^2 y_0 = y_2 - 2y_1 + y_0$$

to show that the Newton forward-form (NFF) and the Lagrange interpolating polynomials for the set of data pairs (x_i, y_i) $(i = 0, 1, 2)$ are simply algebraic rearrangements of the same unique interpolating polynomial. That is, show that

$$P_2(x) = y_0 + \frac{\Delta y_0}{h}(x - x_0) + \frac{\Delta^2 y_0}{2!h^2}(x - x_0)(x - x_1)$$

can be algebraically rearranged in the form

$$P_2(x) = y_0 \frac{(x - x_1)(x - x_2)}{(x_0 - x_1)(x_0 - x_2)} + y_1 \frac{(x - x_0)(x - x_2)}{(x_1 - x_0)(x_1 - x_2)}$$
$$+ y_2 \frac{(x - x_0)(x - x_1)}{(x_2 - x_0)(x_2 - x_1)}.$$

† See Nielsen [19, pp. 33–34].

COMPUTATIONAL EXERCISES

1. Write the quadratic Lagrange interpolating polynomial for the following set of (x_i, y_i): (0.0, 1.0), (0.5, 1.1276), (0.75, 1.2947).

2. Write the cubic Newton forward-form interpolating polynomial for the following set of (x_i, y_i): (0.0, 1.0), (0.5, 1.1276), (1.0, 1.5431), (1.5, 2.3534). Evaluate this polynomial at $x = 1.2$.

3. Use the Aitken-Neville iterated interpolation to evaluate the cubic polynomial for the data of exercise 2 at $x = 1.2$.

4. Use the Newton forward-form interpolation to estimate the value of $\sin 14°$ using values of $\sin 10°, 20°, 30°, 40°$. Check the result against the value in the trig tables for $\sin x$.

5. Use the Lagrange interpolation to estimate the value of $\tan 21°$ using values of $\tan 10°, 20°, 30°$. Check result against value in table.

6. Use the Aitken-Neville iterated interpolation to estimate the value of $\cos 76°$ using values of $\cos 70°, 80°, 90°$. Check result against value in table.

7. Compute the error bound for the interpolated value of $\sin 14°$ using values of $\sin 10°, 20°, 30°$.

8. Compute the error bound for the cubic interpolating polynomial for $\sin x$, using values of $\sin 0°, 10°, 20°, 30°$ to determine this polynomial.

PROGRAMMING EXERCISES—INTERPOLATION

1. Modify the Lagrange flow chart and program to evaluate $P_n(x)$ for each of a set of values $x = X_1, X_2, X_3, \ldots, X_m$.

2. Modify the Lagrange flow chart and program to interpolate several sets of $y_k, \bar{y}_k, \bar{\bar{y}}_k, \ldots$, using the same set of x values.

3. Modify the Lagrange program to conserve storage by storing successive cumulative products $b_k(x)$ in single location rather than in the kth position of the array $B(K)$.

4. Use the Lagrange program to perform inverse interpolation. (Hint: Read x values into y array, and vice versa.)

5. Flow chart and program the Newton backward-form interpolation.

6. Modify the Newton forward-form (NFF) program to conserve storage by storing successive columns of difference table in a single one-dimensional array—like the Romberg program. (Hint: It is necessary to start at the end of each column and work back to the beginning.)

7. If a FORTRAN compiler that permits *zero* subscripts is available, flow chart and program (a) Newton forward-form interpolation, (b) Lagrange interpolation, and (c) iterated linear interpolation, from recursion formulas *before* index shift.

8. Modify the iterated-linear-interpolation program to compute the array by diagonals instead of by columns.

9. Modify the iterated-linear-interpolation program to conserve storage by storing successive columns in a single $1 - D$ array (like the Romberg program). It is possible to work from the beginning of the column forward, since the upper diagonal of array need not be preserved as in the modification of NFF in exercise 6.

10. Modify the iterated-linear-interpolation program of exercise 9 to test the successive elements in the upper diagonal for convergence.

7

Estimation of Parameters
by Least Squares

7.0 Introduction

One of the most extensively used techniques in numerical methods is the estimation of parameters by the principle of least squares. This technique is employed to derive information about the functional relation between x and y, assuming such a relation exists, from a set of data pairs (x_i, y_i) $(i = 0, n)$.

Data is usually obtained by observing physical phenomena and quantitatively measuring variables of interest. For example, an engineer analyzing an electrical circuit may record output voltages at discrete increments of time, obtaining a set of data pairs (x_i, y_i), with x denoting the time variable and y denoting the output-voltage variable. Data so obtained may contain a variety of errors. Some of these errors may be errors of observation (e.g., a meter reader misreading a meter), errors of measurement (e.g., error caused by resolution limits of measuring device), and errors of recording (e.g., truncation or roundoff of the number of significant digits recorded). Additional errors may be introduced in transmission (e.g., data may be transmitted from a remote collection site to a central computing site), or in conversion (e.g., converting from one system of units to another, or even in converting from decimal form to the binary form inside the computer).

The estimation of parameters by least squares causes a *smoothing* of a given set of data and eliminates, to some degree, errors in observation, measurement, recording, transmission, and conversion, as well as other types of random error which may have been introduced in the data.† This is one of the most important features of the principle of least squares, and one which distinguishes it from interpolation. (Recall that an interpolating polynomial exactly fits all of the data points used, so that any errors in the data will be retained by interpolation.)

Categories of Least-Squares Techniques

There are two distinct but related categories of techniques based on the principle of least squares: (1) the estimation of *linear* parameters by least squares; and (2) the estimation of *non-linear* parameters by least squares.

Category 1 is the well-known problem of "fitting" a function of the form

$$(7.0.1) \qquad F(x) = a_0 f_0(x) + a_1 f_1(x) + \cdots + a_m f_m(x)$$

to a set of data pairs (x_i, y_i) $(i = 0, n)$. In this problem, the *unknown parameters* $a_0, a_1, \ldots, a_m$ enter into the functional relation *linearly*. That is, the approximating function $F(x)$ is linear in the parameters $a_0, a_1, \ldots, a_m$.

We will consider two special cases of category 1 for approximating linear parameters.

(a) *Polynomial curve-smoothing by least squares for arbitrary x values.* In this case, the functions $f_k(x) = x^k$, so that function $F(x)$ becomes the mth- degree polynomial

$$(7.0.2) \qquad P_m(x) = a_0 + a_1 x + a_2 x^2 + \cdots + a_m x^m.$$

(b) *Orthogonal-polynomial curve-smoothing by least squares for evenly spaced x values.* In this case, $f_k(x) = O_{kn}(x)$, where $O_{kn}(x)$ is an orthogonal polynomial, so that function $F(x)$ becomes

$$(7.0.3) \qquad P_m(x) = a_0 O_{0n}(x) + a_1 O_{1n}(x) + \cdots + a_m O_{mn}(x).$$

Category 2 is the problem of "fitting" an *explicit function*, such as $F(x) = c_0 e^{c_1 x}$, to a set of data pairs (x_i, y_i) $(i = 0, n)$. In this problem, the unknown parameters c_k enter into the functional relation *non-linearly*. That is, $F(x)$ is non-linear in the c_k.

7.1 Estimation of Linear Parameters by Least Squares

The procedure for the estimation of linear parameters will first be formulated in the generalized notation of equation (7.0.1). Then we will consider the special cases of category 1(a) and category 1(b).

---

† Bias errors (i.e., a fixed offset Δy in each y_i) will *not* be eliminated by the least-squares technique.

Given a set of data pairs (x_i, y_i) $(i = 0, n)$, which can be interpreted as measured values of the coordinates of the points on the graph of $y = f(x)$, let us assume that the unknown function $f(x)$ can be approximated by a *linear combination* of *suitably chosen functions* $f_0(x), f_1(x), \ldots, f_m(x)$ of the form

(7.1.1) $F(x) = a_0 f_0(x) + a_1 f_1(x) + \cdots + a_m f_m(x)$

where the unknown coefficients $a_0, a_1, a_2, \ldots, a_m$ are independent parameters *to be determined*, and $m < n$.

The difference between the approximating function value $F(x_i)$ and the corresponding data value y_i is called a residual r_i and is defined by the relation

(7.1.2) $r_i = F(x_i) - y_i$ $(i = 0, n)$.

We have then a residual r_i for each data pair (x_i, y_i) $(i = 0, n)$.

The function $F(x)$ that *best* approximates the given set of data in a least-squares sense is that linear combination $a_0 f_0(x) + a_1 f_1(x) + \cdots + a_m f_m(x)$ of functions $f_k(x)$ that produces the minimum value of the sum Q of the squared residuals

(7.1.3) $$Q = \sum_i r_i^2 \equiv \sum_i [F(x_i) - y_i]^2.$$

If we write $F(x_i)$ in (7.1.3) in expanded form, we get

(7.1.4) $$Q = \sum_i [a_0 f_0(x_i) + a_1 f_1(x_i) + \cdots + a_m f_m(x_i) - y_i]^2.$$

It is apparent then that the sum Q is a "function" of the independent parameters $a_0, a_1, \ldots, a_m$. That is, if any parameter a_k is varied, a change in Q will result. We can therefore consider the parameters a_k as independent "variables" and write $Q(a_0, a_1, \ldots, a_m)$ to indicate that Q is a "function" of these "variables."

We are now in a position to mathematically formulate the procedure for minimizing the sum $Q(a_0, a_1, \ldots, a_m)$. The reader will recall from the calculus that a minimum or maximum of a function of several variables occurs at a point at which all partial derivatives of the function are simultaneously zero. In the case of the least-squares problem, a *minimum*† is obtained when the $m + 1$ partials of $Q(a_0, a_1, \ldots, a_m)$ with respect to parameters a_k $(k = 0, m)$, simultaneously vanish; i.e., when

(7.1.5) $$\frac{\partial Q}{\partial a_k} \equiv 2 \sum_i [F(x_i) - y_i] \frac{\partial F(x_i)}{\partial a_k} = 0 \qquad (k = 0, m).$$

† See for example Milne [18, pp. 253–255].

The constraints imposed by equations (7.1.5) form a system of $m + 1$ independent algebraic equations (called *normal equations*) which are linear in the $m + 1$ parameters a_k ($k = 0, m$). The solution $(a_0, a_1, \ldots, a_m)$ of this system of normal equations is that set of parameters a_k which produces the minimum sum of squared residuals.

The normal equations can be reduced to a form suitable for computation by the following steps: First, substitute the relation

$$\frac{\partial F(x_i)}{\partial a_k} = f_k(x_i)$$

into (7.1.5) and express $F(x_i)$ in expanded form, obtaining

(7.1.6) $\dfrac{\partial Q}{\partial a_k} = 2 \sum_i [a_0 f_0(x_i) + a_1 f_1(x_i) + \cdots + a_m f_m(x_i) - y_i] f_k(x_i) = 0$

for $k = 0, 1, 2, \ldots, m$. Next, divide each equation by the factor 2, and carry out the indicated multiplication of each term in square brackets by $f_k(x_i)$. The normal equations then appear in the form

(7.1.7) $\sum_i [a_0 f_k(x_i) f_0(x_i) + a_1 f_k(x_i) f_1(x_i) + \cdots + a_m f_k(x_i) f_m(x_i)]$

$$= \sum_i f_k(x_i) y_i \qquad (k = 0, m).$$

Finally, we can rewrite the normal equations in the form

(7.1.8) $a_0 \displaystyle\sum_i f_k(x_i) f_0(x_i) + a_1 \sum_i f_k(x_i) f_1(x_i) + \cdots + a_m \sum_i f_k(x_i) f_m(x_i)$

$$= \sum_i f_k(x_i) y_i \qquad (k = 0, m).$$

The $m + 1$ normal equations of (7.1.8), when evaluated for $k = 0$, $k = 1, k = 2, \ldots, k = m$, can be written as a single matrix equation of the form

$$\begin{bmatrix} \sum [f_0(x_i)]^2 & \sum f_0(x_i)f_1(x_i) & \cdots & \sum f_0(x_i)f_m(x_i) \\ \sum f_1(x_i)f_0(x_i) & \sum [f_1(x_i)]^2 & \cdots & \sum f_1(x_i)f_m(x_i) \\ \cdot \quad \cdot \quad \cdot & \cdot \quad \cdot \quad \cdot & \cdots & \cdot \quad \cdot \quad \cdot \\ \sum f_m(x_i)f_0(x_i) & \sum f_m(x_i)f_1(x_i) & \cdots & \sum [f_m(x_i)]^2 \end{bmatrix} \begin{bmatrix} a_0 \\ a_1 \\ \vdots \\ a_m \end{bmatrix} = \begin{bmatrix} \sum f_0(x_i)y_i \\ \sum f_1(x_i)y_i \\ \vdots \\ \sum f_m(x_i)y_i \end{bmatrix}$$

(7.1.9)

where all summations are over i ($i = 0, n$).

The solution $(a_0, a_1, \ldots, a_m)$ of the matrix form (7.1.9) of the normal equations is the set of parameters a_k ($k = 0, m$) that minimize the sum Q of the squared residuals. The solution can be easily computed by inverting the matrix of coefficients in equation (7.1.9), and multiplying the right-hand column matrix by this inverse matrix.

Now that we have a formulation of the method of obtaining the linear parameters by the principle of least squares, let us consider two special cases of this problem which are commonly encountered in practice, i.e., the problem of curve-smoothing by least-squares using polynomials, and the problem of curve-smoothing by least-squares using orthogonal polynomials.

Curve-Smoothing by Polynomial Least Squares

Let us now consider the special case of the least-squares estimation of linear parameters in which the functions

$$f_k(x) = x^k \qquad (k = 0, m)$$

so that formula (7.1.1) becomes an mth-degree polynomial, $m < n$, denoted $P_m(x)$, of the form

$$P_m(x) = a_0 x^0 + a_1 x^1 + a_2 x^2 + \cdots + a_m x^m.$$

That is, we will approximate function $y = f(x)$ by an mth-degree polynomial $P_m(x)$ over the range of the data pairs (x_i, y_i) $(i = 0, n)$. The parameters $a_0, a_1, \ldots, a_m$ are then determined such that

$$Q = \sum_i r_i{}^2 \equiv \sum_i [P_m(x_i) - y_i]^2$$

is a minimum. That is, we will "fit" an mth-degree polynomial curve to the data in a least-squares sense, as defined earlier. This special case of the estimation of linear parameters is commonly referred to as *polynomial curve-smoothing by least squares*.

The normal equations that determine $a_0, a_1, \ldots, a_m$ for this special case can be obtained directly by substituting $x_i{}^k$ (i.e., x_i to kth power) for $f_k(x_i)$ in equation (7.1.9). This substitution gives us

$$\begin{bmatrix} \sum (x_i{}^0)^2 & \sum x_i{}^0 x_i{}^1 & \ldots & \sum x_i{}^0 x_i{}^m \\ \sum x_i{}^1 x_i{}^0 & \sum (x_i{}^1)^2 & \ldots & \sum x_i{}^1 x_i{}^m \\ \cdot \cdot \cdot \cdot \cdot \cdot \cdot \cdot \cdot \cdot \cdot \cdot \\ \sum x_i{}^m x_i{}^0 & \sum x_i{}^m x_i{}^1 & \ldots & \sum (x_i{}^m)^2 \end{bmatrix} \begin{bmatrix} a_0 \\ a_1 \\ \vdots \\ a_m \end{bmatrix} = \begin{bmatrix} \sum x_i{}^0 y_i \\ \sum x_i{}^1 y_i \\ \vdots \\ \sum x_i{}^m y_i \end{bmatrix}.$$

These normal equations for the least-squares polynomial can then be written as

$$(7.1.10) \quad \begin{bmatrix} n+1 & \sum x_i & \sum x_i{}^2 & \ldots & \sum x_i{}^m \\ \sum x_i & \sum x_i{}^2 & \sum x_i{}^3 & \ldots & \sum x_i{}^{m+1} \\ \cdot & \cdot & \cdot & \cdot & \cdot \\ \sum x_i{}^m & \sum x_i{}^{m+1} & \sum x_i{}^{m+2} & \ldots & \sum x_i{}^{2m} \end{bmatrix} \begin{bmatrix} a_0 \\ a_1 \\ \vdots \\ a_m \end{bmatrix} = \begin{bmatrix} \sum y_i \\ \sum x_i y_i \\ \vdots \\ \sum x_i{}^m y_i \end{bmatrix}.$$

The zero subscripts must be eliminated before the FORTRAN program for the least-squares method can be written. This is accomplished by shifting index i so that the data pairs are (x_i, y_i) $(i = 1, n + 1)$, and shifting the index k so that the mth-degree least-squares polynomial becomes

$$P_m(x) = a_1 x^0 + a_2 x^1 + a_3 x^2 + \cdots + a_m x^{m-1} + a_{m+1} x^m.$$

The normal equations then become

$$
\begin{bmatrix}
n+1 & \sum x_i & \sum x_i^2 & \cdots & \sum x_i^m \\
\sum x_i & \sum x_i^2 & \sum x_i^3 & \cdots & \sum x_i^{m+1} \\
\cdot & \cdot & \cdot & \cdots & \cdot \\
\sum x_i^m & \sum x_i^{m+1} & \sum x_i^{m+2} & \cdots & \sum x_i^{2m}
\end{bmatrix}
\begin{bmatrix}
a_1 \\
a_2 \\
\vdots \\
a_{m+1}
\end{bmatrix}
=
\begin{bmatrix}
\sum y_i \\
\sum y_i x_i \\
\vdots \\
\sum y_i x_i^m
\end{bmatrix}
$$

where all summations are over i from 1 to $n + 1$.

If the matrix of coefficients, the matrix of unknowns, and the matrix on the right are denoted by C, A, and B, respectively, then the normal equations can be written symbolically by the matrix equation

$$CA = B.$$

It should be noted that the determination of the least-squares polynomial of degree m for the set of data pairs (x_i, y_i) $(i = 1, n + 1)$ can be logically subdivided into two distinct problems, as follows: (a) generation of the elements c_{ij} of the matrix of coefficients C, and the elements b_i of the right side matrix B; (b) solution of the matrix equation $CA = B$.

The second of these two problems, i.e., the solution of a system of $m + 1$ linear algebraic equations in $m + 1$ unknowns, has been solved previously, in Chapter 5. We can therefore use the method of inversion by elimination of matrix C followed by matrix multiplication to obtain $A = C^{-1}B$; or, equivalently, we could solve the system of equations by the Gauss-Jordan method, as described in Sec. 5.7.

Computational Summary for Curve-Smoothing by Least Squares Polynomial

Step 1. Input. Read

$$n + 1 = \text{number of data pairs}$$
$$m = \text{degree of least-squares polynomial}$$
$$(x_i, y_i) = \text{data pairs } (i = 1, n + 1).$$

Step 2. Generate terms for constructing normal equations. Calculate

$$X_k = \sum_{i=1}^{n+1} x_i^k \qquad (k = 1, 2m)$$

$$Y = \sum_{i=1}^{n+1} y_i$$

$$YX_k = \sum_{i=1}^{n+1} y_i x_i^k \qquad (k = 1, m)$$

Step 3. Form normal matrix C and column matrix B. Calculate

$$c_{11} = n + 1$$
$$c_{ij} = X_{i+j-2} \qquad (i, j = 1, m + 1), \text{except } c_{11}$$
$$b_1 = Y$$
$$b_i = YX_{i-1} \qquad (i = 2, m + 1)$$

Step 4. Solve normal equations $CA = B$. (a) Invert normal matrix C; (b) form matrix product $A = C^{-1}B$.

Step 5. Output coefficients a_k $(k = 1, m + 1)$ of least-squares polynomial.

A representative flow chart for curve-smoothing by least-squares polynomials is given in Figure 7-1.

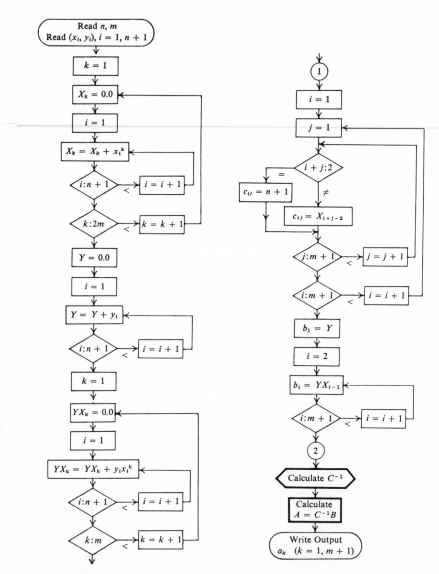

Figure 7-1. Flow chart for curve-smoothing by least squares.

246

```
      PROGRAM LSTSQ
C     MAXIMUM NUMBER OF DATA POINTS = 200
C     DEGREE OF HIGHEST POLYNOMIAL = 20
      DIMENSION X(200),Y(200),XC(40),YX(20),A(20),B(20),C(20,20),
     X D(20,20)
      READ (5,100) N,M
      NP1 = N + 1
      MP1 = M + 1
      M2 = M * 2
      READ (5,101) (X(I), Y(I), I=1,NP1)
C     FORM SUMS OF POWERS OF X(I)
      DO 20 K = 1, M2
      XC(K) = 0.0
      DO 15 I = 1, NP1
   15 XC(K) = XC(K) + X(I)**K
   20 CONTINUE
C     FORM SUM OF Y (I)
      YC = 0.0
      DO 22 I = 1, NP1
   22 YC = YC + Y(I)
C     FORM SUMS OF PRODUCTS  Y(I) * X(I)**K
      DO 30 K = 1, M
      YX(K) = 0.0
      DO 25 I = 1, NP1
   25 YX(K) = YX(K) + Y(I) * X(I)**K
   30 CONTINUE
C     GENERATE NORMAL MATRIX C USING SUMS OF POWERS OF X (I)
      DO 40 I = 1, MP1
      DO 35 J = 1, MP1
      IPJM2 = I + J - 2
      IF (IPJM2) 33, 31, 33
   31 C(1,1) = FLOAT(NP1)
      GO TO 35
   33 C(I,J) = XC(IPJM2)
   35 CONTINUE
   40 CONTINUE
C     GENERATE RIGHT-SIDE MATRIX B
      B(1) = YC
      DO 45 I = 2, MP1
   45 B(I) = YX(I-1)
C     INVERT NORMAL MATRIX C
      CALL INVERSE (C, D, MP1)
C     MATRIX D IS INVERSE OF NORMAL MATRIX
C     FORM MATRIX PRODUCT OF INVERSE AND RIGHT-SIDE MATRIX
      DO 55 I = 1,MP1
      A(I) = 0.0
      DO 54 J = 1,MP1
   54 A(I) = A(I) + D(I,J) * B(J)
   55 CONTINUE
      WRITE (6,111)
      WRITE (6,110) (A(K), K = 1, MP1)
      STOP
  100 FORMAT (I3, I2)
  101 FORMAT (2F14.6)
  110 FORMAT (1X, E14.6)
  111 FORMAT (18H COEFFICIENTS A(K))
```

```
      SUBROUTINE INVERSE (A, B, N)
C     MATRIX INVERSION BY ELIMINATION WITH PARTIAL PIVOTING
C     ORIGINAL MATRIX = A INVERSE MATRIX = B
      DIMENSION A(20,20), B(20,20)
      EPS = 0.0000001
C     CONSTRUCT IDENTITY MATRIX B(I,J) = I
      DO 6 I =1, N
      DO 5 J=1, N
      IF (I-J) 4,3,4
    3 B(I,J) = 1.0
      GO TO 5
    4 B(I, J) = 0.0
    5 CONTINUE
    6 CONTINUE
C     LOCATE MAXIMUM MAGNITUDE A(I, K) ON OR BELOW MAIN DIAGONAL
      DEL = 1.0
      DO 45 K = 1, N
      IF (K-N) 12, 30, 30
   12 IMAX = K
      AMAX = ABS (A(K, K))
      KP1 = K + 1
      DO 20 I = KP1, N
      IF (AMAX - ABS (A(I, K))) 15, 20, 20
   15 IMAX = I
      AMAX = ABS (A(I,K))
   20 CONTINUE
C     INTERCHANGE ROWS IMAX AND K IF IMAX NOT EQUAL TO K
      IF (IMAX - K) 25, 30, 25
   25 DO 29 J = 1, N
      ATMP = A (IMAX, J)
      A (IMAX, J) = A(K,J)
      A(K,J) = ATMP
      BTMP = B(IMAX, J)
      B(IMAX,J) = B(K,J)
   29 B(K,J)= BTMP
      DEL = -DEL
   30 CONTINUE
C     TEST FOR SINGULAR MATRIX
      IF (ABS(A(K,K)) - EPS) 93, 93, 35
   35 DEL = A(K,K) * DEL
C     DIVIDE PIVOT ROW BY ITS MAIN DIAGONAL ELEMENT
      DIV = A(K,K)
      DO 38 J = 1, N
      A(K,J) = A(K,J)/DIV
   38 B(K,J)= B(K,J)/DIV
C     REPLACE EACH ROW BY LINEAR COMBINATION WITH PIVOT ROW
      DO 43 I = 1, N
      AMULT = A(I,K)
      IF (I-K) 39, 43, 39
   39 DO 42 J = 1, N
      A(I,J) = A(I,J) - AMULT * A(K,J)
   42 B(I,J) = B(I,J) - AMULT * B(K,J)
   43 CONTINUE
   45 CONTINUE
      WRITE (6,120)
      WRITE (6,110) ((B(I,J), I=1,N), J=1,N)
      WRITE (6,121)
      WRITE (6,110) DEL
   99 RETURN
   93 WRITE (6,113) K
      GO TO 99
  110 FORMAT (2X, E15,8)
  113 FORMAT (25H SINGULAR MATRIX FOR K =, I2)
  120 FORMAT (20H ELEMENTS OF INVERSE)
  121 FORMAT (21H VALUE OF DETERMINANT)
      END
```

NUMERICAL EXAMPLE. Calculate the third-degree least-sq nomial

$$P_3(x) = a_1 + a_2x + a_3x^2 + a_4x^3$$

for the following set of data pairs:

x	y
0.0	0.0000
0.1	0.1002
0.2	0.2013
0.3	0.3045
0.4	0.4108
0.5	0.5211
0.6	0.6367
0.7	0.7586
0.8	0.8881
0.9	1.0265
1.0	1.1752

The quantities required for constructing the normal equations are:

$$n + 1 = 11 \qquad \qquad \sum y_i = 6.023$$
$$\sum x_i = 5.5 \qquad \qquad \sum y_i x_i = 4.28907$$
$$\sum x_i^2 = 3.85 \qquad \qquad \sum y_i x_i^2 = 3.408437$$
$$\sum x_i^3 = 3.025 \qquad \qquad \sum y_i x_i^3 = 2.8773135$$
$$\sum x_i^4 = 2.5333$$
$$\sum x_i^5 = 2.20825$$
$$\sum x_i^6 = 1.987405$$

The normal equations are then

$$\begin{bmatrix} 11.0 & 5.5 & 3.85 & 3.025 \\ 5.5 & 3.85 & 3.025 & 2.5333 \\ 3.85 & 3.025 & 2.5333 & 2.20825 \\ 3.025 & 2.5333 & 2.20825 & 1.987405 \end{bmatrix} \begin{bmatrix} a_1 \\ a_2 \\ a_3 \\ a_4 \end{bmatrix} = \begin{bmatrix} 6.023 \\ 4.28907 \\ 3.408437 \\ 2.8773135 \end{bmatrix}$$

and the solution of the normal equations is

$$a_1 = -0.000129, \quad a_2 = 1.004383, \quad a_3 = -0.019651, \quad a_4 = 0.190405.$$

The third-degree least-squares polynomial is then

$$P_3(x) = -0.000129 + 1.004383x - 0.019651x^2 + 0.190405x^3.$$

Note. The data for this problem were taken from a table of values for sinh x that has the Taylor-series expansion

$$\sinh x = x + x^3/3! + \cdots.$$

It should be noted that the normal matrix often has a near-zero determinant, and, in general, the higher the order of the least-squares polynomial, the closer the determinant approaches zero. For example, if the x_i are evenly spaced on the interval $(0, 1)$, a segment of the Hilbert matrix appears. The determinant of the Hilbert matrix of order n has the value†

$$H_n = \frac{[1!2!3!\ldots(n-1)!]^3}{n!(n+1)!\ldots(2n-1)!}$$

which approaches zero rapidly as the order n increases.

This difficulty can be avoided for evenly spaced x_i by the use of orthogonal polynomials. One such set of orthogonal polynomials is discussed in the next half of this section.

For arbitrarily spaced x_i, the least-squares polynomials (in powers of x) of high order should be used with caution.

Curve-Smoothing by Use of Orthogonal Polynomials

Another important special case of the estimation of linear parameters uses a linear combination of orthogonal polynomials to fit a smooth curve to a set of points with *evenly spaced abscissas*. A set of orthogonal polynomials commonly used for this purpose are the discrete Legendre polynomials $O_{kn}(x)$, which satisfy the orthogonality relation

$$\sum_{x=0}^{n} O_{jn}(x)O_{kn}(x) = 0 \qquad (j \neq k).$$

One form‡ of the discrete Legendre polynomials for the evenly spaced abscissas $x = 0, 1, 2, \ldots, n$ is a set of $O_{kn}(x)$ as enumerated below.

$$O_{0n}(x) = 1$$

$$O_{1n}(x) = 1 - 2\left(\frac{x}{n}\right)$$

$$O_{2n}(x) = 1 - 6\left(\frac{x}{n}\right) + 6\left(\frac{x}{n}\right)\left(\frac{x-1}{n-1}\right)$$

$$O_{3n}(x) = 1 - 12\left(\frac{x}{n}\right) + 30\left(\frac{x}{n}\right)\left(\frac{x-1}{n-1}\right) - 20\left(\frac{x}{n}\right)\left(\frac{x-1}{n-1}\right)\left(\frac{x-2}{n-2}\right)$$

It should be noted that the *discrete* Legendre polynomials are defined only at the discrete values $x = 0, 1, 2, \ldots, n$.

† See Hamming [5, p. 230] and Arden [1, p.246].

‡ Legendre polynomials of this form are derived in Milne [18, pp. 265–268]. Another form of these Legendre polynomials is given in Nielsen [18, p. 283].

Example. Values of the first four Legendre polynomials $O_{k6}(x)$ are given in Table 7-1. Note that $\sum_{x=0}^{6} O_{j6}(x)O_{k6}(x) = 0$ for $j \neq k$.

Table 7-1

x	0	1	2	3	4	5	6
O_{06}	1	1	1	1	1	1	1
O_{16}	1	4/6	2/6	0	$-2/6$	$-4/6$	-1
O_{26}	1	0	$-3/5$	$-4/5$	$-3/5$	0	1
O_{36}	1	-1	-1	0	1	1	-1

The general formula for the kth-degree ($k \leq n$) discrete Legendre polynomial is

$$O_{kn}(x) = \sum_{j=0}^{k} (-1)^j \binom{k}{j}\binom{k+j}{j} \frac{x^{(j)}}{n^{(j)}}$$

where

$$\binom{k}{j} = \frac{k!}{j!(k-j)!} = \text{the binomial coefficient}$$
$$x^{(j)} = x(x-1)\ldots(x-j+1), \qquad x^{(0)} = 1$$
$$n^{(j)} = n(n-1)\ldots(n-j+1), \qquad n^{(0)} = 1.$$

It is important to note that x has only integral values $0, 1, 2, \ldots, n$.

The coefficients $b_j = (-1)^j \binom{k}{j}\binom{k+j}{j}$ for the Legendre polynomials $O_{kn}(x)$ are given in Table 7-2 for $k = 0, 8$.

Another useful property of the discrete Legendre polynomials is the following:

$$\sum_{x=0}^{n} O_{kn}^2(x) = \frac{(n+k+1)(n+k)^{(k)}}{(2k+1)(n)^{(k)}}$$

where

$$(n+k)^{(k)} = (n+k)(n+k-1)\ldots(n+1)$$
$$(n)^{(k)} = n(n-1)\ldots(n-k+1)$$

EXAMPLE

$$\sum_{x=0}^{6} O_{26}^2(x) = \frac{(6+2+1)(6+2)^{(2)}}{(2\cdot2+1)(6)^{(2)}} = \frac{9[8(7)]}{5[6(5)]}.$$

Applying the Legendre Polynomials to Curve-Smoothing

Now that the discrete Legendre polynomials $O_{kn}(x_i)$ have been defined, we can determine a linear combination of these polynomials that best fits a set of data in the least-squares sense. First, however, it is necessary to transform the data so that the evenly spaced abscissas x_i' take on the values $0, 1, 2, \ldots, n$.

Table 7-2   Coefficients of Discrete Legendre Polynomials

O_{0n}	1								
O_{1n}	1	-2							
O_{2n}	1	-6	6						
O_{3n}	1	-12	30	-20					
O_{4n}	1	-20	90	-140	70				
O_{5n}	1	-30	210	-560	630	-252			
O_{6n}	1	-42	420	-1680	3150	-2772	924		
O_{7n}	1	-56	756	-4200	11550	-16632	12012	-3432	
O_{8n}	1	-72	1260	-9240	34650	-72072	84084	-51480	12870

Let $(x_0', y_0), (x_1', y_1), \ldots, (x_n', y_n)$ be a given set of data where the x_i' are evenly spaced; i.e., such that $x_{i+1}' - x_i' = h$ $(i = 0, n - 1)$. By a substitution of the form

$$x = \frac{x' - x_0'}{h}$$

the abscissas are translated and scaled so that the new abscissas x are the integers $0, 1, 2, \ldots, n$. With this simple transformation the original set of data becomes the set of† (x, y_x) $(x = 0, n)$.

By setting $f_k(x) = O_{kn}(x)$ in expression (7.1.1), we obtain a linear combination of the Legendre polynomials of the form

$$F(x) = a_0 O_{0n}(x) + a_1 O_{1n}(x) + a_2 O_{2n}(x) + \cdots + a_m O_{mn}(x).$$

Again, residuals are defined by the relation

$$r_x = F(x) - y_x \qquad (x = 0, n)$$

and the coefficients $a_0, a_1, \ldots, a_m$ are determined such that the sum of squared residuals $Q = \sum_{x=0}^{n} r_x^2 \equiv \sum_{x=0}^{n} [F(x) - y_x]^2$ is minimized.

The normal equations obtained by setting the partial derivatives $\partial Q/\partial a_0 = \partial Q/\partial a_1 = \cdots = \partial Q/\partial a_m = 0$, reduce to the form

$$
\begin{bmatrix}
\sum_{x=0}^{n} O_{0n}^2(x) & 0 & \cdots & 0 \\
0 & \sum_{x=0}^{n} O_{1n}^2(x) & \cdots & 0 \\
\vdots & \vdots & & \vdots \\
0 & 0 & \cdots & \sum_{x=0}^{n} O_{mn}^2(x)
\end{bmatrix}
\begin{bmatrix}
a_0 \\
a_1 \\
\vdots \\
a_m
\end{bmatrix}
=
\begin{bmatrix}
\sum_{x=0}^{n} O_{0n}(x) \; y_x \\
\sum_{x=0}^{n} O_{1n}(x) \; y_x \\
\vdots \\
\sum_{x=0}^{n} O_{mn}(x) \; y_x
\end{bmatrix}
$$

because of the orthogonality property $\sum_{x=0}^{n} O_{jn}(x) O_{kn}(x) = 0, j \neq k$.

† In the term y_x, x is a subscript (integer) on y.

The coefficients a_k ($k = 0, m$) which produce the minimum Q are the solutions of the normal equations above; these solutions are simply

$$a_k = \frac{\sum\limits_{x=0}^{n} O_{kn}(x)y_x}{\sum\limits_{x=0}^{n} O_{kn}^2(x)} \qquad (k = 0, m).$$

To evaluate the unknown function at points other than the mesh points, each orthogonal polynomial of the linear combination

$$F(x) = a_0 O_{0n}(x) + a_1 O_{1n}(x) + a_2 O_{2n}(x) + \cdots + a_m O_{mn}(x)$$

is replaced by its powers of x representation, giving

$$F(x) = a_0[1] + a_1\left[1 - 2\frac{x}{n}\right] + a_2\left[1 - 6\frac{x}{n} + 6\frac{x(x-1)}{n(n-1)}\right]$$

$$+ \cdots + a_m\left[1 - \binom{m}{1}\binom{m+1}{1}\frac{x}{n} + \binom{m}{2}\binom{m+2}{2}\frac{x(x-1)}{n(n-1)} + \cdots\right].$$

Carrying out the indicated multiplications and collecting terms, we obtain a polynomial in x. This polynomial can then be evaluated for any value of x, whether that value is a mesh point or not, within the interval $(0, n)$.

Numerical Example of Orthogonal-Polynomial Least Squares

Given the set of data (from table of values of $\sinh x$), determine the

x	x_i'	y_x
0	0.0	0.0000
1	0.2	0.2013
2	0.4	0.4108
3	0.6	0.6367
4	0.8	0.8881
5	1.0	1.1752
6	1.2	1.5095

coefficients a_0, a_1, a_2, a_3 such that the linear combination

$$F(x) = a_0 O_{06}(x) + a_1 O_{16}(x) + a_2 O_{26}(x) + a_3 O_{36}(x)$$

of orthogonal polynomials produces the smallest sum of squared residuals.
Solution. Scale the x_i' to obtain $x = 0, 1, 2, 3, 4, 5, 6$.

The values of the orthogonal polynomials and data values, together with the products of orthogonal polynomials and data values, are tabulated in the Table 7-3.

Table 7-3

x	O_{06}	O_{16}	O_{26}	O_{36}	y	$O_{06}y$	$O_{16}y$	$O_{26}y$	$O_{36}y$
0	1.0	1.0	1.0	1.0	0.0000	0.0000	0.0000	0.0000	0.0000
1	1.0	0.666	0.0	-1.0	0.2013	0.2013	0.1342	0.0000	-0.2013
2	1.0	0.333	-0.6	-1.0	0.4108	0.4108	0.1369	-0.2465	-0.4108
3	1.0	0.0	-0.8	0.0	0.6367	0.6367	0.0	-0.5094	0.0000
4	1.0	-0.333	-0.6	1.0	0.8881	0.8881	-0.2960	-0.5329	0.8881
5	1.0	-0.666	0.0	1.0	1.1752	1.1752	-0.7835	0.0000	1.1752
6	1.0	-1.0	1.0	-1.0	1.5095	1.5095	-1.5095	1.5095	-1.5095
						4.8216	-2.3179	0.2207	-0.0583

Then the $\sum_{x=0}^{6} O_{k6}^2$ and $\sum_{x=0}^{6} O_{k6}y_x$ are tabulated:

$$\sum O_{06}^2 = 7.0 \qquad \sum O_{06}y = \quad 4.8216$$
$$\sum O_{16}^2 = 3.1111 \qquad \sum O_{16}y = -2.3179$$
$$\sum O_{26}^2 = 3.36 \qquad \sum O_{26}y = \quad 0.2207$$
$$\sum O_{36}^2 = 6.0 \qquad \sum O_{36}y = -0.0583.$$

Finally, the coefficients a_k ($k = 0, 1, 2, 3$) are calculated:

$$a_0 = 0.6888, \quad a_1 = -0.7450, \quad a_2 = 0.06568, \quad a_3 = -0.009717.$$

The required linear combination is

$$F(x) = 0.6888 O_{06}(x) - 0.7450 O_{16}(x) + 0.06568 O_{26}(x) - 0.009717 O_{36}(x).$$

Substituting the powers of x representations of the orthogonal polynomials into the foregoing expression, we obtain

$$F(x) = 0.6888[1] - 0.7450\left[1 - 2\frac{x}{6}\right] + 0.06568\left[1 - 6\frac{x}{6} + 6\frac{x(x-1)}{6(5)}\right]$$

$$- 0.009717\left[1 - 12\frac{x}{6} + 30\frac{x(x-1)}{6(5)} - 20\frac{x(x-1)(x-2)}{6(5)(4)}\right].$$

Carrying out the indicated multiplications and collecting terms, we get

$$F(x) = 0.6888 - 0.7450 + 0.06568 - 0.009717$$

$$+ x\left[0.7450\left(\frac{2}{6}\right) + 0.06568\left(-1 - \frac{1}{5}\right) - 0.009717\left(2 - 1 - \frac{2}{6}\right)\right]$$

$$+ x^2\left[0.06568\left(\frac{1}{5}\right) - 0.009717\left(1 + \frac{3}{6}\right)\right]$$

$$+ x^3\left[0.009717\left(\frac{1}{6}\right)\right]$$

$$F(x) = -0.000237 + 0.201907x - 0.00144x^2 + 0.0016195x^3.$$

Note that this is a polynomial in the scaled variable x, and not in the original variable x'.

7.2 Estimation of Non-linear Parameters by Least Squares

In many applied problems the form of a function $f(x, c_0, c_1, \ldots, c_m)$ representing a phenomenon is known, whereas the values of the parameters $c_0, c_1, \ldots, c_m$ are to be determined. For example, according to theoretical considerations, a particular phenomenon might be represented by an exponential, logarithmic, trigonometric, or other function. If a set of observations provides empirical values y_i of the function

$$f(x_i, c_0, c_1, \ldots, c_m) \qquad (i = 0, n), n > m,$$

then the parameters c_k $(k = 0, m)$ can often be determined from a set of estimates $c_k{}^0$ by differential-correction techniques based on least squares. Such techniques are used extensively in applied mathematics.

The use of such a technique can be illustrated quite nicely by the following problem:

EXAMPLE. If an electrically charged capacitor is placed in a closed series circuit with a resistor, the capacitor will discharge through the resistor, setting up a transient current in the circuit. This current i can be represented as a function of time t and the parameters q_0 (initial charge), R (resistance), and C (capacitance), and has the form

$$i(t, q_0, R, C) = \frac{-q_0}{RC} e^{-t/RC}.$$

By the substitutions $x = t$, $c_0 = -q_0/RC$, $c_1 = -1/RC$, $i = f$, this equation can be written in the form

$$f(x, c_0, c_1) = c_0 e^{c_1 x}.$$

If values $y_0, y_1, \ldots, y_n$ of the function (current) are known for time values $x_0, x_1, \ldots, x_n$, and if initial estimates $c_0{}^0, c_1{}^0$ of the parameters c_0, c_1 are known, then improved values of c_0, c_1 can be determined by a differential-correction technique based on least squares.

In the following subsection we will develop the formulas for such a differential-correction technique. After this is done, the method will be illustrated by solving the foregoing problem. It should be noted that problems of considerable complexity can be solved by such methods, and that we have selected this simple problem so that the reader can follow an illustration of the method without grinding through the intricacies of a more complex problem.

A Least-Squares Linear-Taylor Differential-Correction Technique

Assume that for a particular phenomenon there exists an explicit function

(7.2.1) $y = f(x, c_0, c_1, \ldots, c_m)$

relating variables x and y, where $c_0, c_1, \ldots, c_m$ are unknown independent parameters that enter non-linearly into the functional expression. Observation of this phenomenon has provided a set of data pairs (x_i, y_i), $i = 0, n$, at distinct, discrete values x_i of the independent variable x. The recorded values y_i may contain errors of observation, measurement, recording, transmission, conversion, and so forth.

If the values of the parameters c_k were known, it would be possible to evaluate $f(x, c_0, c_1, \ldots, c_m)$ for each x_i to obtain a set of "true" residuals

(7.2.2) $r_i = f(x_i, c_0, c_1, \ldots, c_m) - y_i$ $(i = 0, n)$.

A "true" residual then would represent the difference between the actual function value at x_i and the empirical (recorded) value y_i. These "true" residuals cannot be calculated because the actual values of the parameter c_k are unknown.

However, if estimates $c_k{}^0$ of the parameters can be obtained by some means,† then "computed" residuals

(7.2.3) $R_i = f(x_i, c_0{}^0, c_1{}^0, \ldots, c_m{}^0) - y_i$ $(i = 0, n)$

can be calculated. The problem, then, is to obtain improved values of the parameters c_k using the data pairs (x_i, y_i), the estimates $c_k{}^0$, and the "computed" residuals R_i. This can be accomplished by a differential-correction technique based on least squares, provided that the estimates $c_k{}^0$ are sufficiently close to the actual values of the parameters c_k to lead to convergence of the method.

This differential-correction technique can be derived by first expanding the function about $(c_0{}^0, c_1{}^0, \ldots, c_m{}^0)$ using a linear Taylor-series expansion of the form

(7.2.4) $f(x, c_0, c_1, \ldots, c_m) \doteq f(x, c_0{}^0, c_1{}^0, \ldots, c_m{}^0) + \dfrac{\partial f}{\partial c_0}(c_0 - c_0{}^0)$

$$+ \frac{\partial f}{\partial c_1}(c_1 - c_1{}^0) + \cdots + \frac{\partial f}{\partial c_m}(c_m - c_m{}^0)$$

so that a relation between the r_i and R_i can be obtained. This relation can be found by evaluating (7.2.4) at each value of x_i and subtracting y_i from both sides of the equation. Using the definitions

$$\delta c_k = c_k - c_k{}^0$$

† Initial estimates $c_k{}^0$ might be obtained by graphical methods, theoretical considerations, or experience with related problems.

and

$$\frac{\partial f_i}{\partial c_k} = \frac{\partial f}{\partial c_k}\bigg|_{x = x_i, c_k = c_k{}^0}$$

we can write the result in the form

(7.2.5) $f(x_i, c_0, c_1, \ldots, c_m) - y_i \doteq f(x_i, c_0{}^0, c_1{}^0, \ldots, c_m{}^0) - y_i$

$$+ \frac{\partial f_i}{\partial c_0} \delta c_0 + \cdots + \frac{\partial f_i}{\partial c_m} \delta c_m \qquad (i = 0, n).$$

The desired relation between the r_i and R_i can then be found by substituting (7.2.2) and (7.2.3) into (7.2.5); the result is the relation

(7.2.6) $r_i \doteq R_i + \dfrac{\partial f_i}{\partial c_0} \delta c_0 + \cdots + \dfrac{\partial f_i}{\partial c_m} \delta c_m \qquad (i = 0, n).$

Now, let us see how this relation can be used to compute, from $c_k{}^0$, a set of parameters c_k that minimizes the sum of the squares of the "true" residuals r_i; that is, such that the quantity Q, defined by the relation

(7.2.7) $Q \equiv \displaystyle\sum_{i=0}^{n} r_i{}^2 = \sum_{i=0}^{n} \left[R_i + \frac{\partial f_i}{\partial c_0} \delta c_0 + \cdots + \frac{\partial f_i}{\partial c_m} \delta c_m \right]^2$

is minimized.

Equation (7.2.7) indicates that Q is a "function" of the terms δc_k ($k = 0, m$). That is, if any δc_k is varied, then a change in Q will result. We therefore consider Q as a function of the δc_k and write $Q(\delta c_0, \delta c_1, \ldots, \delta c_m)$. This function Q has a minimum value when all of its partials with respect to the δc_k are simultaneously zero; i.e., when

(7.2.8) $\dfrac{\partial Q}{\partial (\delta c_k)} = 2 \displaystyle\sum_{i=0}^{n} \left[R_i + \frac{\partial f_i}{\partial c_0} \delta c_0 + \cdots + \frac{\partial f_i}{\partial c_m} \delta c_m \right] \frac{\partial \Delta}{\partial (\delta c_k)} = 0$

$$(k = 0, m)$$

where

$$\Delta \equiv \frac{\partial f_i}{\partial c_0} \delta c_0 + \cdots + \frac{\partial f_i}{\partial c_m} \delta c_m$$

denotes the total differential of the function f.
Substituting the relation

$$\frac{\partial \Delta}{\partial (\delta c_k)} = \frac{\partial f_i}{\partial c_k}$$

into (7.2.8), and multiplying each term in the brackets by this quantity, we can rewrite (7.2.8) in the form

(7.2.9) $\dfrac{\partial Q}{\partial (\delta c_k)} = 2 \displaystyle\sum_{i=0}^{n} \left[\frac{\partial f_i}{\partial c_k} R_i + \frac{\partial f_i}{\partial c_k} \frac{\partial f_i}{\partial c_0} \delta c_0 + \cdots + \frac{\partial f_i}{\partial c_k} \frac{\partial f_i}{\partial c_m} \delta c_m \right] = 0$

for $k = 0, m$.

Dividing through by 2, and transposing the term containing R_i, we obtain

$$(7.2.10) \quad \delta c_0 \sum_i \frac{\partial f_i}{\partial c_k} \frac{\partial f_i}{\partial c_0} + \cdots + \delta c_m \sum_i \frac{\partial f_i}{\partial c_k} \frac{\partial f_i}{\partial c_m} = -\sum_i \frac{\partial f_i}{\partial c_k} R_i \quad (k = 0, m).$$

Evaluating this equation for each value of k, and writing the result in matrix form, we obtain the normal equations

$$(7.2.11) \quad \begin{bmatrix} \sum_i \left(\frac{\partial f_i}{\partial c_0}\right)^2 & \sum_i \frac{\partial f_i}{\partial c_0} \frac{\partial f_i}{\partial c_1} & \cdots & \sum_i \frac{\partial f_i}{\partial c_0} \frac{\partial f_i}{\partial c_m} \\ \sum_i \frac{\partial f_i}{\partial c_1} \frac{\partial f_i}{\partial c_0} & \sum_i \left(\frac{\partial f_i}{\partial c_1}\right)^2 & \cdots & \sum_i \frac{\partial f_i}{\partial c_1} \frac{\partial f_i}{\partial c_m} \\ \cdot & \cdot & \cdots & \cdot \\ \sum_i \frac{\partial f_i}{\partial c_m} \frac{\partial f_i}{\partial c_0} & \sum_i \frac{\partial f_i}{\partial c_m} \frac{\partial f_i}{\partial c_1} & \cdots & \sum_i \left(\frac{\partial f_i}{\partial c_m}\right)^2 \end{bmatrix} \begin{bmatrix} \delta c_0 \\ \delta c_1 \\ \vdots \\ \delta c_m \end{bmatrix} = \begin{bmatrix} -\sum_i \frac{\partial f_i}{\partial c_0} R_i \\ -\sum_i \frac{\partial f_i}{\partial c_1} R_i \\ \vdots \\ -\sum_i \frac{\partial f_i}{\partial c_m} R_i \end{bmatrix}$$

The solution $(\delta c_0, \delta c_1, \ldots, \delta c_m)$ of this set of normal equations is a first-order approximation of the change in $(c_0{}^0, c_1{}^0, \ldots, c_m{}^0)$ required to obtain the parameters $(c_0, c_1, \ldots, c_m)$. If any $|\delta c_k| > \varepsilon$, we replace $c_k{}^0$ by $c_k{}^0 + \delta c_k$ $(k = 0, m)$, and repeat the entire differential-correction procedure, using these new estimates.

Numerical Example of Differential-Correction Technique

This linear-Taylor differential-correction technique can be illustrated by solving the following problem.

Problem. Determine the values of the parameters c_0, c_1 in the expression

$$f(x, c_0, c_1) = c_0 e^{c_1 x}$$

which represents the transient current established by inserting a charged capacitor (C) in series with the resistor (R) and closing (completing) the circuit at time $x = 0$.

The current is measured at discrete times x_i, and the data pairs (x_i, y_i) are listed in the table below and plotted on the graph (Figure 7-2).

time x_i	current y_i
0.1	1.228
0.2	1.005
0.3	0.823
0.4	0.674
0.5	0.552

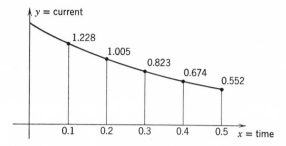

Figure 7-2

Initial estimates $c_0{}^0 = 1.4$ and $c_1{}^0 = -1.0$ are estimated from the graph. The values for the normal equations of the differential-correction technique

$$\begin{bmatrix} \sum_i \left(\dfrac{\partial f_i}{\partial c_0}\right)^2 & \sum_i \dfrac{\partial f_i}{\partial c_0}\dfrac{\partial f_i}{\partial c_1} \\[2mm] \sum_i \dfrac{\partial f_i}{\partial c_0}\dfrac{\partial f_i}{\partial c_1} & \sum_i \left(\dfrac{\partial f_i}{\partial c_1}\right)^2 \end{bmatrix} \begin{bmatrix} \delta c_0 \\[2mm] \delta c_1 \end{bmatrix} = \begin{bmatrix} -\sum_i \dfrac{\partial f_i}{\partial c_0}R_i \\[2mm] -\sum_i \dfrac{\partial f_i}{\partial c_1}R_i \end{bmatrix}$$

are computed and tabulated in Table 7-4.

Table 7-4

x_i	$\dfrac{\partial f_i}{\partial c_0}$	$\dfrac{\partial f_i}{\partial c_1}$	$f(x_i, c_0{}^0, c_1{}^0)$	y_i	R_i	$-\dfrac{\partial f_i}{\partial c_0}R_i$	$-\dfrac{\partial f_i}{\partial c_1}R_i$
0.1	0.905	0.127	1.267	1.228	0.039	−0.0353	−0.00495
0.2	0.819	0.229	1.147	1.005	0.142	−0.1163	−0.03252
0.3	0.741	0.311	1.037	0.823	0.214	−0.1586	−0.06655
0.4	0.670	0.375	0.938	0.674	0.264	−0.1769	−0.09900
0.5	0.607	0.425	0.850	0.552	0.298	−0.1809	−0.12665

The resulting normal equations for this first iteration are

$$\begin{bmatrix} 2.856 & 1.042 \\ 1.042 & 0.4864 \end{bmatrix} \begin{bmatrix} \delta c_0 \\ \delta c_1 \end{bmatrix} = \begin{bmatrix} -0.668 \\ -0.3297 \end{bmatrix}$$

with solution

$$\delta c_0 = 0.06141$$
$$\delta c_1 = -0.809$$

The improved estimates $c_0{}^1$, $c_1{}^1$ are obtained by adding these differential corrections to the previous estimates, i.e.,

$$c_0{}^1 = c_0{}^0 + \delta c_0 = 1.4 + 0.06141 = 1.46141$$
$$c_1{}^1 = c_1{}^0 + \delta c_1 = -1.0 - 0.809 = -1.809$$

Table 7-5

x_i	$\dfrac{\partial f_i}{\partial c_0}$	$\dfrac{\partial f_i}{\partial c_1}$	$f(x_i, c_0^1, c_1^1)$	y_i	R_i	$-\dfrac{\partial f_i}{\partial c_0} R_i$	$-\dfrac{\partial f_i}{\partial c_1} R_i$
0.1	0.835	0.122	1.219	1.228	-0.009	0.00752	0.00110
0.2	0.698	0.204	1.019	1.005	0.014	-0.00977	-0.00286
0.3	0.583	0.255	0.851	0.813	0.028	-0.01632	-0.00714
0.4	0.487	0.284	0.711	0.674	0.037	-0.01802	-0.01051
0.5	0.407	0.297	0.594	0.552	0.042	-0.01709	-0.01247

For the second iteration, these estimates are truncated to the values $c_0^1 = 1.46$ and $c_1^1 = -1.8$ to save computation time. Again, the values required for the normal equations are computed and tabulated in Table 7-5.

The resulting normal equations are

$$\begin{bmatrix} 1.927 & 0.652 \\ 0.652 & 0.2904 \end{bmatrix} \begin{bmatrix} \delta c_0 \\ \delta c_1 \end{bmatrix} = \begin{bmatrix} -0.05368 \\ -0.03188 \end{bmatrix}$$

with solutions

$$\delta c_0 = 0.0386$$
$$\delta c_1 = -0.1965$$

and improved estimates are

$$c_0^2 = c_0^1 + \delta c_0 = 1.46 + 0.0386 = 1.4986$$
$$c_1^2 = c_1^1 + \delta c_1 = -1.8 - 0.1965 = -1.9965$$

Another iteration should be performed since the $|\delta c_k|$ are not negligible. However, it can be shown† that the true parameter values are $c_0 = 1.500$, $c_1 = -2.000$, so that two iterations of the differential-correction technique have produced values of c_0 and c_1, which agree to three decimal places with the true values.

COMPUTATIONAL EXERCISES

1. Fit a linear least-squares polynomial to the following data:

x	y
-2	-3
-1	-1
0	1
1	3
2	5
3	7

† Evaluate $1.5e^{-2.0x}$. This function was used to obtain the data values y_i.

2. Fit a quadratic least-squares polynomial to the following data:

x	y
-3	-1
-2	-4
-1	-5
0	-4
1	-1
2	4
3	9

3. Fit a quadratic least-squares polynomial to the following data:

x	y
0.0	1.0000
0.1	1.1052
0.2	1.2214
0.3	1.3499
0.4	1.4918
0.5	1.6487

4. Determine coefficients a_0, a_1, a_2 such that the linear combination

$$F(x) = a_0 O_{05}(x) + a_1 O_{15}(x) + a_2 O_{25}(x)$$

of orthogonal polynomials fits the data of exercise 2 with the smallest sum of squared residuals.

5. The same as exercise 4, using the data of exercise 3.

PROGRAMMING EXERCISES

1. Flow chart the SHARE routine SDA 3289 that fits least-squares polynomials of degree 1 to 100 to a given set of data points.

2. Prepare a flow chart for curve-smoothing, using orthogonal polynomials.

3. Write a FORTRAN program for curve-smoothing, using orthogonal polynomials.

4. Flow chart the SHARE routine SDA 3094 for the least-squares estimation of non-linear parameters. The theory for the algorithm used in this routine is given in D. W. Marquardt, "An Algorithm for Least-Squares Estimation of Nonlinear Parameters" *J. SIAM*, Vol. 11, No. 2, 1963, pp. 431–441.

8

Numerical Integration
(Quadrature)

8.0 Introduction

In the integral calculus we learned that the definite integral,

$$(8.0.1) \qquad I = \int_a^b f(x)\, dx$$

where $f(x)$ is a continuous function over interval (a, b), can be interpreted geometrically as the area under the graph† of $y = f(x)$ between $x = a$ and $x = b$. By dividing the interval (a, b) into n equal subintervals (x_i, x_{i+1}), each of length h, we obtain a set of rectangles of width h, height $f(x_i)$, and area $f(x_i)h$. Note that $x_0 = a$ and $x_n = b$. The area under the graph of $f(x)$ can then be approximated by the sum of the areas of these rectangles. Also, the definite integral is *defined* by the relation

$$(8.0.2) \qquad I = \lim_{n \to \infty} \sum_{i=0}^{n-1} f(x_i)h.$$

In classical mathematics, the definite integral is usually evaluated by determining the anti-derivative

$$(8.0.3) \qquad F(x) = \int f(x)\, dx$$

† Throughout this chapter, $y \equiv f(x)$ and y_i denotes $f(x_i)$.

262

i.e., a function $F(x)$ with derivative $F'(x) \equiv f(x)$, and evaluating $F(x)$ at $x = a$ and $x = b$. The result is written

$$(8.0.4) \qquad \int_a^b f(x)\, dx = F(b) - F(a).$$

If the anti-derivative $F(x)$ cannot be found directly, we can try classical techniques such as trigonometric substitution or integration by parts, or can search a table of integrals for the particular function to be integrated.

However, if such approaches are unsuccessful or if we know beforehand that the definite integral cannot be evaluated by classical methods, we must resort to numerical methods for approximating the value of the definite integral.

The definition (8.0.2) of the definite integral leads us to one obvious method of approximating I. That is, we can compute the sum

$$(8.0.5) \qquad S_n = \sum_{i=0}^{n-1} f(x_i) h$$

where n is a large but finite integer. We must choose n as finite, since if it were infinite, an infinite length of time would be needed to compute the sum S_n by formula (8.0.5).

Figure 8-1 illustrates the fact that the area I under the graph of $f(x)$ between $x = a$ and $x = b$ could be obtained from the sum S_n by adding the sum of the shaded areas and subtracting the sum of the cross-hatched areas.

Intuitively, it would appear that a better numerical approximation of the definite integral could be obtained by computing the sum of the

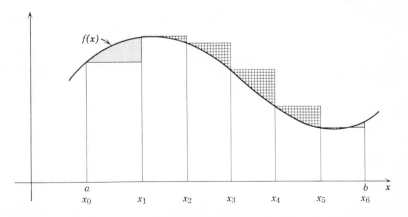

Figure 8-1

trapezoidal areas depicted in Figure 8-2, where the area of the ith trapezoid is

(8.0.6) $$t_i = \frac{h}{2}[f(x_i) + f(x_i + h)].$$

The sum T of these trapezoidal areas is then

(8.0.7) $$T = \sum_{i=0}^{n-1} \frac{h}{2}[f(x_i) + f(x_i + h)].$$

Writing the above formula in expanded form, we obtain

(8.0.8) $$T = \frac{h}{2} ([f(x_0) + f(x_0 + h)] + [f(x_1) + f(x_1 + h)]$$
$$+ \cdots + [f(x_{n-1}) + f(x_{n-1} + h)]).$$

Substituting the relation $x_{i+1} = x_i + h$ into the above, and collecting terms, we obtain the classical form of the Trapezoid Rule:

(8.0.9) $$T = \frac{h}{2}[f(x_0) + 2f(x_1) + 2f(x_2) + \cdots + 2f(x_{n-1}) + f(x_n)].$$

Examining formulas (8.0.5) and (8.0.9), we see that either of these simple numerical methods for approximating a definite integral can be written in the form of a weighted sum of the ordinates

(8.0.10) $$I \doteq \sum_i A_i f(x_i)$$

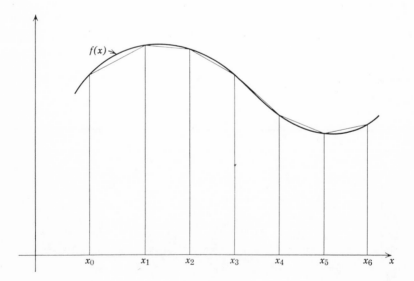

Figure 8-2

where the A_i are suitable constants (weights), and the $f(x_i)$ are suitably chosen ordinates of the function. As we will see in this chapter, *various numerical methods for approximating a definite integral can be represented in the form of (8.0.10).*

In this chapter we will develop representative numerical quadrature methods for approximating the definite integral. The methods developed herein can be categorized into the following distinct types.

Type 1. For *arbitrary distinct* x_i, determine constants A_i such that the error E defined by the relation†

$$(8.0.11) \qquad E = \int_a^b f(x)\,dx - \sum_{i=0}^{n} A_i f(x_i)$$

is zero when $f(x)$ is a polynomial of max degree n; that is, a polynomial of degree k, where $k \le n$.

Type 2. For *equally spaced* x_i, determine constants A_i such that the error E defined by the relation

$$(8.0.12) \qquad E = \int_a^b f(x)\,dx - \sum_{i=0}^{n} A_i f(x_i)$$

is zero when $f(x)$ is a polynomial of max degree n for n odd, and when $f(x)$ is a polynomial of max degree $n + 1$ for n even.

Type 3. For *properly selected* x_i, determine constants A_i such that the error E defined by the relation

$$(8.0.13) \qquad E = \int_a^b f(x)\,dx - \sum_{i=0}^{n} A_i f(x_i)$$

is zero when $f(x)$ is a polynomial of max degree $2n + 1$.

Type 4. For x_i obtained by *successive halving of the interval* (a, b), an iterative procedure is established for approximating the definite integral $\int_a^b f(x)\,dx$ by computing trapezoidal sums for $h_0 = b - a$, $h_1 = h_0/2$, $h_2 = h_0/2^2$, $h_3 = h_0/2^3, \ldots$, and applying repeated (iterated) linear interpolation in variable h^2.

8.1   Numerical Integration by Method of Undetermined Coefficients

In this section, we will determine a set of constants A_i such that the error E defined in the relation

$$(8.1.1) \quad E = \int_a^b f(x)\,dx - [A_0 f(x_0) + A_1 f(x_1) + \cdots + A_n f(x_n)]$$

is zero *for arbitrary distinct* x_i *when* $f(x)$ is any polynomial $p_n(x)$ of degree not exceeding n.

† $\sum A_i f(x_i) \equiv \sum A_i y_i$, since $y_i \equiv f(x_i)$.

Now, if error E is zero when $f(x)$ is *any* polynomial of degree not exceeding n, then it must be zero when $f(x) = 1, x, x^2, \ldots, x^n$. By substituting successively $1, x, x^2, \ldots, x^n$ for $f(x)$ in (8.1.1) we obtain the following system of equations, which are linear in the unknown constants A_i (imposing the constraint that $E = 0$ in each case):

For $f(x) = 1$,

$$(8.1.2a) \qquad \int_a^b 1 \, dx = A_0 1 \quad + A_1 1 \quad + \cdots + A_n 1$$

for $f(x) = x$,

$$(8.1.2b) \qquad \int_a^b x \, dx = A_0 x_0 \quad + A_1 x_1 \quad + \cdots + A_n x_n$$

for $f(x) = x^2$,

$$(8.1.2c) \qquad \int_a^b x^2 \, dx = A_0 x_0{}^2 + A_1 x_1{}^2 + \cdots + A_n x_n{}^2$$

$$\cdot \quad \cdot \quad \cdot \quad \cdot \quad \cdot \quad \cdot \quad \cdot \quad \cdot \quad \cdot \quad \cdot \quad \cdot \quad \cdot$$

for $f(x) = x^n$,

$$(8.1.2d) \qquad \int_a^b x^n \, dx = A_0 x_0{}^n + A_1 x_1{}^n + \cdots + A_n x_n{}^n$$

Equations (8.1.2) constitute $n + 1$ conditions for the determination of the $n + 1$ unknown constants A_i. Evaluating the integrals $\int_a^b x^k \, dx$ in (8.1.2) and writing the above equations in matrix form, we obtain

$$(8.1.3) \qquad \begin{bmatrix} 1 & 1 & \ldots & 1 \\ x_0 & x_1 & \ldots & x_n \\ x_0{}^2 & x_1{}^2 & \ldots & x_n{}^2 \\ \cdot & \cdot & \cdot & \cdot \\ x_0{}^n & x_1{}^n & \ldots & x_n{}^n \end{bmatrix} \begin{bmatrix} A_0 \\ A_1 \\ A_2 \\ \vdots \\ A_n \end{bmatrix} = \begin{bmatrix} b - a \\ (b^2 - a^2)/2 \\ (b^3 - a^3)/3 \\ \vdots \\ (b^{n+1} - a^{n+1})/(n + 1) \end{bmatrix}$$

The determinant of the matrix of coefficients in (8.1.3) is the well-known Vandermonde determinant, which is non-zero for distinct x_i. Since the x_i were assumed to be distinct, there exists a unique solution $(A_0, A_1, A_2, \ldots, A_n)$ of equation (8.1.3). This solution is the set of unique A_i $(i = 0, n)$, which make error E in equation (8.1.1) zero when $f(x) = 1, x, x^2, \ldots, x^n$.

To show that error E in (8.1.1) is zero when $f(x)$ is any polynomial $p_n(x)$ of degree not exceeding n, we need only demonstrate that

$$(8.1.4) \quad \int_a^b p_n(x) \, dx = A_0 p_n(x_0) + A_1 p_n(x_1) + \cdots + A_n p_n(x_n)$$

where $f(x) = p_n(x) \equiv a_0 + a_1 x + a_2 x^2 + \cdots + a_n x^n$ is an arbitrary polynomial of degree not exceeding n, and $(A_0, A_1, A_2, \ldots, A_n)$ is the solution of matrix equation (8.1.3).

First, rewrite equations (8.1.2) in summation notation, as follows:

$$\int_a^b 1 \, dx = \sum_{i=0}^n A_i$$

$$\int_a^b x \, dx = \sum_{i=0}^n A_i x_i$$

(8.1.5)

$$\int_a^b x^2 \, dx = \sum_{i=0}^n A_i x_i^2$$

$$\vdots$$

$$\int_a^b x^n \, dx = \sum_{i=0}^n A_i x_i^n.$$

Now

$$\int_a^b p_n(x) \, dx = \int_a^b (a_0 + a_1 x + a_2 x^2 + \cdots + a_n x^n) \, dx$$

$$= a_0 \int_a^b dx + a_1 \int_a^b x \, dx + a_2 \int_a^b x^2 \, dx + \cdots + a_n \int_a^b x^n \, dx.$$

(8.1.6)

Substituting (8.1.5) into the last line of (8.1.6) we obtain

(8.1.7) $$\int_a^b p_n(x) \, dx = a_0 \sum A_i + a_1 \sum A_i x_i + a_2 \sum A_i x_i^2$$

$$+ \cdots + a_n \sum A_i x_i^n$$

$$= \sum_{i=0}^n A_i (a_0 + a_1 x_i + a_2 x_i^2 + \cdots + a_n x_i^n)$$

$$= \sum_{i=0}^n A_i p_n(x_i).$$

That is,

(8.1.8) $$\int_a^b p_n(x) \, dx = A_0 p_n(x_0) + A_1 p_n(x_1) + A_2 p_n(x_2) + \cdots + A_n p_n(x_n).$$

Equation (8.1.8) states that error E in (8.1.1) is zero when $f(x)$ is any polynomial $p_n(x)$ of degree not exceeding n, for a unique set of A_i ($i = 0, n$), which satisfy equation (8.1.3).

EXAMPLE (Method of Undetermined Coefficients). Given the points $(x_0, f(x_0))$, $(x_1, f(x_1))$, $(x_2, f(x_2))$ on the graph of $y = f(x)$, where x_0, x_1, $x_2 \in (a, b)$, evaluate $\int_a^b f(x) \, dx$ by the formula

$$\int_a^b f(x) \, dx = A_0 f(x_0) + A_1 f(x_1) + A_2 f(x_2)$$

where the coefficients A_0, A_1, A_2 constitute the solution of the matrix equation

$$\begin{bmatrix} 1 & 1 & 1 \\ x_0 & x_1 & x_2 \\ x_0{}^2 & x_1{}^2 & x_2{}^2 \end{bmatrix} \begin{bmatrix} A_0 \\ A_1 \\ A_2 \end{bmatrix} = \begin{bmatrix} b - a \\ (b^2 - a^2)/2 \\ (b^3 - a^3)/3 \end{bmatrix}$$

NUMERICAL EXAMPLE. Given the points $(0.0, 1.0)$, $(1.0, 1.5431)$, $(2.0, 3.7622)$ from the graph of $f(x) = \cosh x$, calculate the value of

$$\int_0^2 \cosh x \, dx = A_0 f(0) + A_1 f(1) + A_2 f(2)$$

where

$$\begin{bmatrix} 1 & 1 & 1 \\ 0.0 & 1.0 & 2.0 \\ 0.0 & 1.0 & 4.0 \end{bmatrix} \begin{bmatrix} A_0 \\ A_1 \\ A_2 \end{bmatrix} = \begin{bmatrix} 2.0 - 0.0 \\ (4.0 - 0.0)/2 \\ (8.0 - 0.0)/3 \end{bmatrix}$$

The solution $(A_0, A_1, A_2) = (1/3, 4/3, 1/3)$. Substituting these values into the integral equation we obtain

$$\int_0^2 \cosh x \, dx = \frac{1}{3}(1.0) + \frac{4}{3}(1.5431) + \frac{1}{3}(3.7622) = 3.6449.$$

The exact value of

$$\int_0^2 \cosh x \, dx = \sinh x \, |_0^2 = 3.6269.$$

The error

$$E = \int_0^2 \cosh x \, dx - \sum A_i y_i = 3.6269 - 3.6449 = -0.0180.$$

8.2 Newton-Cotes Quadrature Formulas (Closed)

The well-known Trapezoid Rule, Simpson's 1/3 Rule, and Newton's 3/8 Rule belong to the Newton-Cotes (closed) class of numerical integration (quadrature) methods for approximating $I \equiv \int_a^b f(x) \, dx$. These quadrature formulas are denoted by Q_{nn} and are of the general form

$$(8.2.1) \qquad Q_{nn} = \sum_{i=0}^{N} A_i f(x_i)$$

where the x_i are *evenly spaced* values of interval $(x_0, x_N) \equiv (a, b)$, with subinterval length $h = x_{i+1} - x_i$ $(i = 0, N - 1)$.

The Newton-Cotes (closed) quadrature formulas Q_{nn} *are* derived by approximating $f(x)$ by interpolating polynomials of degree n over each set of n subintervals† of (x_0, x_N), and computing the integral of the interpolating polynomial over its interval of definition. It should be noted that the Newton-Cotes quadrature formulas *could* be derived by the method of undetermined coefficients, for the special case of evenly spaced x_i instead of arbitrary x_i as described in Sec. 8.1.

The constants A_i are determined so that the error E defined in the relation

$$(8.2.2) \qquad E = \int_{x_0}^{x_N} f(x)\,dx - \sum_{i=0}^{N} A_i f(x_i)$$

is zero when $f(x)$ is a polynomial of max degree n for n odd, and when $f(x)$ is a polynomial of max degree $n + 1$ for n even.

In the next two subsections we will derive the formulas for Q_{11} (Trapezoid Rule) and Q_{22} (Simpson's 1/3 Rule). The method and notation used in these derivations can be directly extended for the derivation of the higher-order Newton-Cotes quadrature formulas $Q_{33}, Q_{44}, \ldots$.

The truncation error E will also be analyzed for the case when $f(x)$ is a non-polynomial function. This analysis can be accomplished by investigating the integral of the error inherent in approximation by interpolating polynomials (see Sec. 6.1), but is difficult for all cases except the Trapezoid Rule. However, the Taylor-series expansion provides a straightforward means for evaluating this error for all the Newton-Cotes quadrature formulas.

The Trapezoid Rule (Q_{11})

The Trapezoid Rule, described briefly in Sec. 8.0, can be derived by first writing

$$(8.2.3) \qquad I = \int_{x_0}^{x_N} f(x)\,dx = \sum_{i=0}^{N-1} \int_{x_i}^{x_{i+1}} f(x)\,dx$$

and then approximating $f(x)$ in interval (x_i, x_{i+1}) by the *linear* interpolating polynomial (Newton forward form)

$$(8.2.4) \qquad P_1{}^i(x) = y_i + \frac{\Delta y_i}{h}(x - x_i)$$

where $y_i \equiv f(x_i)$, and superscript i denotes interval (x_i, x_{i+1}) of definition for the linear interpolating polynomial. This interpolating polynomial is then integrated over its range of definition (x_i, x_{i+1}), i.e.,

$$(8.2.5) \qquad I_i = \int_{x_i}^{x_{i+1}} P_1{}^i(x)\,dx = \int_{x_i}^{x_{i+1}} \left[y_i + \frac{\Delta y_i}{h}(x - x_i) \right] dx$$

To simplify the integration, we introduce variable $u = (x - x_i)/h$, where

† Interval (x_0, x_N) is divided into N equal subintervals each of width $h = (x_N - x_0)/N$.

also $du = dx/h$. The range of integration (x_i, x_{i+1}) in x becomes $(0, 1)$ in variable u. We can then rewrite (8.2.5) in the form

$$(8.2.6) \qquad I_i = h \int_0^1 [y_i + u \, \Delta y_i] \, du$$

$$= h \left[y_i u + \frac{u^2}{2} \Delta y_i \right]_0^1 = h \left[y_i(1) + \frac{1}{2} \Delta y_i \right].$$

Substituting the relation $y_{i+1} - y_i = \Delta y_i$ into the above, we obtain

$$(8.2.7) \qquad I_i = h \left[y_i + \frac{1}{2} (y_{i+1} - y_i) \right]$$

$$= \frac{h}{2} [y_i + y_{i+1}].$$

Now, since $P_1{}^i(x)$ approximates $f(x)$ in (x_i, x_{i+1}), we obtain the relation

$$(8.2.8) \qquad \sum_{i=0}^{N-1} \int_{x_i}^{x_{i+1}} f(x) \, dx \doteq \sum_{i=0}^{N-1} \int_{x_i}^{x_{i+1}} P_1{}^i(x) \, dx$$

It follows then that

$$(8.2.9) \qquad \int_{x_0}^{x_N} f(x) \, dx \doteq \sum_{i=0}^{N-1} I_i = \sum_{i=0}^{N-1} \frac{h}{2} [y_i + y_{i+1}]$$

$$\doteq \frac{h}{2} [y_0 + 2y_1 + 2y_2 + \cdots + 2y_{N-1} + y_N] \equiv Q_{11}$$

This is the well-known form of the Trapezoid Rule.

Simpson's 1/3 Rule (Q_{22})

The Simpson's 1/3 Rule (which requires that N be even)

$$(8.2.10) \quad Q_{22} = \frac{h}{3} [y_0 + 4y_1 + 2y_2 + 4y_3 + \cdots + 2y_{N-2} + 4y_{N-1} + y_N]$$

for approximating the definite integral $\int_{x_0}^{x_N} f(x) \, dx$ is obtained by first writing

$$(8.2.11) \qquad \int_a^b f(x) \, dx = \sum_{i=0}^{N/2-1} \int_{x_{2i}}^{x_{2i+2}} f(x) \, dx$$

and then approximating $f(x)$ by a second-degree interpolating polynomial $P_2{}^i(x)$ over each of the subinterval pairs (x_{2i}, x_{2i+2}); that is, over (x_0, x_2), (x_2, x_4), (x_4, x_6), ... Superscript i in $P_2{}^i(x)$ denotes the ith pair of subintervals.

Hence, for the subinterval pair (x_{2i}, x_{2i+2}), we approximate the integral of $f(x)$ by

$$I_i = \int_{x_{2i}}^{x_{2i+2}} P_2{}^i(x)\, dx$$

$$= \int_{x_{2i}}^{x_{2i+2}} \left[y_{2i} + \frac{\Delta y_{2i}}{h}(x - x_{2i}) + \frac{\Delta^2 y_{2i}}{2h^2}(x - x_{2i})(x - x_{2i+1}) \right] dx$$

(8.2.12)

using a second-degree Newton forward-form interpolating polynomial with range of definition (x_{2i}, x_{2i+2}).

To simplify the integration, we introduce variable $u = (x - x_{2i})/h$, where $du = dx/h$, and interval (x_{2i}, x_{2i+2}) in x becomes $(0, 2)$ in u. With this substitution, we can rewrite (8.2.12) as

$$(8.2.13) \qquad I_i = h \int_0^2 \left[y_{2i} + u\,\Delta y_{2i} + u(u - 1)\frac{\Delta^2 y_{2i}}{2} \right] du$$

$$= h\left[u y_{2i} + \frac{u^2}{2}\Delta y_{2i} + \left(\frac{u^3}{3} - \frac{u^2}{2} \right)\frac{\Delta^2 y_{2i}}{2} \right]_0^2$$

$$= h\left[2 y_{2i} + 2\Delta y_{2i} + \frac{1}{3}\Delta^2 y_{2i} \right].$$

Substituting the relations

$$\Delta y_{2i} = y_{2i+1} - y_{2i}$$
$$\Delta^2 y_{2i} = y_{2i+2} - 2y_{2i+1} + y_{2i}$$

into (8.2.13), and collecting terms, we obtain

$$(8.2.14) \qquad I_i = \frac{h}{3}[y_{2i} + 4y_{2i+1} + y_{2i+2}].$$

Now, since $P_2{}^i(x)$ approximates $f(x)$ over (x_{2i}, x_{2i+2}), and also

$$\int_{x_{2i}}^{x_{2i+2}} f(x)\, dx \doteq \int_{x_{2i}}^{x_{2i+2}} P_2{}^i(x)\, dx$$

we can write

$$\int_{x_0}^{x_N} f(x)\, dx \doteq \sum_{i=0}^{N/2-1} \int_{x_{2i}}^{x_{2i+2}} P_2{}^i(x)\, dx = \sum_{i=0}^{N/2-1} I_i$$

$$\doteq \frac{h}{3}[y_0 + 4y_1 + 2y_2 + 4y_3 + \cdots + 4y_{N-1} + y_N] = Q_{22}$$

(8.2.15)

which is the desired form of Simpson's 1/3 Rule for approximating I.

Derivation of Higher-Order Newton-Cotes Formulas

Higher-order Newton-Cotes formulas Q_{nn} $(n = 3, 4, \ldots)$ can be derived by directly extending the method and notation used in the foregoing pages to derive Q_{11} and Q_{22}. For example, to derive Q_{33} (for which N must be a multiple of 3), we divide (x_0, x_N) into trios of subintervals (x_{3i}, x_{3i+3}), and approximate $f(x)$ over each such triple interval by a third-degree interpolating polynomial $P_3{}^i(x)$.

The integration of

$$I_i = \int_{x_{3i}}^{x_{3i+3}} P_3{}^i(x)\, dx$$

is simplified by introducing variable $u = (x - x_{3i})/h$, where $du = dx/h$ and interval (x_{3i}, x_{3i+3}) in x becomes $(0, 3)$ in variable u. After some calculation and substitution of the ordinate-difference relations, we obtain

$$I_i = \left(\frac{3h}{8}\right)(y_{3i} + 3y_{3i+1} + 3y_{3i+2} + y_{3i+3})$$

so that we obtain Newton's 3/8 Rule

$$\int_{x_0}^{x_N} f(x)\, dx \doteq \frac{3h}{8}[y_0 + 3y_1 + 3y_2 + 2y_3 + 3y_4 + \cdots + 3y_{N-1} + y_N] = Q_{33}$$
(8.2.16)

The derivation of Q_{44}, Q_{55}, $\ldots$ can be accomplished in the same way, by simply extending the notation employed above.

Numerical Examples of Newton-Cotes Quadrature

EXAMPLE 1. Evaluate

$$I \equiv \int_1^2 \frac{dx}{x^2}$$

by the Trapezoid Rule

$$I \doteq \frac{h}{2}[y_0 + 2y_1 + 2y_2 + \cdots + 2y_{N-1} + y_N]$$

using: (a) $h = 1.0$; (b) $h = 0.5$; (c) $h = 0.25$.

Solution:

(a) $I \doteq \dfrac{1.0}{2}\left[\dfrac{1}{1^2} + \dfrac{1}{2^2}\right] = 0.625$

(b) $I \doteq \dfrac{0.5}{2}\left[\dfrac{1}{1^2} + \dfrac{2}{1.5^2} + \dfrac{1}{2^2}\right] = 0.534722$

(c) $I = \dfrac{0.25}{2}\left[\dfrac{1}{1^2} + \dfrac{2}{1.25^2} + \dfrac{2}{1.5^2} + \dfrac{2}{1.75^2} + \dfrac{1}{2^2}\right] = 0.508993.$

EXAMPLE 2. Evaluate

$$I \equiv \int_1^2 \frac{dx}{x^2}$$

by Simpson's 1/3 Rule

$$I \doteq \frac{h}{3} [y_0 + 4y_1 + 2y_2 + 4y_3 + \cdots + 4y_{N-1} + y_N]$$

using (a) $h = 0.5$; (b) $h = 0.25$.
Solution:

(a) $I \doteq \dfrac{0.5}{3} \left[\dfrac{1}{1^2} + \dfrac{4}{1.5^2} + \dfrac{1}{2^2} \right] = 0.504630$

(b) $I \doteq \dfrac{0.25}{3} \left[\dfrac{1}{1^2} + \dfrac{4}{1.25^2} + \dfrac{2}{1.5^2} + \dfrac{4}{1.75^2} + \dfrac{1}{2^2} \right] = 0.500418.$

Note. Closed form solution

$$\int_1^2 \frac{dx}{x^2} = -x^{-1} \Big]_1^2 = 0.500000.$$

These examples give some indication of the accuracy of the Trapezoid Rule and Simpson's Rule. Note that the latter is more accurate than the former, and the accuracy of either is improved by reducing the size of h.†

An analysis of the error in the Newton-Cotes quadrature formulas is presented in the following subsection.

Error in Newton-Cotes (Closed) Quadrature Formulas

Truncation Error in Trapezoid Rule

If $f(x)$ is a linear function or a piece-wise linear function with vertices at (x_i, y_i), then $f(x)$ coincides with the linear interpolating polynomials $P_1{}^i(x)$ throughout each subinterval (x_i, x_{i+1}), i.e., $f(x) - P_1{}^i(x) \equiv 0$. It follows then that

$$E_1{}^i = \int_{x_i}^{x_{i+1}} [f(x) - P_1{}^i(x)] \, dx = 0$$

and hence the total truncation error for the Trapezoid Rule for (x_0, x_N) is zero for this case; i.e.,

(8.2.17) $$E_1 = \sum_{i=0}^{N-1} E_1{}^i = 0.$$

If $f(x)$ is a non-linear function, then we would expect that the Trapezoid Rule truncation error

$$E_1 = \int_{x_0}^{x_N} f(x) \, dx - \frac{h}{2} [y_0 + 2y_1 + 2y_2 + \cdots + 2y_{N-1} + y_N]$$

† Because of roundoff, there is a point of diminishing returns on the reduction of the size of h. (See p. 275.)

to be non-zero. That this is so can be demonstrated by investigating the local truncation error [i.e., the error in interval (x_i, x_{i+1})]

(8.2.18) $$\int_{x_i}^{x_{i+1}} [f(x) - P_1{}^i(x)]\, dx \equiv \int_{x_i}^{x_{i+1}} e_1{}^i(x)\, dx$$

where†

$$e_1{}^i(x) \equiv f(x) - P_1{}^i(x) = \frac{f''(x^*)}{2} (x - x_i)(x - x_{i+1}).$$

This method of analyzing the error is feasible for the Trapezoid Rule, and an excellent presentation is given in Macon [14, pp. 114–116]. However, for Simpson's 1/3 Rule and higher-order Newton-Cotes formulas this approach is difficult. For this reason, we seek another approach to truncation error analysis that can be applied to each of the Newton-Cotes formulas. The approach we will use is based on the Taylor-series expansion of the function $f(x)$ and its anti-derivative $F(x) = \int f(x)\, dx$. Note also that $F'(x) = f(x)$, $F''(x) = f'(x)$, $F'''(x) = f''(x)$, and so forth.

Assuming that the required derivatives exist, we can investigate the Trapezoid Rule truncation error in interval (x_i, x_{i+1}) defined by the relation

(8.2.19) $$E_1{}^i = \int_{x_i}^{x_{i+1}} f(x)\, dx - \frac{h}{2} [y_i + y_{i+1}].$$

Using the relations $y \equiv f(x)$ and $F(x) = \int f(x)\, dx$, we can write the preceding equation in the form

(8.2.20) $$E_1{}^i = [F(x_{i+1}) - F(x_i)] - \frac{h}{2} [f(x_i) + f(x_{i+1})]$$

$$= F(x_i + h) - F(x_i) - \frac{h}{2} [f(x_i) + f(x_i + h)].$$

We can now expand each term of the above expression in a Taylor series at $x = x_i$ in powers of h, as follows:

(8.2.21)
$$F(x_i + h) = F(x_i) + hf(x_i) + \frac{h^2}{2!} f'(x_i) + \frac{h^3}{3!} f''(x_i) + \cdots$$
$$F(x_i) \qquad = F(x_i)$$
$$f(x_i) \qquad = f(x_i)$$
$$f(x_i + h) = f(x_i) + hf'(x_i) + \frac{h^2}{2!} f''(x_i) + \cdots.$$

Substituting (8.2.21) into (8.2.20) and collecting terms, we find that

$$E_1{}^i = -\frac{h^3}{12} f''(x_i) + \cdots.$$

† See Sec. 6.1.

If higher-order terms in the expansion of $E_1{}^i$ are neglected, we find that the total error in the Trapezoid Rule for (x_0, x_N) is

$$(8.2.22) \qquad E_1 = \sum_{i=0}^{N-1} E_1{}^i \doteq N\left(-\frac{h^3}{12}f''(x^*)\right) = Nh\left(\frac{-h^2}{12}f''(x^*)\right)$$

$$= -\frac{(x_N - x_0)h^2 f''(x^*)}{12}$$

where x^* is some value of $x \in (x_0, x_N)$, such that

$$f''(x^*) = \frac{1}{N}\sum_{i=0}^{N-1} f''(x_i).$$

Roundoff Error in the Trapezoid Rule

The formula (8.2.9) for the Trapezoid Rule can be written in the form (for $N = 4$)

$$Q_{11} = \left[\frac{y_0}{2} + y_1 + y_2 + y_3 + \frac{y_4}{2}\right]\cdot h.$$

Let r_k denote† the relative error inherent in y_k, α_k the relative roundoff in the k addition, and μ the relative roundoff in the multiplication by h. The corresponding relative-error expression can then be written as

$$r_Q = \left(\left\{\left[\left(r_0\frac{y_0/2}{S_1} + r_1\frac{y_1}{S_1} + \alpha_1\right)\frac{S_1}{S_2} + r_2\frac{y_2}{S_2} + \alpha_2\right]\frac{S_2}{S_3} + r_3\frac{y_3}{S_3} + \alpha_3\right\}\frac{S_3}{S_4} + r_4\frac{y_4/2}{S_4} + \alpha_4\right)1 + r_h\ 1 + \mu$$

$$\underbrace{}_{\begin{array}{cccc} N & O & NO & R \end{array}}$$

$$\underbrace{}_{\begin{array}{cccc} N & O & NO & R \end{array}}$$

$$\underbrace{}_{\begin{array}{cccc} N & O & NO & R \end{array}}$$

$$\underbrace{}_{\begin{array}{cccc} N & O & NO & R \end{array}}$$

$$\underbrace{}_{\begin{array}{cccc} N & O & NO & R \end{array}}$$

$$\underbrace{}_{N}$$

where the S_k denote the partial sums

$$S_1 = \frac{y_0}{2} + y_1$$

$$S_2 = \frac{y_0}{2} + y_1 + y_2$$

$$S_3 = \frac{y_0}{2} + y_1 + y_2 + y_3$$

$$S_4 = \frac{y_0}{2} + y_1 + y_2 + y_3 + \frac{y_4}{2}.$$

† As written here r_0 and r_4 denote the relative errors in $y_0/2$ and $y_4/2$, respectively, while r_1, r_2, r_3 denote the relative errors in y_1, y_2, y_3, respectively.

Multiplying α_k by S_k/S_k, the relative-error expression can be written in the form

$$r_Q = \left(\left\{ \left[\left(\frac{r_0 y_0/2 + r_1 y_1 + \alpha_1 S_1}{S_1} \right) \frac{S_1}{S_2} + \frac{r_2 y_2 + \alpha_2 S_2}{S_2} \right] \frac{S_2}{S_3} \right. \right.$$
$$\left. \left. + \frac{r_3 y_3 + \alpha_3 S_3}{S_3} \right\} \frac{S_3}{S_4} + \frac{r_4 y_4/2 + \alpha_4 S_4}{S_4} \right) + \frac{r_h S_4 + \mu S_4}{S_4}.$$

Now, $S_4 \cdot h = Q_{11}$, and $\varepsilon_Q = Q_{11} r_Q$; collecting terms and multiplying through by $S_4 h$, we obtain

$$\varepsilon_Q = h\left(r_0 \frac{y_0}{2} + r_1 y + \alpha_1 S_1 + r_2 y_2 + \alpha_2 S_2 + r_3 y_3 + \alpha_3 S_3 + r_4 \frac{y_4}{2} \right)$$
$$+ \alpha_4 S_4 + r_h S_4 + \mu S_4$$

$$\varepsilon_Q = h\left(\left[r_0 \frac{y_0}{2} + r_1 y_1 + r_2 y_2 + r_3 y_3 + r_4 \frac{y_4}{2} \right] \right.$$
$$+ \alpha_1 \left[\frac{y_0}{2} + y_1 \right] + \alpha_2 \left[\frac{y_0}{2} + y_1 + y_2 \right] + \alpha_3 \left[\frac{y_0}{2} + y_1 + y_2 + y_3 \right]$$
$$\left. + (\alpha_4 + r_h + \mu)\left[\frac{y_0}{2} + y_1 + y_2 + y_3 + \frac{y_4}{2} \right] \right).$$

This last expression gives us the error in the Trapezoid Rule for $N = 4$. For the general case, the error in the Trapezoid Rule

$$Q_{11} = \left[\frac{y_0}{2} + y_1 + y_2 + y_3 + \cdots + y_{N-1} + \frac{y_N}{2} \right] \cdot h$$

can be expressed in the form

$$\varepsilon_Q = h\left[\sum_{k=1}^{N-1} r_k y_k + r_0\left(\frac{y_0}{2}\right) + r_N\left(\frac{y_N}{2}\right) \right]$$
$$+ h\left(\frac{y_0}{2}\right)[\alpha_1 + \alpha_2 + \alpha_3 + \cdots + \alpha_{N-1}]$$
$$+ h[\alpha_1 y_1 + \alpha_2(y_1 + y_2) + \cdots + \alpha_{N-1}(y_1 + y_2 + \cdots + y_{N-1})]$$
$$+ h(\alpha_N + r_h + \mu)\left[\left(\frac{y_0}{2}\right) + y_1 + y_2 + \cdots + y_{N-1} + \left(\frac{y_N}{2}\right) \right].$$

If the $|r_k|$ are bounded by $K \cdot 10^{-d}$ and the $|\alpha_k|$ and $|\mu|$ by $5 \cdot 10^{-d}$, a very conservative† bound for $|\varepsilon_Q|$ can be obtained by replacing y_k by Y, where $Y = \max_k |y_k|$, and assuming that $r_h = 0$ since h can be measured quite accurately. Under these assumptions, we obtain

$$|\varepsilon_Q| \leq hY\left[K(N+1) + 5\left(\frac{1}{2}(N-1) + \sum_{k=1}^{N-1} k + 2N\right) \right]10^{-d}$$
$$\leq \frac{hY}{2}[2K(N+1) + 5(N-1 + N(N-1) + 4N)]10^{-d}$$
$$\leq \frac{hY}{2}[2K(N+1) + 5(N^2 + 4N - 1)]10^{-d}$$

Since $N = (b - a)/h$, where (a, b) is the interval of integration that is divided into N subintervals of length h, we find that

$$|\varepsilon_Q| \leq \frac{Y}{2}\left[2K(b - a) + 5\frac{(b - a)^2}{h} + 5(4)(b - a) + (2K - 5)h\right]10^{-d}.$$

For very small h, the second term in the brackets predominates, so that

$$|\varepsilon_Q| \doteq \left[\frac{Y}{2}(b - a)^2 5 \; 10^{-d}\right]\left(\frac{1}{h}\right).$$

This indicates that the roundoff error increases approximately as $1/h$ for small h. Since the truncation error is proportional to h^2, the truncation error can be made extremely small by reducing h, which unfortunately increases the roundoff error. Hence, we cannot make h too small, or else the roundoff error will dominate the truncation error, which is highly undesirable.

Truncation Error in Simpson's 1/3 Rule

If $f(x)$ is a linear or second-degree polynomial function, or a function whose graph consists of line segments or parabolas that contain the points (x_{2i}, y_{2i}), (x_{2i+1}, y_{2i+1}), (x_{2i+2}, y_{2i+2}), then $f(x)$ coincides with the *quadratic* interpolating polynomials $P_2{}^i(x)$ throughout (x_{2i}, x_{2i+2}), i.e., $e_2{}^i(x) \equiv f(x) - P_2{}^i(x) \equiv 0$, so that

$$(8.2.23) \qquad E_2{}^i = \int_{x_{2i}}^{x_{2i+2}} [f(x) - P_2{}^i(x)]\,dx = 0.$$

It follows then that the total truncation error for Simpson's 1/3 Rule for (x_0, x_N) is zero for polynomial functions of max degree 2; i.e.,

$$(8.2.24) \qquad E_2 = \sum_{i=0}^{N/2-1} E_2{}^i = 0.$$

It is interesting to note that Simpson's Rule is also *exact* when $f(x)$ is a cubic polynomial function. To show this, we note that the error inherent in approximating $f(x)$ by a *quadratic* interpolating polynomial over (x_{2i}, x_{2i+2}) is (see Sec. 6.1)

$$(8.2.25) \qquad e_2{}^i(x) = \frac{f'''(x^*)}{6}(x - x_{2i})(x - x_{2i+1})(x - x_{2i+2}).$$

---

† A smaller bound is given in McCracken and Dorn [16] by replacing y_k by an average $\bar{y}$ and neglecting small quantities. Note also that the form of the error expression differs slightly because of the difference in which Q_{11} was formed.

Now, if $f(x)$ is a *cubic* polynomial function, it follows that $f'''(x) = K$, where K is a constant. It follows then that

$$(8.2.26) \qquad e_2{}^i(x) = \frac{K}{6}(x - x_{2i})(x - x_{2i+1})(x - x_{2i+2}).$$

Function $e_2{}^i(x)$ is an *odd-symmetric* function with x_{2i+1} as the point of symmetry. Therefore

$$(8.2.27) \quad E_2{}^i = \int_{x_{2i}}^{x_{2i+2}} \frac{K}{6}(x - x_{2i})(x - x_{2i+1})(x - x_{2i+2})\, dx = 0.$$

We see by (8.2.24) and (8.2.27) that the truncation error in Simpson's Rule is zero when $f(x)$ is a polynomial function of max degree 3.

The odd-symmetry property of $e_2{}^i(x)$ illustrated in Figure 8-3 is characteristic of the property of the error function inherent in approximating (for n even) a polynomial of degree $n + 1$ by an interpolating polynomial of degree n. The fact that the integral of such an error function is zero over (x_{ni}, x_{ni+n}) for n even causes the Newton-Cotes formulas of even degree to be exact for polynomial functions of degree $n + 1$.

If $f(x)$ is a polynomial function of degree exceeding 3, or if $f(x)$ is a non-polynomial function, then we would expect that the error in Simpson's 1/3 Rule is non-zero. This error can be expressed as

$$E_2 = \int_{x_0}^{x_N} f(x)\, dx - \frac{h}{3}[y_0 + 4y_1 + 2y_2 + 4y_3 + \cdots + 4y_{N-1} + y_N].$$

$$(8.2.28)$$

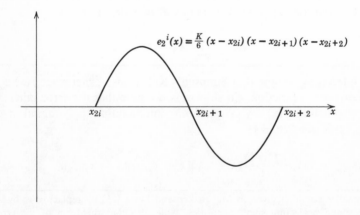

Figure 8-3

As indicated in the foregoing discussion, the error in (x_{2i}, x_{2i+2}) could be determined by calculating the integral of the error inherent in approximating $f(x)$ by a second-degree interpolating polynomial $P_2{}^i(x)$; that is,

$$(8.2.29) \quad E_2{}^i = \int_{x_{2i}}^{x_{2i+2}} [f(x) - P_2{}^i(x)] \, dx$$

$$= \int_{x_{2i}}^{x_{2i+2}} \frac{f'''(x^*)}{6} (x - x_{2i})(x - x_{2i+1})(x - x_{2i+2}) \, dx.$$

This approach is difficult because the function

$$(x - x_{2i})(x - x_{2i+1})(x - x_{2i+2})$$

changes sign in (x_{2i}, x_{2i+2}). (See Jennings [12, p. 120].) For this reason, we will use the Taylor-series approach employed in the preceding section.

The local truncation error in Simpson's 1/3 Rule is

$$(8.2.30) \quad E_2{}^i = \int_{x_{2i}}^{x_{2i+2}} f(x) \, dx - \frac{h}{3} [y_{2i} + 4y_{2i+1} + y_{2i+2}].$$

Evaluating the anti-derivative $F(x) = \int f(x) \, dx$ at x_{2i} and x_{2i+2}, we obtain

$$(8.2.31) \quad E_2{}^i = [F(x_{2i} + 2h) - F(x_{2i})]$$

$$- \frac{h}{3} [f(x_{2i}) + 4f(x_{2i} + h) + f(x_{2i} + 2h)].$$

Expanding each term of the above in a Taylor series at $x = x_{2i}$, we find:

$$F(x_{2i} + 2h) = F(x_{2i}) + f(x_{2i})2h + f'(x_{2i}) \frac{(2h)^2}{2} + f''(x_{2i}) \frac{(2h)^3}{6}$$

$$+ f'''(x_{2i}) \frac{(2h)^4}{24} + f^{(4)}(x_{2i}) \frac{(2h)^5}{120} + \cdots$$

$$F(x_{2i}) = F(x_{2i})$$

$$f(x_{2i}) = f(x_{2i})$$

$$f(x_{2i} + h) = f(x_{2i}) + f'(x_{2i})h + f''(x_{2i}) \frac{h^2}{2} + f'''(x_{2i}) \frac{h^3}{6}$$

$$+ f^{(4)}(x_{2i}) \frac{h^4}{24} + \cdots$$

$$f(x_{2i} + 2h) = f(x_{2i}) + f'(x_{2i})(2h) + f''(x_{2i}) \frac{(2h)^2}{2} + f'''(x_{2i}) \frac{(2h)^3}{6}$$

$$+ f^{(4)}(x_{2i}) \frac{(2h)^4}{24} + \cdots.$$

Substituting these into (8.2.31) and collecting terms, we get

$$E_2{}^i = -\frac{h^5}{90} f^{(4)} + \cdots.$$

Then

$$E_2 = \sum_{i=0}^{N/2-1} E_2{}^i \doteq -\frac{(x_N - x_0)h^4 f^{(4)}}{180}.$$

8.3   Gauss-Legendre Quadrature

The Gauss-Legendre method of numerical integration determines a şet of constants A_i such that the error

$$(8.3.1) \qquad E^+ = \int_a^b f^+(X) \, dX - \sum_{i=0}^{n} A_i{}^+ f^+(X_i)$$

is zero when $f^+(X)$ is a polynomial of max degree $2n + 1$, *provided* that the X_i are *properly selected*. As we will show, the X_i are translated and scaled values corresponding to the zeros x_i of the Legendre orthogonal polynomials $0_{n+1}(x)$. In this discussion, we will use Legendre polynomials with orthogonality range $(0, 1)$ and which have zeros x_i in interval $(0, 1)$.

To derive the Gauss-Legendre quadrature formulas we must first transform the range of integration in (8.3.1) from (a, b) to $(0, 1)$. This is accomplished by substituting variable x for X, where

$$(8.3.2) \qquad x = \frac{X - a}{b - a}, \quad \text{and} \quad dx = \frac{dX}{b - a}.$$

We can then write

$$(8.3.3) \qquad \int_a^b f^+(X) \, dX = (b - a) \int_0^1 f(x) \, dx$$

where

$$(8.3.4) \qquad \begin{aligned} X &= (b - a)x + a \\ f(x) &= f^+([b - a]x + a). \end{aligned}$$

Our problem then is to determine a set of A_i such that the error

$$(8.3.5) \qquad E = \int_0^1 f(x) \, dx - \sum_{i=0}^{n} A_i f(x_i), \quad \text{where} \quad A_i = \frac{A_i{}^+}{b - a}$$

is zero when $f(x)$ is a polynomial of max degree $2n + 1$, where the x_i are properly selected.

Let us assume that $f^+(X)$ is a polynomial of max degree $2n + 1$ in X. It follows then that $f(x)$ is a polynomial of max degree $2n + 1$ in x. In the following discussion, we will assume that $f(x)$ is an arbitrary polynomial $P_{2n+1}(x)$ of max degree $2n + 1$.

Suppose now that we select $n + 1$ distinct points (x_i, f_i) $(i = 0, n)$ on the graph of $f(x)$. These $n + 1$ points determine a unique polynomial $P_n(x)$ of max degree n, which can be written in Lagrangian form as

$$P_n(x) = \sum_{i=0}^{n} f_i \frac{(x - x_0)\ldots(x - x_{i-1})(x - x_{i+1})\ldots(x - x_n)}{(x_i - x_0)\ldots(x_i - x_{i-1})(x_i - x_{i+1})\ldots(x_i - x_n)}.$$

Now, polynomial $P_{2n+1}(x)$ coincides with $P_n(x)$ at the $n + 1$ points (x_i, f_i) $(i = 0, n)$, because $P_n(x_i) = f_i$ $(i = 0, n)$ by construction, and $P_{2n+1}(x)$ is the function $f(x)$.

It is a fact that polynomial $P_{2n+1}(x)$, which coincides with $P_n(x)$ at values $x_0, x_1, \ldots, x_n$, can be expressed in the form

(8.3.6) $\qquad P_{2n+1}(x) = P_n(x) + (x - x_0)(x - x_1)\ldots(x - x_n)Q_n(x)$

where $Q_n(x)$ is some polynomial of max degree n. Integrating (8.3.6) over $(0, 1)$ we obtain

(8.3.7) $\qquad \displaystyle\int_0^1 P_{2n+1}(x)\, dx = \int_0^1 P_n(x)\, dx + \int_0^1 (x - x_0)\ldots(x - x_n)Q_n(x)\, dx.$

If we can determine a set of values $x_0, x_1, \ldots, x_n$ such that

(8.3.8) $\qquad \displaystyle\int_0^1 (x - x_0)(x - x_1)\ldots(x - x_n)Q_n(x)\, dx = 0$

then equation (8.3.7) will reduce to

(8.3.9) $\qquad \displaystyle\int_0^1 P_{2n+1}(x)\, dx = \int_0^1 P_n(x)\, dx.$

And, from Sec. 8.1, we know that if $x_0, x_1, \ldots, x_n$ are distinct values of x, then a set of constants A_i can be uniquely determined such that

(8.3.10) $\qquad \displaystyle\int_0^1 P_n(x)\, dx = \sum_{i=0}^{n} A_i P_n(x_i).$

Equating (8.3.9) and (8.3.10) we see that

(8.3.11) $\qquad \displaystyle\int_0^1 P_{2n+1}(x)\, dx = \sum_{i=0}^{n} A_i P_n(x_i)$

provided that a set of distinct $x_0, x_1, \ldots, x_n$ can be determined such that (8.3.8) is satisfied. Further, since $P_{2n+1}(x_i) = P_n(x_i)$ $(i = 0, n)$, we can write

(8.3.12) $\qquad \displaystyle\int_0^1 P_{2n+1}(x)\, dx = \sum_{i=0}^{n} A_i P_{2n+1}(x_i).$

Therefore, for a properly selected set of distinct $x_0, x_1, \ldots, x_n$ the error

$$(8.3.13) \qquad E = \int_0^1 f(x)\, dx - \sum_{i=0}^n A_i f(x_i)$$

will be zero when $f(x)$ is a polynomial $P_{2n+1}(x)$ of max degree $2n + 1$. Our problem then is to determine a set of values $x_0, x_1, \ldots, x_n$ such that $\int_0^1 (x - x_0)(x - x_1)\ldots(x - x_n)Q_n(x)\, dx = 0$ for an arbitrary polynomial $Q_n(x)$. The zeros of Legendre polynomial $O_{n+1}(x)$ are such a set of values, as we will show in the following paragraphs.

Orthogonal Polynomials

If a set of polynomials $O_0(x), O_1(x), \ldots, O_n(x)$ satisfy the condition

$$(8.3.14) \qquad \int_\alpha^\beta O_i(x)O_k(x)\, dx = 0 \qquad (i \neq k)$$

then the set of polynomials $O_i(x)$, $i = 1, 2, \ldots$, constitute a set of simply orthogonal polynomials with range of orthogonality (α, β).

Lanczos [36, p. 400] shows that in general the orthogonality of a set of polynomials involves a weight factor $w(x)$ in the integrand, and that *only* in the special case of the Legendre polynomials is $w(x) = 1$, i.e., the case of the simply orthogonal polynomials defined in (8.3.14). The Legendre polynomials, which we denote by $O_i(x)$ in the remainder of this section, satisfy the orthogonality conditions

$$(8.3.15) \qquad \int_0^1 O_i(x)O_k(x)\, dx = 0, \quad i \neq k$$

and

$$(8.3.16) \qquad \int_0^1 O_{n+1}(x)Q_n(x)\, dx = 0$$

where $Q_n(x)$ is any arbitrary polynomial of max degree n. The interval $(0, 1)$ is the range of orthogonality of these "shifted" Legendre polynomials. (Some authors [e.g., Nielsen, 19, p. 127] use range $(-1, 1)$ or $(-1/2, 1/2)$ "unshifted" Legendre polynomials.)

The Legendre "shifted" polynomials, as derived in Milne [18, p. 257], are seen in Table 8-1. These Legendre polynomials are orthogonal on the interval $(0, 1)$, satisfying the relations

$$\int_0^1 O_m(x)O_n(x)\, dx = \begin{cases} 0 & , \text{ if } m \neq n \\ \dfrac{1}{2m + 1}, & \text{ if } m = n. \end{cases}$$

It should be noted that these are Legendre polynomials in the *continuous* variable x, and are not to be confused with the discrete-variable Legendre polynomials used in Chapter 7.

Table 8-1

n	$O_{n+1}(x)$
0	$O_1(x) = 1 - 2x$
1	$O_2(x) = 1 - 6x + 6x^2$
2	$O_3(x) = 1 - 12x + 30x^2 - 20x^3$
3	$O_4(x) = 1 - 20x + 90x^2 - 140x^3 + 70x^4$
4	$O_5(x) = 1 - 30x + 210x^2 - 560x^3 + 630x^4 - 252x^5$
$\vdots$	$\vdots$
$m - 1$	$O_m(x) = \sum_{k=0}^{m} (-1)^k \binom{m}{k} \binom{m+k}{k} x^k$

Now, if $x_0, x_1, \ldots, x_n$ are the zeros of Legendre polynomial $O_{n+1}(x)$ of degree $n + 1$, then we can express this polynomial in the form

(8.3.17) $O_{n+1}(x) = c(x - x_0)(x - x_1)\ldots(x - x_n).$

And, multiplying both sides of this equation by polynomial $Q_n(x)$, we get

(8.3.18) $O_{n+1}(x)Q_n(x) = c(x - x_0)(x - x_1)\ldots(x - x_n)Q_n(x).$

It then follows that

$$\int_0^1 O_{n+1}(x)Q_n(x)\, dx = c \int_0^1 (x - x_0)(x - x_1)\ldots(x - x_n)Q_n(x)\, dx.$$

(8.3.19)

Now, the integral on the left is zero by orthogonality property (8.3.16). Therefore, we obtain the desired result (8.3.8)

$$\int_0^1 (x - x_0)(x - x_1)\ldots(x - x_n)Q_n(x)\, dx = 0$$

which implies (8.3.12)

$$\int_0^1 P_{2n+1}(x)\, dx = \sum_{i=0}^{n} A_i P_{2n+1}(x_i)$$

by choosing $x_0, x_1, \ldots, x_n$ as the zeros of Legendre polynomial $O_{n+1}(x)$. That is, the equation

(8.3.20) $$\int_0^1 f(x)\, dx = \sum_{i=0}^{n} A_i f(x_i)$$

is exact when $f(x)$ is an arbitrary polynomial $P_{2n+1}(x)$ of max degree $2n + 1$.

Determination of the constants A_i. Since equation (8.3.20) is exact when $f(x)$ is an arbitrary polynomial $P_{2n+1}(x)$, it must be exact when $f(x) = 1$, $x, x^2, \ldots, x^n, \ldots, x^{2n+1}$. The $n + 1$ constants A_i can be determined by $n + 1$ independent equations linear in A_i. Such a set of equations can be obtained by successively substituting $1, x, x^2, \ldots, x^n$ for $f(x)$ in (8.3.20). The resulting set of equations is

$$\int_0^1 1 \, dx = A_0 \cdot 1 + A_1 \cdot 1 + \cdots + A_n \cdot 1, \quad \text{when } f(x) = 1$$

$$\int_0^1 x \, dx = A_0 x_0 + A_1 x_1 + \cdots + A_n x_n, \quad \text{when } f(x) = x$$

(8.3.21) $$\int_0^1 x^2 \, dx = A_0 x_0^2 + A_1 x_1^2 + \cdots + A_n x_n^2, \quad \text{when } f(x) = x^2$$

$$\cdot \quad \cdot \quad \cdot \quad \cdot \quad \cdot \quad \cdot \quad \cdot \quad \cdot \quad \cdot \quad \cdot \quad \cdot \quad \cdot \quad \cdot \quad \cdot$$

$$\int_0^1 x^n \, dx = A_0 x_0^n + A_1 x_1^n + \cdots + A_n x_n^n, \quad \text{when } f(x) = x^n$$

Evaluating the integrals $\int_0^1 x^k \, dx$, and writing the above system of linear equations in matrix form, we obtain

(8.3.22) $$\begin{bmatrix} 1 & 1 & 1 & \ldots & 1 \\ x_0 & x_1 & x_2 & \ldots & x_n \\ x_0^2 & x_1^2 & x_2^2 & \ldots & x_n^2 \\ \cdot & \cdot & \cdot & \cdot & \cdot \\ x_0^n & x_1^n & x_2^n & \ldots & x_n^n \end{bmatrix} \begin{bmatrix} A_0 \\ A_1 \\ A_2 \\ \cdot \\ A_n \end{bmatrix} = \begin{bmatrix} 1 \\ 1/2 \\ 1/3 \\ \cdot \\ 1/(n+1) \end{bmatrix}$$

This matrix equation has a unique solution $(A_0, A_1, A_2, \ldots, A_n)$ provided the $x_0, x_1, x_2, \ldots, x_n$ are distinct, since the determinant of coefficients is the well-known Vandermonde determinant which is non-zero for distinct x_i. This condition is satisfied because the zeros of the "shifted" Legendre polynomials are real, distinct, and lie in the interval $(0, 1)$. A proof of this fact is given in Milne [18, p. 262].

Alternate determination of the Constants A_i. The polynomial $P_n(x)$ determined by the $n + 1$ points (x_i, f_i) $(i = 0, n)$, where the x_i are the zeros of Legendre polynomial $0_{n+1}(x)$, can be expressed in Lagrangian form as

(8.2.23) $$P_n(x) = \sum_{i=0}^{n} f_i \frac{(x - x_0) \ldots (x - x_{i-1})(x - x_{i+1}) \ldots (x - x_n)}{(x_i - x_0) \ldots (x_i - x_{i-1})(x_i - x_{i+1}) \ldots (x_i - x_n)}$$

i.e., as

$$P_n(x) = \sum_{i=0}^{n} f_i \frac{L_i(x)}{L_i(x_i)}.$$

Integrating the foregoing expression, we find

$$\int_0^1 P_n(x)\,dx = \int_0^1 \left(\sum_{i=0}^n f_i \frac{L_i(x)}{L_i(x_i)} \right) dx$$

$$= \sum_{i=0}^n f_i \frac{1}{L_i(x_i)} \int_0^1 L_i(x)\,dx$$

and since

$$\int_0^1 P_n(x)\,dx = \sum_{i=0}^n A_i f_i$$

it follows that (equating the above two equations)

$$(8.3.24) \qquad A_i = \frac{1}{L_i(x_i)} \int_0^1 L_i(x)\,dx.$$

We see then that the constants A_i can be calculated either by equation (8.3.24) or by solving system (8.3.22). In practice, however, constants A_i are precalculated and tabulated together with the zeros $x_0, x_1, \ldots, x_n$ of the Legendre polynomials $0_{n+1}(x)$. Such a tabulation is given in Table 8-2 for $n = 0, 1, 2, 3, 4, 5$. (See Milne [18, p. 288].)

Table 8-2

n		$i = 0$	$i = 1$	$i = 2$	$i = 3$	$i = 4$	$i = 5$
0	x_i	0.5					
	A_i	1.0					
1	x_i	0.2113249	0.7886751				
	A_i	0.5	0.5				
2	x_i	0.1127017	0.5000000	0.8872983			
	A_i	0.2777778	0.4444444	0.2777778			
3	x_i	0.0694318	0.3300095	0.6699905	0.9305682		
	A_i	0.1739274	0.3260726	0.3260726	0.1739274		
4	x_i	0.0469101	0.2307653	0.5000000	0.7692347	0.9530899	
	A_i	0.1184634	0.2393144	0.2844444	0.2393144	0.1184634	
5	x_i	0.0337652	0.1693953	0.3806904	0.6193096	0.8306047	0.9662348
	A_i	0.0856622	0.1803808	0.2339570	0.2339570	0.1803808	0.0856622

Numerical Examples of Gauss-Legendre Quadrature

EXAMPLE 1. Evaluate

$$\int_1^2 \frac{dX}{X^2}$$

by Gauss-Legendre quadrature, using $n = 1$.

Solution:

$$\int_1^2 f^+(X)\,dX = (2-1)\int_0^1 f(x)\,dx \doteq (2-1)[A_0 f(x_0) + A_1 f(x_1)]$$

where

$$X = (b-a)x + a = (2-1)x + 1 = x + 1$$

$$f(x) = f^+([b-a]x + a) = \frac{1}{(x+1)^2}$$

$$x_0 = 0.2113249 \qquad x_1 = 0.7886751$$

$$A_0 = 0.5 \qquad A_1 = 0.5$$

$$f(x_0) = 0.681520171 \qquad f(x_1) = 0.312562649$$

$$\int_0^1 f(x)\,dx \doteq A_0 f(x_0) + A_1 f(x_1) = \underline{0.497041410}$$

$$\int_1^2 f^+(X)\,dX = (2-1)\int_0^1 f(x)\,dx = (2-1)0.497041410 = \underline{0.497041410.}$$

Note. Closed-form solution:

$$\int_1^2 \frac{dX}{X^2} = [-X^{-1}]_1^2 = 0.5000000.$$

EXAMPLE 2. Same problem, using $n = 2$.

$$x_0 = 0.1127017 \qquad x_1 = 0.5000000 \qquad x_2 = 0.8872983$$

$$A_0 = 0.2777778 \qquad A_1 = 0.4444444 \qquad A_2 = 0.2777778$$

$$f(x_0) = 0.807685891, \quad f(x_1) = 0.444444444, \quad f(x_2) = 0.280749442$$

$$\int_0^1 f(x)\,dx \doteq A_0 f(x_0) + A_1 f(x_1) + A_2 f(x_2) \qquad\qquad = 0.499874014$$

$$\int_1^2 f^+(X)\,dX = (2-1)\int_0^1 f(x)\,dx \doteq (2-1)0.499874014 = 0.499874014$$

The Gauss quadrature truncation error E^+ defined in (8.3.1) can be expressed as

$$E^+ = \frac{(b-a)^{2n+3}}{2n+3}\left[\frac{(n+1)!}{(n+2)(n+3)\ldots(2n+2)}\right]^2 \frac{f^{(2n+2)}(\bar{x})}{(2n+2)!}$$

for some $\bar{x}$ in the interval (a, b).

Other expressions for E^+ are given throughout the literature. (See, for example, Scarborough [25, p. 167] or Hildebrand [10, Chapter 7].) An excellent treatment of the Gauss-Legendre quadrature for interval $(-1, 1)$, as well as other types of the Gauss quadrature, is given in Chapter 7 of Hildebrand.

8.4   Romberg's Method of Numerical Integration

Romberg's method is a very efficient algorithm for the computation by digital computer of successive approximations of the definite integral $\int_a^b f(x) \, dx$. This method can be derived in the following two-step procedure.

1. Compute trapezoid sums for $\int_a^b f(x) \, dx$, using successively the intervals $h_0 = b - a$, $h_1 = h_0/2$, $h_2 = h_0/2^2$, ..., $h_k = h_0/2^k$.

2. Apply repeated (iterated) linear interpolation in variable h^2, starting with points (h_0^2, T_0^0), (h_1^2, T_0^1), (h_2^2, T_0^2), ..., (h_k^2, T_0^k), where the T_0^k are the trapezoid sum using interval h_k $(k = 0, 1, 2, \ldots)$.

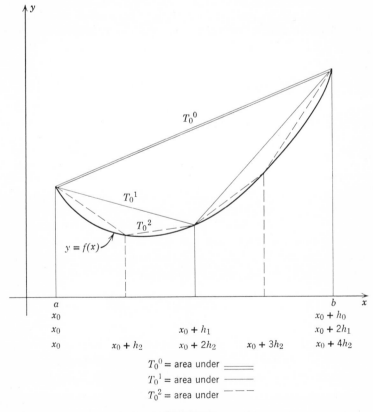

T_0^0 = area under _____
T_0^1 = area under ————
T_0^2 = area under – – –

Figure 8-4

Each of these steps will be explained in detail in the following subsection.

Computing Successive Approximations by Trapezoid Rule

Consider the following sequential method of computing successive trapezoid sums which approximate $\int_a^b f(x)\,dx$, as illustrated in Figure 8-4. First, compute the area $T_0{}^0$ of the single trapezoid for the interval $(a, b) \equiv (x_0, x_0 + h_0)$, that is, where $h_0 = b - a$:

$$T_0{}^0 = \frac{h_0}{2}\left[f(x_0) + f(x_0 + h_0)\right].$$

Next, halve the interval h_0, obtaining $h_1 = h_0/2$, and compute the sum $T_0{}^1$ of the areas of the *two* trapezoids for intervals $(x_0, x_0 + h_1)$ and $(x_0 + h_1, x_0 + 2h_1)$:

$$T_0{}^1 = \frac{h_1}{2}\left[f(x_0) + 2f(x_0 + h_1) + f(x_0 + 2h_1)\right].$$

Again, halve the intervals of length h_1, obtaining $h_2 = h_1/2 = h_0/2^2$, and compute the sum $T_0{}^2$ of the areas of the *four* trapezoids for intervals $(x_0, x_0 + h_2)$, $(x_0 + h_2, x_0 + 2h_2)$, $(x_0 + 2h_2, x_0 + 3h_2)$, $(x_0 + 3h_2, x_0 + 4h_2)$:

$$T_0{}^2 = \frac{h_2}{2}\left[f(x_0) + 2\sum_{j=1}^{2^2-1} f(x_0 + jh_2) + f(x_0 + 2^2 h_2)\right].$$

Continue this process of successive halving of the intervals, obtaining for the kth step of the procedure the sum $T_0{}^k$ of the areas of the 2^k trapezoids of width $h_k = h_0/2^k$

$$(8.4.1) \qquad T_0{}^k = \frac{h_k}{2}\left[f(x_0) + 2\sum_{j=1}^{2^k-1} f(x_0 + jh_k) + f(x_0 + 2^k h_k)\right]$$

until h_k is a "sufficiently small" subdivision of the original interval h_0.

The computational procedure just outlined generates a sequence of trapezoid sums $T_0{}^0, T_0{}^1, T_0{}^2, \ldots, T_0{}^k$ that converge to the value of the definite integral $I = \int_a^b f(x)\,dx$, i.e., such that

$$\lim_{k \to \infty} T_0{}^k = \int_{x_0}^{x_0 + h_0} f(x)\,dx$$

provided that $f(x)$ is analytic throughout $(x_0, x_0 + h_0)$.

This method of successive approximations by computing trapezoid sum with successive halving of the intervals is a practical method for the manual computation of the definite integral $\int_a^b f(x)\,dx$. However, as we shall see, this principle can be combined with the principle of repeated (iterated) linear interpolation (in variable h^2) to produce a very efficient digital-computer method of numerical integration, known as Romberg's method. To show this, we will now discuss the application of repeated linear interpolation to the points $(h_0{}^2, T_0{}^0), (h_1{}^2, T_0{}^1), (h_2{}^2, T_0{}^2), \ldots,$ $(h_k{}^2, T_0{}^k)$.

Repeated Linear Interpolation in Variable h^2

In Sec. 8.2 we showed that the error in numerical integration by the Trapezoid Rule was of the order h^2, where h was the interval size of each trapezoid in (a, b). Therefore, using interval size h_k, the Trapezoid Rule error E_k can be written

$$E_k = I - T_0{}^k \doteq \frac{-(b-a)}{12} f''(\bar{x})h_k{}^2, \quad \text{where} \quad \bar{x} \in (a, b).$$

Similarly, using interval size $h_{k+1} = h_k/2$, the error is

$$E_{k+1} = I - T_0^{k+1} \doteq \frac{-(b-a)}{12} f''(\bar{\bar{x}})h_{k+1}^2, \quad \text{where} \quad \bar{\bar{x}} \in (a, b).$$

Now, $f''(\bar{x})$ and $f''(\bar{\bar{x}})$ are averages of values of $f''(x)$ in the intervals of size h_k and h_{k+1}, respectively, and since each is the average of values over the entire interval (a, b), it would seem reasonable that $f''(\bar{x})$ and $f''(\bar{\bar{x}})$ are approximately equal.

If we assume then that $f''(\bar{x})$ and $f''(\bar{\bar{x}})$ are equal, then we can write the above two approximate relations in the form

$$T_0{}^k \doteq I + g(x)\, h_k{}^2$$

$$T_0^{k+1} \doteq I + g(x)h_{k+1}^2$$

where

$$g(x) = \frac{(b-a)}{12} f''(\bar{x}) = \frac{(b-a)}{12} f''(\bar{\bar{x}}).$$

We can solve for I in the two approximate relations above by first multiplying the first relation by h_{k+1}^2 and the second relation by $h_k{}^2$, obtaining

$$T_0{}^k h_{k+1}^2 \doteq I h_{k+1}^2 + g(x)h_k{}^2 h_{k+1}^2$$

$$T_0^{k+1} h_k{}^2 \doteq I h_k{}^2 + g(x)h_{k+1}^2 h_k{}^2$$

and then subtracting the first relation above from the second, and solving for I:

(8.4.2)
$$I \doteq \frac{T_0^{k+1} h_k{}^2 - T_0{}^k h_{k+1}^2}{h_k{}^2 - h_{k+1}^2}.$$

It should be noted that formula (8.4.2) is really just an *approximation* for $I = \int_a^b f(x)\, dx$, because the Trapezoid Rule error was obtained by neglecting terms of higher order in the Taylor-series expansion for the error, and further because we assumed $f''(\bar{x}) = f''(\bar{\bar{x}})$.

Let us now consider formula (8.4.2) from a geometric viewpoint. It should be recalled that in Chapter 6, equation (6.0.3), we showed that the equation of a line through points (x_k, y_k) and (x_{k+1}, y_{k+1}) can be written in the form

$$y(x) = \frac{y_k(x - x_{k+1}) - y_{k+1}(x - x_k)}{x_k - x_{k+1}}.$$

Now, if we extrapolate this line to $x = 0$, we obtain the y-axis intercept

(8.4.3)
$$y(0) = \frac{y_{k+1}x_k - y_k x_{k+1}}{x_k - x_{k+1}}.$$

If we consider $(h_k{}^2, T_0{}^k)$ and (h_{k+1}^2, T_0^{k+1}) as points on a graph (Figure 8-5) with h^2 as abscissa and T_0 as ordinate, then we see that (8.4.2) can be interpreted as the T_0-axis intercept of "line" (linear equation in h^2) through $(h_k{}^2, T_0{}^k)$ and (h_{k+1}^2, T_0^{k+1}) extrapolated to $h^2 = 0$. That is, equation (8.4.2) is of the same form as (8.4.3), with h^2 replacing x and T_0 replacing y.

Therefore, if we plot the trapezoid sums $T_0{}^k$ as functions of a variable h^2, we obtain new approximations to $I = \int_a^b f(x)\, dx$ by determining the

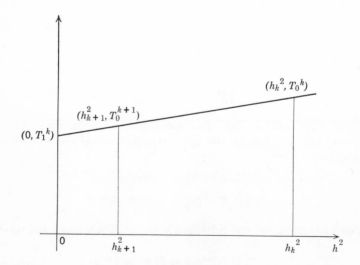

Figure 8-5

T_0-axis intercept of the "line" determined by points $(h_k{}^2, T_0{}^k)$ and (h_{k+1}^2, T_0^{k+1}). This approximation for each value of k is computed by formula (8.4.2). For notation purposes, we call this approximation $T_1{}^k$; i.e., formula (8.4.2) is written as

(8.4.4)
$$T_1{}^k = \frac{T_0^{k+1}h_k{}^2 - T_0{}^k h_{k+1}^2}{h_k{}^2 - h_{k+1}^2}.$$

Using the fact that $h_{k+1} = h_k/2$, we can reduce the equation above to the form

(8.4.5)
$$T_1{}^k = \frac{4T_0^{k+1} - T_0{}^k}{3}.$$

It can be shown that formula (8.4.5) is equivalent to Simpson's Rule of interval h_{k+1}. Simpson's Rule has an error of order h^4; i.e.,

$$E_k = I - T_1{}^k \doteq \frac{-(b-a)}{180} f^{(4)}(\bar{x}) h_{k+1}^4, \qquad \bar{x} \in (a, b).$$

and

$$E_{k+1} = I - T_1^{k+1} \doteq \frac{-(b-a)}{180} f^{(4)}(\bar{\bar{x}}) h_{k+2}^4, \qquad \bar{\bar{x}} \in (a, b).$$

These error equations for Simpson's Rule for h_{k+1} and h_{k+2} can be manipulated in a manner analogous to the manipulation of the Trapezoid Rule error equations that produced formula (8.4.2) for approximating I. That is, since $f^{(4)}(\bar{x})$ and $f^{(4)}(\bar{\bar{x}})$ are averages of values of $f^{(4)}(x)$ for h_{k+1} and h_{k+2}, we assume that

$$f^{(4)}(\bar{x}) = f^{(4)}\bar{\bar{x}} = g_1(x)\frac{180}{(b-a)}$$

and rewrite the error equations for Simpson's Rule in the form

$$T_1{}^k \doteq I + g_1(x) h_{k+1}^4$$

$$T_1^{k+1} \doteq I + g_1(x) h_{k+2}^4.$$

These two approximate relations can be solved for I by multiplying the first by h_{k+2}^4 and the second by h_{k+1}^4; i.e.,

$$T_1{}^k h_{k+2}^4 \doteq I h_{k+2}^4 + g_1(x) h_{k+1}^4 h_{k+2}^4$$

$$T_1^{k+1} h_{k+1}^4 \doteq I h_{k+1}^4 + g_1(x) h_{k+2}^4 h_{k+1}^4.$$

Subtracting the resulting first relation from the second, and solving for I, we obtain

(8.4.6)
$$I \doteq \frac{T_1^{k+1} h_{k+1}^4 - T_1{}^k h_{k+2}^4}{h_{k+1}^4 - h_{k+2}^4} = T_2{}^k.$$

Again, this is but an approximation of I which we denote by $T_2{}^k$. Using the fact that $h_{k+2} = h_{k+1}/2$, formula (8.4.6) can be reduced to the form

(8.4.7) $$T_2{}^k = \frac{16T_1^{k+1} - T_1{}^k}{15}$$

and it can be shown that this formula is equivalent to Newton-Cotes (closed) quadrature formula Q_{44} for interval h_{k+2}.

Formula (8.4.6) can be interpreted geometrically as the linear interpolation between two "Simpson points" $(h_{k+1}^4, T_1{}^k)$ and (h_{k+2}^4, T_1^{k+1}) *extrapolated* to $h^4 = 0$. And since the Simpson values $T_1{}^k$ were obtained by linear interpolation (extrapolated to $h^2 = 0$) between two "trapezoid points" $(h_k{}^2, T_0{}^k)$ and (h_{k+1}^2, T_0^{k+1}), we see that formula (8.4.6) can be obtained by *two repeated linear interpolations* (extrapolated to $h^2 = 0$) on the pairs of trapezoid points.

This process of repeated (iterated) linear interpolation can be directly extended to obtain approximations of I of increasing accuracy. To do this, recall that the formula (6.4.10*) for repeated linear interpolation on the set of pairs (x_k, y_k) is

$$y_m{}^k(x) = \frac{y_{m-1}^k(x)[x - x_{k+m}] - y_{m-1}^{k+1}(x)[x - x_k]}{x_k - x_{k+m}}$$

which, evaluated at $x = 0$, reduces to the form

(8.4.8) $$y_m{}^k(0) = \frac{y_{m-1}^{k+1}(0)x_k - y_{m-1}^k(0)x_{k+m}}{x_k - x_{k+m}}.$$

An analogous formula for repeated linear interpolation of the trapezoid points $(h_k{}^2, T_0{}^k)$ and (h_{k+1}^2, T_0^{k+1}) in variable h^2 and extrapolated to $h^2 = 0$, can be obtained from (8.4.8) by simply replacing x_k by $h_k{}^2$ and x_{k+m} by h_{k+m}^2, replacing $y_{m-1}^k(0)$ by T_{m-1}^k and $y_{m-1}^{k+1}(0)$ by T_{m-1}^{k+1}, and replacing $y_m{}^k(0)$ by $T_m{}^k$. The result of these substitutions is the formula

(8.4.9) $$T_m{}^k = \frac{T_{m-1}^{k+1}h_k{}^2 - T_{m-1}^k h_{k+m}^2}{h_k{}^2 - h_{k+m}^2}.$$

This formula is the basic recursion formula of the Romberg algorithm for numerical integration. Formula (8.4.9) can be reduced by using the relation $h_{k+m} = h_k/2^m$ to the form

(8.4.10) $$T_m{}^k = \frac{4^m T_{m-1}^{k+1} - T_{m-1}^k}{4^m - 1}.$$

The elements of Romberg's table are generated by this form of the basic recursion formula. The order of generation is explained in the following subsection, together with the summary explanation of Romberg's method, which uses in combination the Trapezoid Rule with successive halving of the interval *and* repeated linear interpolation, as explained in detail earlier.

Bauer et al.† prove that each column of the Romberg table converges to $\int_a^b f(x)\,dx$, assuming that no roundoff error occurs. That is, each of the sequences $\{T_m^0, T_m^1, \ldots, T_m^k, \ldots\}$ $(m = 0, 1, 2, \ldots)$ converges to $\int_a^b f(x)\,dx$ so that the truncation error can be made as small as desired by continuing the Romberg table until two successive elements in the same column differ by less than some positive ε.

It can also be shown that for some functions $f(x)$, the column of trapezoid sums $\{T_0^0, T_0^1, \ldots, T_0^k, \ldots\}$ converges faster to $\int_a^b f(x)\,dx$ than do succeeding columns in the Romberg table. For this reason, Bauer et al. recommend that the column of trapezoid sums through T_0^k, where h_k is sufficiently small, be computed first. If the sequence $\{T_0^k\}$ converges, it is unnecessary then to complete the Romberg table.

† *Proceedings of Symposia of Applied Mathematics*, Vol. XV, Amer. Math. Soc., 1963.

Computational Summary for Romberg Integration

Step 0. Define $f(x)$. Input and Initialization. Read $a, b =$ limit of integration; $\varepsilon =$ convergence term; $\delta =$ smallest subdivision of $b - a$ required. Set

$$x_0 = a$$
$$h_0 = b - a.$$

Step 1. Compute trapezoid sums for $h = h_0, h_0/2, h_0/2^2, \ldots$

(a) $\quad T_0{}^n = \dfrac{h}{2}\left[f(x_0) + 2 \displaystyle\sum_{j=1}^{2^n-1} f(x_0 + jh) + f(x_0 + 2^n h) \right], \quad$ for $h = h_0/2^n$.

Store trapezoid sums in one-dimensional array T

$$T_0{}^0 \;\rightarrow\; T_1$$
$$T_0{}^1 \;\rightarrow\; T_2$$
$$T_0{}^2 \;\rightarrow\; T_3$$

Continue computation of trapezoid sums until convergence or until subinterval size is "sufficiently" small.

(b) if $|T_{n+1} - T_n| \le \varepsilon$, go to Step 3; if $|T_{n+1} - T_n| > \varepsilon$, go to Step 1c.
(c) If current $h > \delta$, let $h = h/2$, return to Step 1a; if current $h \le \delta$, set *nmax* $= n$, go to Step 2.

Step 2. Generate elements $T_m{}^k$ $(m = 1, 2, \ldots, n)$ $(k = n - 1, n - 2, \ldots, 0)$ of Romberg table. Starting with iteration $n = 1$, compute elements in row n to the right of the vertical line, storing in one-dimensional array T, as shown at the right:

		$m = 1$	$m = 2$	$m = 3$			$m = 1$	$m = 2$	$m = 3$
	h_0 $T_0{}^0$					h_0 T_1			
$n = 1$	h_1 $T_0{}^1$	$T_1{}^0$			$n = 1$	h_1 T_2	T_1		
$n = 2$	h_2 $T_0{}^2$	$T_1{}^1$	$T_2{}^0$		$n = 2$	h_2 T_3	T_2	T_1	
$n = 3$	h_3 $T_0{}^3$	$T_1{}^2$	$T_2{}^1$	$T_3{}^0$	$n = 3$	h_3 T_4	T_3	T_2	T_1

In iteration n, compute all elements in row n *unless* convergence occurs before row n is completed. Elements are generated by recursion formula

$$T_K = [4^m T_{K+1} - T_K]/[4^m - 1], \quad (K = n, n - 1, \ldots, 1), \quad (m = 1, 2, \ldots).$$

Convergence is achieved when $|T_K - T_{K-1}| \le \varepsilon$, where T_K and T_{K-1} are elements in the same column m. When converged, go to Step 3. If n exceeds *nmax* before convergence, halve δ and return to Step 1a.

Step 3. Output T_{n+1} if the trapezoid sum converged. Output T_K if the Romberg table converged.

A representative flow chart of Romberg Integration is given in Figure 8-6.

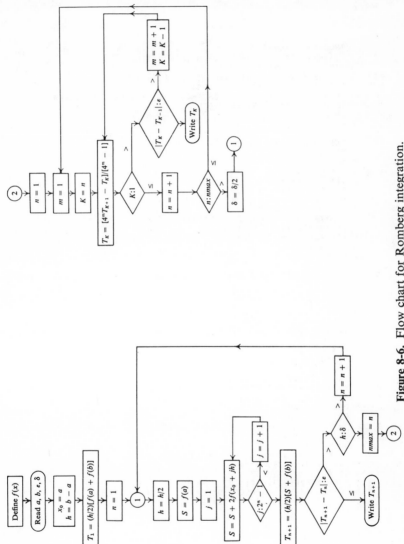

Figure 8-6. Flow chart for Romberg integration.

```
        PROGRAM ROMBERG
C       FORTRAN PROGRAM FOR ROMBERG INTEGRATION
        DIMENSION T(50)
C       DEFINE FUNCTION FCT (ARG) TO BE INTEGRATED
        FCT(ARG) = ARG**2 * EXP(ARG)
C       INPUT LIMITS A, B, CONVERGENCE EPS, SMALLEST SUBINTERVAL DEL
        READ (5,100) A, B, EPS, DEL
        X0 = A
        H = B - A
C       GENERATE TRAPEZOID SUMS (THRU STATEMENT 22)
        T(1) = (H/2.) * (FCT(A) + FCT(B))
        N = 1
      5 H = H/2.
        S = FCT(A)
        LIM = 2**N - 1
        DO 10 J = 1, LIM
        X = X0 + FLOAT(J) * H
     10 S = S + 2.0 * FCT(X)
        T(N+1) = (H/2.) * (S + FCT(B))
        IF(ABS(T(N+1) - T(N)) - EPS) 60, 60, 15
     15 IF(H-DEL) 25, 25, 20
     20 N = N +1
     22 GO TO 5
     25 NMAX = N
C       GENERATE ROMBERG ELEMENTS ON LINE N
        N = 1
     26 M = 1
        K = N
     28 T(K) = (4.**M * T(K+1) - T(K))/(4.**M - 1.)
        IF (K - 1) 40,40,30
     30 IF(ABS(T(K) - T(K-1)) - EPS) 50, 50, 35
     35 M = M + 1
        K = K - 1
        GO TO 28
     40 N = N + 1
        IF(N - NMAX) 26, 26, 45
C       ROMBERG HAS NOT CONVERGED FOR ORIGINAL DEL
C       HALVE DEL, AND REPEAT ENTIRE PROCESS
     45 DEL = DEL/2.
        GO TO 5
C       OUTPUT FOR CONVERGED ROMBERG ELEMENTS
     50 WRITE (6,110) T(K)
        GO TO 65
C       OUTPUT FOR CONVERGED TRAPEZOID SUM
     60 WRITE (6, 111) T(N+1)
     65 STOP
    100 FORMAT (2F15.7, 2F10.9)
    110 FORMAT (29H CONVERGED ROMBERG ELEMENT = , E15.7)
    111 FORMAT (27H CONVERGED TRAPEZOID SUM = , E15.7)
        END
```

Numerical Example of Romberg Integration

Use Romberg's method to evaluate the definite integral

$$\int_0^1 x^2 e^x \, dx$$

using $\varepsilon = 0.00005$, and $\delta = 0.125$.

First, tabulate values of the integrand at endpoints of subintervals of $(0, 1)$, as follows:

x	$f(x) = x^2 e^x$
0.000	0.0
0.125	0.017 705 445
0.250	0.080 251 589
0.375	0.204 608 168
0.500	0.412 180 318
0.625	0.729 783 577
0.750	1.190 812 510
0.875	1.836 638 897
1.000	2.718 281 828

Then compute the trapezoid sums using $h = 1, 0.5, 0.25, 0.125$, i.e., until $h \leq \delta = 0.125$. These trapezoid sums constitute the first column of Table 8-3. Then generate the elements of the Romberg table by repeated linear interpolation (extrapolated to $h^2 = 0$) on the trapezoid sum points.

Table 8-3

h^2	Trapezoid sums	Elements generated by repeated linear interpolation		
1.0^2	1.359 140 914			
0.5^2	0.885 660 616	0.727 833 850		
0.25^2	0.760 596 332	0.718 908 237	0.718 313 196	
0.125^2	0.728 890 177	0.718 321 458	0.718 282 339	0.718 281 849
		$\longrightarrow$		

The Romberg method converges at the point indicated by the arrow, i.e., when two successive elements in same column differ by less than ε.
Note. The closed-form solution of this problem is

$$\int_0^1 x^2 e^x \, dx = [e^x(x^2 - 2x + 2)]_0^1 = 0.718\ 281\ 828$$

Summary

The method of undetermined coefficients (for arbitrary x values) is *not* recommended as a method practical for numerical integration; it was introduced primarily to show that a definite integral can be accurately represented by a weighted sum of functional values of the form $\sum A_i f(x_i)$. However, in the event that the function to be integrated is represented *only* by a set of arbitrary (x_i, y_i) values, recourse must be made to numerical integration by this method.

The Newton-Cotes quadrature formulas (especially the Trapezoid Rule and Simpson's Rule) can be recommended for manual computation of definite integrals because of their simplicity and reasonable accuracy. The accuracy of these methods does not in general warrant their use for automatic computation. The Newton-Cotes formulas of higher degree are to be used with caution; Henrici [8] points out that there exist functions for which the sequences of the integrals of the interpolating polynomials do not converge toward the integral of the function.

The Gauss-Legendre quadrature formulas are difficult to apply in manual computations, primarily because the abscissas are usually irrational. However, for automatic computation these abscissas and the corresponding weights can be precomputed and stored, so that these formulas are well suited for automatic computation. In general, the Gauss formulas are considerably more accurate than the Newton-Cotes formulas. Other variations of Gauss quadrature, such as Gauss-Chebyshev, Gauss-Laguerre, and Gauss-Hermite, are also recommended for automatic computations. A number of SHARE routines (SDA 3292 through SDA 3303) exist for these different forms. The Gauss formulas do not suffer from the instability that characterizes the high-order Newton-Cotes formulas in some instances.

The Romberg method of numerical integration can be especially recommended for automatic computation because of the simplicity of its recursion formulas for programming, its computational efficiency, and its automatic accuracy determination (i.e., its ability to automatically determine the integration increment for a specified accuracy). For some functions, the trapezoid sums converge faster than other columns in the Romberg table (Bauer *et al.*), so the program has been written to test convergence of the trapezoid sums before completing the Romberg table.

COMPUTATIONAL EXERCISES

1. Use the method of undetermined coefficients to approximate

$$\int_0^1 (1 + x^3)\, dx$$

using $x_0 = 0.0$, $x_1 = 0.5$, $x_2 = 1.0$.

2. Approximate $\int_0^2 x^2 e^x \, dx$ using

(a) Trapezoid Rule with $h = 0.5$,
(b) Trapezoid Rule with $h = 0.25$,
(c) Trapezoid Rule with $h = 0.125$.

Compare your results with the closed-form solution obtained by integration by parts or by using a table of integrals.

3. Approximate $\int_1^4 \dfrac{dx}{x}$ by Gauss-Legendre integration, using $n = 2$.

4. Approximate $\int_0^1 \dfrac{dx}{1 + x}$ by Gauss-Legendre integration, using $n = 2$.

5. Approximate $\int_0^2 x^2 e^x \, dx$ by Romberg integration, using the trapezoid sums computed in exercise 2 for the first column of the Romberg table.

6. Approximate $\int_0^\pi \dfrac{\sin x}{x} \, dx$ by Romberg integration, using $\varepsilon = 5(10^{-5})$ for the convergence criterion.

7. Approximate $\int_0^1 \dfrac{dx}{1 + x^2}$ by Romberg integration, with $\varepsilon = 5(10^{-5})$.

PROGRAMMING EXERCISES

1. Write a FORTRAN program for the Romberg integration, using Henrici's recursion formulas [7, p. 260] to generate elements A_{mn} of the Romberg table. These formulas include a recursion formula for successive trapezoid sums and therefore reduce the amount of computations as compared with the program in this text.

2. Flow chart the SHARE routine SDA 3292 for six-point Gauss integration.

3. Convert Arden's MAD program [1, p. 196] to FORTRAN. Note that Arden uses the interval $(-1, 1)$ instead of $(0, 1)$ for the Gauss-Legendre quadrature.

4. Modify the Romberg integration program presented in this text to incorporate the improvement suggested in "Remark on Romberg Quadrature," *Communications of The ACM*, Vol. 8, No. 4 (Apr., 1965), p. 236, by A. M. Krasun and W. Prager.

RECOMMENDED READING

Henrici, P., *Elements of Numerical Analysis*, pp. 259–261 (on the Romberg integration), and pp. 255–258 (Trapezoidal Rule with end correction).

Arden, B. W., *Introduction to Digital Computing*, pp. 187–197 (Gaussian quadrature using interval $(-1, 1)$).

F. L. Bauer, H. Rutishauser, and E. Stiefel, "New Aspects in Numerical Integration," *Proceedings of Symposia of Applied Mathematics*, Vol. XV, Amer. Math. Soc., 1963.

Communications of the ACM, Vol. 8, No. 4 (Apr., 1965), pp. 236–237 ("Remark on Romberg Quadrature"), A. M. Krasun and W. Prager.

9

Solution of Ordinary
Differential Equations

9.0 Introduction

A linear first-order ordinary differential equation of the form†

(9.0.1) $$\frac{dy}{dx} = f(x, y)$$

can be used as the mathematical model for a variety of phenomena, either physical or non-physical, in any of a number of scientific and non-scientific disciplines. Examples of such phenomena include the following: heat-flow problems (thermodynamics), simple electrical circuits (electrical engineering), force problems (mechanics), rate of bacterial growth (biological science), rate of decomposition of radioactive material (atomic physics), crystallization rate of a chemical compound (chemistry), and rate of population growth (statistics). Powell and Wells [23] illustrate in lucid detail the formulation of the mathematical model and the *classical* solution for a number of problems of this type.

A typical elementary differential equations text presents several general classes of methods for solving a linear first-order differential equation. The principal classes of methods are (1) variables-separable, or reduction thereto; (2) exact equations, or reduction thereto; and (3) solution by infinite series.

† dy/dx is also denoted $y'(x)$ throughout this chapter.

In an introductory course in differential equations, the student is taught to apply the general method that appears best for the solution of the particular differential equation. For example, the linear first-order differential equation

$$(9.0.2) \qquad\qquad \frac{dy}{dx} = xy$$

can be easily solved by the variables-separable method. This is accomplished by rewriting the equation in the form

$$(9.0.3) \qquad\qquad \frac{dy}{y} = x \, dx$$

and integrating both sides to obtain

$$(9.0.4) \qquad\qquad \ln y = \frac{x^2}{2} + c$$

where c is an arbitrary constant of integration.

The general solution of equation (9.0.2) can then be written as

$$(9.0.5) \qquad\qquad y = c_1 e^{x^2/2}$$

where $c_1 = e^c$.

The *general solution* of such a linear first-order differential equation consists of a family of curves, called integral curves. The family of integral curves $y = c_1 e^{x^2/2}$ that constitute the solution of equation (9.0.2) is illustrated in Figure 9-1. For each value of c_1, a particular member of this family of curves is determined.

A *particular solution* of equation (9.0.2) can be determined if a condition on the solution curve is specified. For example, if we require that the particular solution curve pass through the point (0, 1), then we obtain the particular solution $y = e^{x^2/2}$.

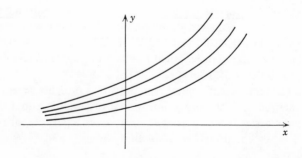

Figure 9-1

The determination of a particular solution $y(x)$ that satisfies the linear first-order ordinary differential equation

$$\frac{dy}{dx} = f(x, y)$$

as well as the condition that the integral curve $y(x)$ pass through the point (x_0, y_0) is commonly referred to as the *initial-value problem*.

In real-life problems, we encounter many differential equations that either cannot be solved by elementary classical methods or for which the evaluation of the analytic solution is quite difficult. In such instances, we must resort to *numerical methods* for obtaining one or more particular solutions of the initial-value problem. A number of representative methods for the numerical solution of the initial-value problem are developed in this chapter.

The initial-value problem can be restated in simple terms as follows:

Given a point (x_0, y_0) and the differential equation $y'(x) = f(x, y)$, *find* the (unknown) function $y(x)$ that passes through the point (x_0, y_0) and has the derivative $y'(x) = f(x, y)$.

The solution $y(x)$ is the integral curve that passes through the point (x_0, y_0); its derivative $y'(x) = f(x, y)$ can be interpreted geometrically as the slope of the integral curve at any point (x, y) on the curve. Let us assume (for the moment) that the graph of $y(x)$ is smooth, i.e., that the derivative $f(x, y)$ is changing slowly as we move along the integral curve.

If it is assumed that $f(x, y)$ is constant over small intervals (x_i, x_{i+1}), a simple numerical procedure can be derived for calculating numerical-solution values $y_1, y_2, y_3, \ldots, y_n$ that approximate the true-solution values $y(x_1), y(x_2), y(x_3), \ldots, y(x_n)$, respectively.

That is, if we replace the integrand in the relation †

$$y(x_{i+1}) = y(x_i) + \int_{x_i}^{x_{i+1}} f(x, y)\, dx$$

by the constant value $f(x_i, y_i)$, we obtain an algorithm of the form

$$y_{i+1} = y_i + f(x_i, y_i)(x_{i+1} - x_i), \quad \text{where} \quad y_0 = y(x_0)$$

for generating numerical-solution values y_i that approximate true-solution values $y(x_i)$, $(i = 0, 1, 2, \ldots)$.

† Where $f(x, y)$ denotes $y'(x)$.

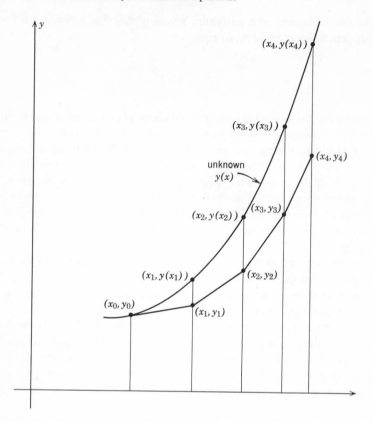

Figure 9-2. Solution of initial-value problem by Euler's algorithm.

The line segments (x_0, y_0), (x_1, y_1), (x_2, y_1), ..., (x_n, y_n) in general form a "broken" line which approximates the graph of the true solution $y(x)$. See Figure 9-2.

This procedure constitutes the simplest numerical method, known as Euler's algorithm (page 307), for solving the initial-value problem. However, it nicely illustrates the techniques of solving the initial-value problem by a step-by-step numerical method.

Numerical Methods for Solving the Initial-Value Problem

A numerical solution of the initial-value problem

$$\frac{dy}{dx} = f(x, y), \quad \text{where} \quad y(x_0) = y_0$$

consists of a number of values $y_1, y_2, y_3, \ldots, y_N$ that approximate, respectively, the values $y(x_1), y(x_2), y(x_3), \ldots, y(x_N)$ on the integral curve $y(x)$ which passes through point (x_0, y_0). The values y_i are referred to as numerical-solution values, while the values $y(x_i)$ are referred to as true-solution values. Usually, the numerical solution values are computed for evenly spaced discrete values of the abscissa; i.e., $x_1, x_2, x_3, \ldots, x_N$, where $x_{i+1} = x_i + h$. The magnitude of the abscissa increment h is called the step size of the numerical method.

Outline

   In this chapter numerical methods for solving the initial-value problem are organized into three categories, as follows:

   1. *One-Step Methods*—those algorithms that can compute y_{n+1} when only one point (x_n, y_n) and step size h are known. These methods are generally derived in either of two ways:

   (a) By using a Taylor-series expansion of the form

$$y(x_{n+1}) = y(x_n) + hy'(x_n) + \frac{h^2}{2} y''(x_n) + \cdots + \frac{h^p}{p!} y^{(p)}(x_n) + T_{n+1}$$

   (b) By using the definition of the definite integral

$$y(x_{n+1}) - y(x_n) = \int_{x_n}^{x_{n+1}} y'(x) \, dx$$

where $y'(x)$ is assumed to be constant or a linear function of x in the interval (x_n, x_{n+1}).

   2. *Multi-Step Methods*—those algorithms that require step size h and more than one point $(x_n, y_n), (x_{n-1}, y_{n-1}), \ldots$ in order to compute y_{n+1}. Hence, these methods must be provided with starting values $y_n, y_{n-1}, \cdots$ by a one-step method. Multi-step methods are generally derived by using the definition of the definite integral

$$y(x_{n+1}) - y(x_{n-k}) = \int_{x_{n-k}}^{x_{n+1}} y'(x) \, dx$$

where $y'(x)$ is approximated by an interpolating polynomial.

   3. *Iterated One-Step Method* with successive interval halving—Technically, this method can be considered as a one-step method since it requires only one point (x_n, y_n) and h for computing y_{n+1}.

   One-step methods and multi-step methods can be further categorized as either (1) *explicit* algorithms or (2) *implicit* algorithms.

   One-step *explicit* algorithms are of the form (in which y_{n+1} appears explicitly as a function of $x_n, y_n,$ and h):

$$y_{n+1} = y_n + hg(x_n, y_n, h).$$

One-step *implicit* algorithms are of the form

$$y_{n+1}^c = y_n + hg(x_n, y_n, x_{n+1}, y_{n+1}^p, h)$$

where the unknown appears in an implicit relation; i.e., on both sides of the equation. The y_{n+1}^p on the right denotes a *predicted* value of y_{n+1}, which is initially computed by an explicit formula. The y_{n+1}^c on the left denotes a *corrected* value of y_{n+1} computed by the implicit formula.

Multi-step *explicit* algorithms are derived from the relation

$$y(x_{n+1}) - y(x_{n-k}) = \int_{x_{n-k}}^{x_{n+1}} y'(x)\, dx$$

where $y'(x)$ is approximated by a polynomial defined by the interpolating points $x_n, x_{n-1}, \ldots$, (i.e., excluding x_{n+1}).

Multi-step *implicit* algorithms are derived from the relation

$$y(x_{n+1}) - y(x_{n-m}) = \int_{x_{n-m}}^{x_{n+1}} y'(x)\, dx$$

where $y'(x)$ is approximated by a polynomial defined by the interpolating points $x_{n+1}, x_n, x_{n-1}, \ldots$ (i.e., *including* x_{n+1}). Since (x_{n+1}, y_{n+1}) appears on the right side of this relation, the formula is implicit.

Algorithms that use both an explicit formula and an implicit formula are called predictor-corrector methods. The solution approximation computed by the explicit formula is denoted y_{n+1}^p and is called a predictor. This predictor is then used initially in the right side of an implicit formula that computes a corrector y_{n+1}^c. The implicit formula can be repeatedly applied, using the y_{n+1}^c from the preceding iteration in the right side and computing a new y_{n+1}^c on the left.

9.1   One-Step Methods

Given the differential equation

(9.1.1) $$y' = f(x, y)$$

and a numerical-solution point (x_n, y_n), we seek to calculate an approximation y_{n+1} to true-solution value $y(x_{n+1})$, where $x_{n+1} - x_n = h$.

The Taylor-series expansion for the solution function $y(x)$ can be expressed in powers of h in the form†

(9.1.2) $$y(x_{n+1}) = y(x_n) + hy'(x_n) + \frac{h^2}{2} y''(x_n) + \cdots + \frac{h^p}{p!} y^{(p)}(x_n) + T_{n+1}$$

† Where

$$T_{n+1} = \frac{h^{p+1} y^{(p+1)}(\bar{x}_n)}{(p+1)!}, \qquad x_n < \bar{x}_n < x_{n+1}$$

denotes the truncation error in the value of $y(x_{n+1})$ resulting from truncation of the Taylor series after term of order p.

or in the equivalent form

(9.1.3) $y(x_{n+1}) = y(x_n) + hf(x_n, y(x_n)) + \dfrac{h^2}{2} f'(x_n, y(x_n)) + \cdots$

$$+ \dfrac{h^p}{p!} f^{(p-1)}(x_n, y(x_n)) + T_{n+1}$$

obtained by substituting $f(x_n, y(x_n))$ and derivatives for $y'(x_n)$ and its derivatives.

It should be noted that since $y'(x) = f(x, y)$ is a function of both x and y, the derivatives of $f(x, y)$ are obtained by implicit differentiation, using the operator $D \equiv \partial/\partial x + \partial/\partial y \, dy/dx$. If we denote the partial derivatives with respect to x and y by the subscripts x and y, respectively, then we can write the derivatives as follows

(9.1.4)
$$\begin{aligned}
y' &= f \\
y'' &= f' = f_x + f_y f \\
y''' &= f'' = f_{xx} + 2ff_{xy} + f^2 f_{yy} + f_y[f_x + f_y f] \\
&\vdots \\
y^{(p)} &= f^{(p-1)}.
\end{aligned}$$

It is apparent that the computation of $y(x_{n+1})$ by direct use of Taylor series would be quite difficult for high-order p because of the involved form of the derivatives (9.1.4). As we will see later, there are methods that indirectly use the Taylor-series expansion without computing the higher-order derivatives of $f(x, y)$.

Euler's Method

The Taylor-series expansion of first degree is obtained by evaluating (9.1.3) for $p = 1$. The result is

(9.1.5) $y(x_{n+1}) = y(x_n) + hf(x_n, y(x_n)) + T_{n+1}.$

Euler's algorithm uses this linear Taylor-series form to calculate an approximation y_{n+1} by the formula

(9.1.6) $y_{n+1} = y_n + hf(x_n, y_n).$

Note that this is an *explicit formula* of the *one-step category*.

Euler's method is the simplest of all algorithms for solving ordinary differential equations. However, this simplicity allows us to explain properties characteristic of this and other methods of solving ordinary differential equations.

Given the differential equation $y'(x) = f(x, y)$ and the initial value $y(x_0) = y_0$, we can compute approximate solution values $y_1, y_2, y_3, \ldots,$

by using (9.1.6) with $n = 0, 1, 2, \ldots$, respectively. It is assumed that the initial-value point (x_0, y_0) lies on the graph of the solution curve $y(x)$ as illustrated in Figure 9-2.

Succeeding solution points $(x_1, y_1), (x_2, y_2), (x_3, y_3), \ldots$, however, will probably not lie on the solution curve because we have truncated the Taylor-series expansion of $y(x)$ after terms of degree one. As a result, a truncation error

$$T_{n+1} = -\frac{h^2}{2} y''(\bar{x}_n), \qquad (x_n \leqslant \bar{x}_n \leqslant x_{n+1})$$

occurs at each step of the solution. Furthermore, a roundoff error R_{n+1} occurs at each step. The magnitude of the roundoff error is dependent on the number of digits carried in the computation. And further, these errors are propagated from step to step in the solution.

Macon [14, pp. 131–132] shows that the total error ε_{n+1} between the computed solution y_{n+1} and the exact solution $y(x_{n+1})$ is the sum of the propagated error, the local truncation error, and the local roundoff error, and can be expressed in the form

(9.1.7) $\varepsilon_{n+1} = \varepsilon_n[1 + hf_y(x_n, \bar{y}_n)] + T_{n+1} + R_{n+1}$

for some $\bar{y}_n$ between y_n and $y(x_n)$.

Taylor's Algorithm of Order p

The following algorithm for computing an approximation y_{n+1} when only one point (x_n, y_n) and h are known

$$(9.1.8) \quad y_{n+1} = y_n + hf(x_n, y_n) + \frac{h^2}{2} f'(x_n, y_n) + \cdots + \frac{h^p}{p!} f^{(p-1)}(x_n, y_n)$$

is obtained by replacing $y(x_n)$ and $y(x_{n+1})$ by approximations y_n and y_{n+1} in the Taylor-series expansion (9.1.3), and neglecting the error term of order h^{p+1}. Algorithm (9.1.8) is therefore referred to as the Taylor algorithm of order p. The truncation error of this algorithm is

$$T_{n+1} = -\frac{h^{p+1}}{[p+1]!} y^{(p+1)}(\bar{x}_n), \qquad x_n \leqslant \bar{x}_n \leqslant x_{n+1}.$$

The computation of y_{n+1} by the Taylor algorithm for high-order p can be quite difficult because of the involved form of the derivatives $f', f'', \ldots,$ $f^{(p-1)}$, as indicated in (9.1.4). For this reason, methods of the so-called Runge-Kutta type have been devised, which *indirectly* use the Taylor algorithm.

In general, the Runge-Kutta methods evaluate $f(x, y)$ at more than one point in the neighborhood of (x_n, y_n) *instead* of evaluating the derivatives of $f(x, y)$, which would be required for direct use of the Taylor-series algorithm (9.1.8).

A Runge-Kutta method of order p is an algorithm that is equivalent to the Taylor algorithm of order p. The derivation of these methods is accomplished by assuming an algorithm of a particular form with certain *undetermined constants*. These constants are then solved for by equating this Runge-Kutta formula of order p to the Taylor algorithm of order p.

In the following subsection we will derive formulas for Runge-Kutta methods of order 2 and simply state the formulas for a Runge-Kutta method of order 4. The fourth-order Runge-Kutta methods are among the most widely used one-step methods for starting the solution of the initial value problem in ordinary differential equations. Predictor-corrector methods for continuing the solution can be combined with the Runge-Kutta starter to make a very efficient digital-computer method for solving the initial-value problem.

Second-Order Runge-Kutta Methods

The second-order (quadratic) Runge-Kutta methods are derived by assuming that y_{n+1} can be computed by a formula of the form

(9.1.9) $y_{n+1} = y_n + h[k_1 f(x_n, y_n) + k_2 f(x_n + \alpha h, y_n + \beta h f(x_n, y_n))]$

which is equivalent to a quadratic Taylor algorithm that we write in the form

(9.1.10) $$y_{n+1} = y_n + h\left[f(x_n, y_n) + \frac{h}{2} f'(x_n, y_n) \right].$$

The undetermined constants k_1, k_2, α, β are then solved for by equating (9.1.9) to (9.1.10). We find that

(9.1.11) $k_1 f(x_n, y_n) + k_2 f(x_n + \alpha h, y_n + \beta h f(x_n, y_n))$

$$= f(x_n, y_n) + \frac{h}{2} f'(x_n, y_n).$$

Expanding the second term on the left side in a linear Taylor expansion, we obtain

(9.1.12) $f(x_n + \alpha h, y_n + \beta h f(x_n, y_n))$

$$= f(x_n, y_n) + \alpha h f_x(x_n, y_n) + \beta h f(x_n, y_n) f_y(x_n, y_n).$$

Substituting (9.1.12) into (9.1.11), we obtain

$k_1 f(x_n, y_n) + k_2 [f(x_n, y_n) + \alpha h f_x(x_n, y_n) + \beta h f(x_n, y_n) f_y(x_n, y_n)]$

$$= f(x_n, y_n) + \frac{h}{2} [f_x(x_n, y_n) + f(x_n, y_n) f_y(x_n, y_n)]$$

where the new right side is obtained by replacing f' by $f_x + ff_y$.

Equating coefficients of like terms in the above expression, we get three equations in k_1, k_2, α, β, as follows:

$$k_1 + k_2 = 1$$

(9.1.13)
$$k_2\alpha = \frac{1}{2}$$

$$k_2\beta = \frac{1}{2}.$$

Hence, we have a family of solutions in terms of one of the parameters k_1, k_2, α, β. If we let $k_2 \neq 0$ be a free parameter, then we obtain the following solutions for k_1, α, β:

$$k_1 = 1 - k_2$$

(9.1.14)
$$\alpha = \frac{1}{2k_2}$$

$$\beta = \frac{1}{2k_2}.$$

As we will show in the next subsection, particular algorithms are obtained by selecting certain values for k_2.

Improved Euler Method

If we choose $k_2 = 1/2$, formulas (9.1.14) become $k_1 = 1/2, \alpha = 1, \beta = 1$. If these values are then substituted into the general quadratic Runge-Kutta formula (9.1.9), we obtain

(9.1.15) $$y_{n+1} = y_n + \frac{h}{2} [f(x_n, y_n) + f(x_n + h, y_n + hf(x_n, y_n))].$$

Denoting $x_n + h$ by x_{n+1} and $f(x_n, y_n)$ by f_n, we can write this formula as

(9.1.16) $$y_{n+1} = y_n + \frac{h}{2} [f_n + f(x_{n+1}, y_n + hf_n)].$$

This special case of the quadratic Runge-Kutta formula is usually referred to as the Improved Euler method. Although this method is only as accurate as a quadratic Taylor expansion, we can illustrate the step-by-step solution of the initial-value problem by setting up the computational summary and flow chart (Figure 9-3) for this simple method.

The Improved Euler Predictor-Corrector Method

A simple iterative method, called a predictor-corrector method can be derived using *both* the *explicit* Euler formula (9.1.6) and the *implicit*

Computational Summary	Flow Chart
	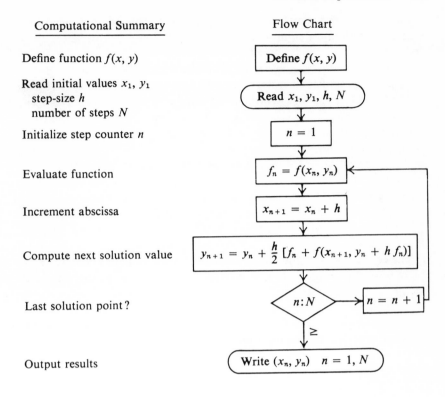
Define function $f(x, y)$	

Read initial values x_1, y_1
  step-size h
  number of steps N	

Initialize step counter n	
Evaluate function	
Increment abscissa	
Compute next solution value	
Last solution point?	
Output results	

Figure 9-3

† To eliminate zero subscripts we use (x_1, y_1) as the initial point instead of (x_0, y_0). This "slipping the index" will be done on each of the numerical methods for solving the initial-value problem so that the FORTRAN processor can handle the indices.

improved Euler formula (9.1.16). To show this, we write Euler's formula as

$$(9.1.17) \qquad y^p_{n+1} = y_n + h f_n$$

where y^p_{n+1} denotes a predicted value approximating $y(x_{n+1})$, computed by a *linear* Taylor algorithm. If we substitute (9.1.17) into (9.1.16) we obtain

$$(9.1.18) \qquad y^c_{n+1} = y_n + \frac{h}{2} [f_n + f(x_{n+1}, y^p_{n+1})]$$

where y^c_{n+1} denotes a corrected approximation of $y(x_{n+1})$ computed by a *quadratic* algorithm. Note that formula (9.1.18) is an implicit formula.

Given a solution point (x_n, y_n) and step size h, an iterative predictor-corrector method for computing numerical solution y_{n+1} can be set up as follows:

Step 1. Compute y_{n+1}^p by predictor formula (9.1.17).

Step 2. Compute correctors y_{n+1}^c by corrector formula (9.1.18) as follows: (a) For the first iteration, use value computed in Step 1 as term y_{n+1}^p in (9.1.18). (b) For succeeding iterations, use value y_{n+1}^c computed in preceding iteration as term y_{n+1}^p in (9.1.18). Usually, two or three iterations of the corrector formula will suffice.

When the corrector formula converges, step index n is incremented; i.e., n is replaced by $n + 1$, and the process of Steps 1 and 2 is repeated to compute the next numerical solution value. This procedure is continued until all solution values through y_N are computed.

Note that the predictor formula is used only *once* for each step, whereas the corrector is iterated until it converges for each step.

The convergence of this simple predictor-corrector method can be tested by comparing $|y_{n+1}^c - y_{n+1}^p|$ with some positive convergence term ε. If $|y_{n+1}^c - y_{n+1}^p| \leq \varepsilon$, we say that convergence has been achieved for step n of the solution; index n is then incremented (replace n by $n + 1$), and we proceed to calculate a solution approximation for the next x increment. If $|y_{n+1}^c - y_{n+1}^p| > \varepsilon$, we must replace y_{n+1}^p by y_{n+1}^c and repeat the entire process (for the same value of n), by computing new $f(x_{n+1}, y_{n+1}^p)$ and calculating new y_{n+1}^c by formula (9.1.18), until convergence for step n is achieved.

NUMERICAL EXAMPLE. Calculate approximate solution values y_1, y_2, y_3, y_4 for the initial-value problem $y' \equiv f(x, y) = x^2 y$, where $y(0) = 1$. Use Euler's improved predictor-corrector method, with $h = 0.1$, $\varepsilon = 0.000001$.

Solution. The approximate solution values are calculated by a single application of the predictor formula

$$y_{n+1}^p = y_n + hf(x_n, y_n)$$

followed by repeated application of the corrector formula

$$y_{n+1}^c = y_n + \frac{h}{2}(f_n + f_{n+1}^p)$$

until $|y_{n+1}^c - y_{n+1}^p| \leq \varepsilon$, with the old corrector becoming the new predictor (i.e., the corrector replaces the predictor in each successive iteration, for a given step n of the solution). The actual calculations are as in Table 9-1.

It would appear that the accuracy of this method for $h_4 = 0.1$ does not warrant carrying 9 significant places or use of such a small epsilon. The student should reduce h and note the improvement in accuracy.

Table 9-1†

n	x_n	$f(x_n, y_n)$	x_{n+1}	y_{n+1}^p	f_{n+1}^p	y_{n+1}^c
0	0.0	0.0	0.1	1.0	0.01	1.000500000
					0.010005000	1.000500250 $= y_1$
1	0.1	0.010005003	0.2	1.001500750	0.040060003	1.003003500
					0.040120140	1.003006507 $= y_2$
2	0.2	0.040120261	0.3	1.007126774	0.090641410	1.009544597
					0.090859014	1.009555477
					0.090859993	1.009555526 $= y_3$
3	0.3	0.090859997	0.4	1.019415523	0.163106484	1.022253850
					0.163560616	1.022276557
					0.163564249	1.022276738 $= y_4$

† Closed-form solution $y(x) = \exp(x^3/3)$ has the following values: $y(0.1) = 1.000333$, $y(0.2) = 1.0026701$, $y(0.3) = 1.0090406$, $y(0.4) = 1.0215825$.

A Fourth-Order Runge-Kutta Method

Fourth-order Runge-Kutta methods are widely used for starting the solution of the initial-value problem. Once a sufficient number of starting values is generated, the solution can be continued by more efficient and more accurate multi-step methods, described in Sec. 9.2.

Given a solution point (x_n, y_n) $n = 0, 1, 2, \ldots$, of the numerical solution of the initial-value problem

$$y' = f(x, y), \quad \text{with} \quad y(x_0) = y_0$$

a typical quartic Runge-Kutta method calculates solution value y_{n+1} using the formula

(9.1.19) $$y_{n+1} = y_n + \frac{k_1 + 2k_2 + 2k_3 + k_4}{6}$$

where

$$k_1 = hf(x_n, y_n)$$

$$k_2 = hf\left(x_n + \frac{h}{2}, y_n + \frac{k_1}{2}\right)$$

(9.1.20)

$$k_3 = hf\left(x_n + \frac{h}{2}, y_n + \frac{k_2}{2}\right)$$

$$k_4 = hf(x_n + h, y_n + k_3).$$

The quartic Runge-Kutta methods are based on a Taylor's expansion of the form of (9.1.2), truncated after terms of the fourth order. A derivation of this method is given in Ralston and Wilf [24, pp. 110–120].

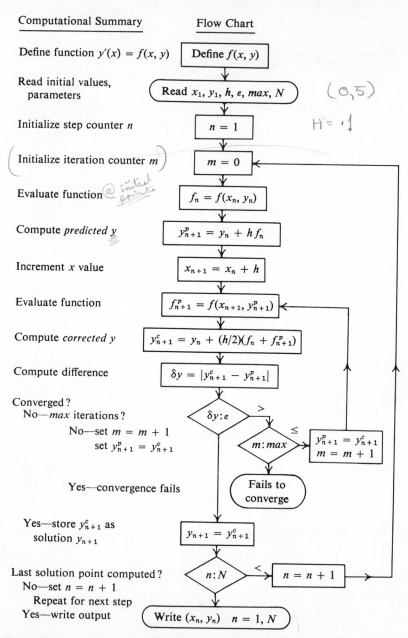

Computational Summary	Flow Chart

Define function $y'(x) = f(x, y)$ — Define $f(x, y)$

Read initial values, parameters — Read $x_1, y_1, h, \varepsilon, max, N$ $(0, 5)$

Initialize step counter n — $n = 1$ $H = .1$

Initialize iteration counter m — $m = 0$

Evaluate function @ initial point — $f_n = f(x_n, y_n)$

Compute *predicted y* — $y_{n+1}^p = y_n + h f_n$

Increment x value — $x_{n+1} = x_n + h$

Evaluate function — $f_{n+1}^p = f(x_{n+1}, y_{n+1}^p)$

Compute *corrected y* — $y_{n+1}^c = y_n + (h/2)(f_n + f_{n+1}^p)$

Compute difference — $\delta y = |y_{n+1}^c - y_{n+1}^p|$

Converged?
  No—*max* iterations?
    No—set $m = m + 1$
      set $y_{n+1}^p = y_{n+1}^c$ — $\delta y : \varepsilon$ $>$ $\quad \leq \quad$ $m : max$ — $y_{n+1}^p = y_{n+1}^c$ $m = m + 1$

    Yes—convergence fails — Fails to converge

  Yes—store y_{n+1}^c as solution y_{n+1} — $y_{n+1} = y_{n+1}^c$

Last solution point computed?
  No—set $n = n + 1$
    Repeat for next step — $n : N$ $<$ $n = n + 1$
  Yes—write output — Write (x_n, y_n) $n = 1, N$

Figure 9-4. Euler's improved method—predictor-corrector. Given $y'(x) = f(x, y)$, with initial value $y(x_1) = y_1$, the improved Euler predictor-corrector method computes succeeding numerical solution values $y_2, y_3, \ldots, y_N$ which are approximations of the true solution values $y(x_2), y(x_3), \ldots, y(x_N)$.

```
      PROGRAM EULIMP
C     FORTRAN PROGRAM - EULER IMPROVED METHOD PREDICTOR CORRECTOR
      DIMENSION X(100), Y(100)
C     DEFINE FUNCTION IN DIFFERENTIAL EQUATION
      FCT(ARG1,ARG2) = ARG1**2 * ARG2
C     INPUT INITIAL VALUES, STEP SIZE, CONVERGENCE TERM,
C     MAXIMUM NUMER OF ITERATIONS, NUMBER OF SOLUTION POINTS
      READ (5, 100) X (1), Y(1), H, EPS, MAX, NC
      DO 40 N = 1, NC
    5 M = 0
      FN = FCT (X(N), Y(N))
      YP = Y(N) + H * FN
      X(N+1) = X(N) + H
   10 FP = FCT (X(N+1),YP)
      YC = Y(N) + (H/2.) * (FN + FP)
      DELY = ABS (YC-YP)
      IF (DELY - EPS) 30, 30, 15
   15 IF(M-MAX) 20, 20, 98
   20 YP = YC
      M = M + 1
      GO TO 10
   30 Y(N+1) = YC
   40 CONTINUE
      WRITE (6,110) (X(N), Y(N), N = 1, NC)
   99 STOP
   98 NLAST = N
      WRITE (6,113) NLAST
      WRITE (6,110) (X(N), Y(N), N = 1, NLAST)
      GO TO 99
  100 FORMAT (2F15.7, 2F10.8, 2I2)
  110 FORMAT (1X, 2E18.8)
  113 FORMAT (29H FAILURE TO CONVERGE FOR N = , I2)
      END
```

To solve another first-order differential equation, simply redefine FCT (ARG1, ARG2), and input appropriate initial values and parameters.

The quartic Runge-Kutta method using formulas (9.1.19) and (9.1.20) is sometimes called the Kutta-Simpson method because it reduces to Simpson's Rule if $y'(x)$ is independent of y. That is, if $y'(x) = f(x)$, then

$$k_1 = hf(x_n)$$

$$k_2 = hf\left(x_n + \frac{h}{2}\right)$$

(9.1.21)

$$k_3 = hf\left(x_n + \frac{h}{2}\right)$$

$$k_4 = hf(x_n + h).$$

Substituting these values into (9.1.19), we find

$$(9.1.22) \quad y_{n+1} = y_n + \frac{h}{6}\left[f(x_n) + 4f\left(x_n + \frac{h}{2}\right) + f(x_n + h)\right].$$

Now, if we use step size $h^* = \frac{h}{2}$, equation (9.1.22) can be written

$$(9.1.23) \qquad y_{n+1} = y_n + \frac{h^*}{3}\left[f(x_n) + 4f(x_n + h^*) + f(x_n + 2h^*)\right]$$

which gives solution values $y_1, y_2, \ldots, y_n$ of the initial-value problem by numerical integration, using Simpson's Rule.

This quartic Runge-Kutta method could be used to compute all the desired numerical solution values $y_1, y_2, y_3, \ldots, y_N$, if the step size h is kept sufficiently small. The local truncation error of the quartic Runge-Kutta method is of the order of h^5 since it is equivalent to the Taylor algorithm of order 4. The principal advantage of this method is that it is self-starting (i.e., a one-step method). This fact, coupled with its reasonable accuracy, makes the method one of the most widely used for solving the initial-value problem. A disadvantage of the method is that it requires four evaluations of $f(x, y)$ for each step of the solution.

In Sec. 9.2 more accurate multi-step predictor-corrector methods are developed; such methods require several points $(x_n, y_n), (x_{n-1}, y_{n-1}), \ldots$ in order to compute y_{n+1}. The quartic Runge-Kutta method is often used to compute the starting values required for the more efficient multi-step methods. If the truncation error of the multi-step method is of order higher than h^5, the Runge-Kutta method can use a step size, which is some integer divisor of the step size used by the multi-step method.

An example of a quartic Runge-Kutta subroutine is given on p. 338.

The Propagation of Error in Representative One-Step Methods

The propagation of error in Euler's algorithm can be analyzed by determining the relation of the error at step $n + 1$ to the error at step n.

To do this, let Y_n denote the true solution values $y(x_n)$ of the initial-value problem $y'(x) = f(x, y)$, $y(x_0) = y_0$. Then the total solution error ε_n at step n is defined by

$$\varepsilon_n = Y_n - y_n.$$

The numerical-solution values computed by Euler's algorithm (9.1.6) satisfy the relation

$$y_{n+1} = y_n + hf(x_n, y_n) - R_{n+1}$$

where R_{n+1} denotes the roundoff error resulting from the evaluation of (9.1.6). Similarly, the true solution values satisfy the relation

$$Y_{n+1} = Y_n + hf(x_n, Y_n) + T_{n+1}$$

where T_{n+1} denotes the local truncation† error in Euler's algorithm. Subtracting these last two relations, we find that

$$Y_{n+1} - y_{n+1} = Y_n - y_n + h[f(x_n, Y_n) - f(x_n, y_n)] + T_{n+1} + R_{n+1}$$

i.e.,

$$\varepsilon_{n+1} = \varepsilon_n + h[f(x_n, Y_n) - f(x_n, y_n)] + E_{n+1}$$

where $E_{n+1} = T_{n+1} + R_{n+1}$. By the mean value theorem, this relation between successive errors can be written as

$$\varepsilon_{n+1} = \varepsilon_n + h[Y_n - y_n]f_y(x_n, \bar{y}_n) + E_{n+1}$$

i.e.,

$$\varepsilon_{n+1} = \varepsilon_n[1 + hf_y(x_n, \bar{y}_n)] + E_{n+1}$$

where f_y denotes $\partial f / \partial y$, and $\bar{y}_n$ lies between y_n and Y_n.

If $|f_y(x, y)| \leq C$ and $|E_{n+1}| \leq E$, where C and E are both positive constants, the error expression above can be replaced by the related first-order difference equation‡

$$e_{n+1} = e_n[1 + hC] + E.$$

---

† Actually T_{n+1} denotes the *rounded* truncation error of the algorithm in step $n + 1$. The true (exact) truncation error $\bar{T}_{n+1}$ is equal to the value of the quantity

$$Y_{n+1} - [Y_n + hf(x_n, Y_n)]$$

computed without roundoff error. Wherever T_{n+1} is used (in a computational sense) in the remainder of this chapter, it will denote the rounded truncation error (round-trunc error) of the particular algorithm under consideration.

‡ Let $F = F_n(y, z)$ be a sequence of functions defined on a set I of consecutive integers and for all y and z belonging to some set of real numbers S. If

$$x_n \in S, \quad x_{n-1} \in S, \quad \text{for all } n \in I$$

and

$$F_n(x_n, x_{n-1}) = 0$$

then a sequence $\{x_n\}$ satisfying these conditions is called a solution of the first-order difference equation $F_n(x_n, x_{n-1}) = 0$. [See Henrici (7, p. 46).]

Under the stated assumptions, it follows from (9.1.7) that

$$|\varepsilon_{n+1}| \leq |\varepsilon_n|[1 + hC] + E.$$

Now, if $|\varepsilon_n| \leq e_n$, it follows that

$$|\varepsilon_{n+1}| \leq e_n[1 + hC] + E = e_{n+1}$$

i.e.,

$$|\varepsilon_{n+1}| \leq e_{n+1}.$$

Therefore $|\varepsilon_0| \leq e_0 \Rightarrow |\varepsilon_1| \leq e_1 \Rightarrow |\varepsilon_2| \leq e_2 \Rightarrow \cdots$, (where the symbol $\Rightarrow$ means "implies that").

That is, the propagated error is bounded by the solution of the related first-order difference equation. Now, since $Y_0 - y_0 = 0$, the condition that $|\varepsilon_0| \leq e_0$ will be satisfied by setting $e_0 = 0$, which is the initial condition on the difference equation.

Defining $G = 1 + hC$, we can write the difference equation in the form

$$e_{n+1} = Ge_n + E, \quad \text{with initial value } e_0 = 0.$$

The solution of this first-order difference equation can be found by successive substitution as follows:

$$
\begin{aligned}
e_1 &= Ge_0 &&+ E = E \\
e_2 &= Ge_1 &&+ E = (G + 1)E \\
e_3 &= Ge_2 &&+ E = (G^2 + G + 1)E \\
&\ \vdots && \quad \vdots \\
e_n &= Ge_{n-1} &&+ E = (G^{n-1} + G^{n-2} + \cdots + G + 1)E \\
&\ \vdots && \quad \vdots
\end{aligned}
$$

The solution e_n of the difference equation is therefore

$$e_n = (G^{n-1} + G^{n-2} + \cdots + G + 1)E$$

which can be written, using the fact that the solution is a geometric series, as

$$e_n = E\left(\frac{G^n - 1}{G - 1}\right).$$

It follows then that the propagated error in Euler's algorithm is bounded by the expression

$$|\varepsilon_{n+1}| \leq E\left(\frac{(1 + hC)^{n+1} - 1}{hC}\right).$$

An excellent treatment of the propagation of error in Euler's algorithm, together with numerical examples, is given in Macon [14, p. 133].

Propagation of Error in Runge-Kutta Methods

Let Y_n denote the true-solution value $y(x_n)$ of the initial-value problem $y'(x) = f(x, y)$, $y(x_0) = y_0$, and let y_n denote the corresponding numerical solution calculated by a Runge-Kutta method of order p. Then the total solution error at step n, due to truncation in the integration formula, roundoff in the computations, and propagated error from preceding stages, is defined as

(9.1.24) $\varepsilon_n = Y_n - y_n.$

The propagation of error in the Runge-Kutta methods can be estimated by determining the relation between ε_{n+1} and ε_n. Let us examine the error propagation in a quadratic Runge-Kutta method, and then show how the results can be extended to methods of higher order.

The solution value generated by the quadratic Runge-Kutta method (9.1.9) satisfies the relation

$$(9.1.25) \quad y_{n+1} = y_n + hk_1 f(x_n, y_n) \\ + hk_2 f(x_n + \alpha h, y_n + \beta h f(x_n, y_n)) - R_{n+1}$$

where R_{n+1} denotes the roundoff error in evaluating the general quadratic Runge-Kutta formula.

Similarly, the true-solution values Y_n satisfy the relation

$$(9.1.26) \quad Y_{n+1} = Y_n + hk_1 f(x_n, Y_n) \\ + hk_2 f(x_n + \alpha h, Y_n + \beta h f(x_n, Y_n)) + T_{n+1}$$

where T_{n+1} denotes the local truncation error in the integration formula.

Subtracting (9.1.25) from (9.1.26), and using the error definition (9.1.24), we find that

$$\varepsilon_{n+1} = \varepsilon_n + hk_1[f(x_n, Y_n) - f(x_n, y_n)] \\ + hk_2[f(x_n + \alpha h, Y_n + \beta h f(x_n, Y_n)) \\ - f(x_n + \alpha h, y_n + \beta h f(x_n, y_n))] + E_{n+1}$$

where

$$E_{n+1} = R_{n+1} + T_{n+1}.$$

Using the mean value theorem, we can write the expression for ε_{n+1} as

$$\varepsilon_{n+1} = \varepsilon_n + hk_1[Y_n - y_n]\bar{f}_y \\ + hk_2[Y_n + \beta h f(x_n, Y_n) - y_n - \beta h f(x_n, y_n)]\bar{\bar{f}}_y + E_{n+1}$$

where†

$$\bar{f}_y = \frac{\partial f}{\partial y}(x_n, \bar{y}_n), \quad \text{and} \quad \bar{\bar{f}}_y = \frac{\partial f}{\partial y}(x_n + \alpha h, \bar{\bar{y}}_n).$$

† $\bar{y}_n$ lies between y_n and Y_n, and $\bar{\bar{y}}_n$ lies between $y_n + \beta h f(x_n, y_n)$ and $Y_n + \beta h f(x_n, Y_n)$.

Collecting terms, we obtain

$$\varepsilon_{n+1} = \varepsilon_n + hk_1[Y_n - y_n]\vec{f}_y + hk_2[Y_n - y_n]\vec{\bar{f}}_y$$
$$+ hk_2\beta h[f(x_n, Y_n) - f(x_n, y_n)]\vec{\bar{f}}_y + E_{n+1}.$$

A second application of the mean value theorem gives us

$$\varepsilon_{n+1} = \varepsilon_n + hk_1[Y_n - y_n]\vec{f}_y + hk_2[Y_n - y_n]\vec{\bar{f}}_y$$
$$+ h^2k_2\beta[Y_n - y_n]\vec{\bar{f}}_y\vec{f}_y + E_{n+1}.$$

Finally, substituting in the error definition, we obtain

$$(9.1.27) \quad \varepsilon_{n+1} = \varepsilon_n + hk_1\varepsilon_n\vec{f}_y + hk_2\varepsilon_n\vec{\bar{f}}_y + h^2k_2\beta\varepsilon_n\vec{\bar{f}}_y\vec{f}_y + E_{n+1}.$$

If $|f_y(x, y)| \le C$ in the neighborhood of values considered, and if $|E_{n+1}| \le E$, where C and E are constants, it follows that

$$(9.1.28) \quad |\varepsilon_{n+1}| \le |\varepsilon_n|[1 + hC(|k_1| + |k_2|) + h^2C^2|\beta k_2|] + E.$$

Under these assumptions, (9.1.28) can be replaced by the related first-order difference equation

$$(9.1.29) \quad e_{n+1} = e_n[1 + hC(|k_1| + |k_2|) + h^2C^2|\beta k_2|] + E.$$

Further, if $|\varepsilon_n| \le e_n$, it follows that

$$|\varepsilon_{n+1}| \le e_n[1 + hC(|k_1| + |k_2|) + h^2C^2|\beta k_2|] + E = e_{n+1}$$

that is,

$$|\varepsilon_{n+1}| \le e_{n+1}$$

so that

$$|\varepsilon_0| \le e_0 \Rightarrow |\varepsilon_1| \le e_1 \Rightarrow |\varepsilon_2| \le e_2 \Rightarrow \cdots$$

where the symbol $\Rightarrow$ means "implies that."

Let $G = 1 + hC(|k_1| + |k_2|) + h^2C^2|\beta k_2|$. Then the difference equation (9.1.29) can be written in the form

$$(9.1.30) \qquad\qquad e_{n+1} = Ge_n + E.$$

Now, since $Y_0 = y_0$, it follows that $\varepsilon_0 = 0$. Therefore we can let $e_0 = 0$, and the condition that $|\varepsilon_0| \le e_0$ is satisfied.

The solution of difference equation (9.1.30), satisfying the initial value $e_0 = 0$, can be obtained by successive substitution as follows:

$$e_1 = Ge_0 \quad + E = E$$
$$e_2 = Ge_1 \quad + E = GE + E$$
$$e_3 = Ge_2 \quad + E = G(GE + E) + E = G^2E + GE + E$$
$$e_4 = Ge_3 \quad + E = G(G^2E + GE + E) + E = G^3E + G^2E + GE + E$$
$$\vdots$$
$$e_n = Ge_{n-1} + E = G(G^{n-2}E + \cdots) + E = G^{n-1}E + \cdots + GE + E$$
$$(9.1.31)$$

The solution of the difference equation is then

$$e_n = E(G^{n-1} + G^{n-2} + \cdots + G + 1)$$

$$= E\left(\frac{G^n - 1}{G - 1}\right).$$

It follows then that

(9.1.32) $$|\varepsilon_{n+1}| \le E\left(\frac{G^{n+1} - 1}{G - 1}\right)$$

where $G = 1 + hC(|k_1| + |k_2|) + h^2 C^2 |\beta k_2|$.

EXAMPLE. For $k_1 = k_2 = 1/2$ and $\alpha = \beta = 1$, the quadratic Runge-Kutta formula (9.1.9) becomes the improved Euler formula (9.1.15), which can be written as

$$y_{n+1} = y_n + \frac{h}{2} f(x_n, y_n) + \frac{h}{2} f(x_n + h, y_n + h f(x_n, y_n)).$$

The total solution errors $\varepsilon_n = Y_n - y_n$ are related by the expression

$$\varepsilon_{n+1} = \varepsilon_n + \frac{h}{2} \varepsilon_n \bar{f}_y + \frac{h}{2} \varepsilon_n \bar{\bar{f}}_y + \frac{h^2}{2} \varepsilon_n \bar{f}_y \bar{\bar{f}}_y + E_{n+1}$$

which has the related difference equation

$$e_{n+1} = e_n \left[1 + hC + \frac{h^2 C^2}{2}\right] + E$$

provided that $|f_y(x, y)| \le C, |E_{n+1}| \le E$.

By defining $G = 1 + hC + h^2 C^2 / 2$, we can rewrite the difference equation as

$$e_{n+1} = G e_n + E.$$

If $e_0 = 0$, the condition that $|\varepsilon_0| \le e_0$ will be satisfied since $\varepsilon_0 \equiv Y_0 - y_0 = 0$. The solution of the difference equation having initial value $e_0 = 0$ is

$$e_n = E\left(\frac{G^n - 1}{G - 1}\right)$$

i.e.,

$$e_n = E\left(\frac{(1 + hC + h^2 C^2 / 2)^n - 1}{hC + h^2 C^2 / 2}\right)$$

and the total solution error at x_{n+1} is bounded by the expression

$$|\varepsilon_{n+1}| \le E\left(\frac{(1 + hC + h^2 C^2 / 2)^{n+1} - 1}{hC + h^2 C^2 / 2}\right).$$

9.2 Multi-Step Predictor-Corrector Methods

The numerical-solution values $y_1, y_2, y_3, \ldots, y_n, y_{n+1}, \ldots, y_N$ of the initial-value problem

$$(9.2.1) \qquad y'(x) = f(x, y), \qquad y(x_0) = y_0$$

could be computed by an appropriate *one-step* method such as the quartic Runge-Kutta method. However, once we have computed $y_1, y_2, \ldots, y_n$ by an appropriate one-step method, we can compute the remaining values $y_{n+1}, y_{n+2}, \ldots, y_N$ by multi-step methods that are in general more efficient than the one-step methods.

Assume then that for $x_1, x_2, \ldots, x_n$ that we have computed numerical-solution values $y_1, y_2, \ldots, y_n$ by some one-step method. We can then evaluate $f(x_i, y_i)$ $(i = 0, n)$ to obtain $f_0, f_1, f_2, \ldots, f_n$. From these values we can construct the finite-difference table of backward differences

$$(9.2.2)$$

$$
\begin{array}{lll}
x_0 & y_0 & f_0 \\
& & \qquad \nabla f_1 \\
x_1 & y_1 & f_1 \qquad\qquad \nabla^2 f_2 \quad . \\
& & \qquad \nabla f_2 \qquad\qquad\qquad . \\
x_2 & y_2 & f_2 \qquad\qquad \vdots \qquad\qquad . \quad \nabla^n f_n \\
\vdots & \vdots & \vdots \qquad\qquad\qquad\qquad\qquad . \\
& & \qquad\qquad\qquad \nabla^2 f_n \quad . \\
& & \qquad \nabla f_n \\
x_n & y_n & f_n
\end{array}
$$

which will be used in the construction of a multi-step method for continuing the solution, i.e., to compute y_{n+1} using preceding values.

A multi-step method can be based on the relation

$$(9.2.3) \qquad y(x_{n+1}) - y(x_{n-M}) = \int_{x_{n-M}}^{x_{n+1}} y'(x)\, dx$$

where M is some non-negative integer.

Now, since we have values of $f_n, f_{n-1}, \ldots$, we can approximate $y' \equiv f(x, y)$ by a polynomial $P(x)$ using interpolating points $x_n, x_{n-1}, \ldots$. Substituting this interpolating polynomial $P(x)$ for $y'(x)$ as the integrand in (9.2.3), we obtain the relation

$$(9.2.4) \qquad y_{n+1} = y_{n-M} + \int_{x_{n-M}}^{x_{n+1}} P(x)\, dx$$

where y_{n-M} is an approximation of $y(x_{n-M})$ and y_{n+1} is an approximation of $y(x_{n+1})$. Note that (9.2.4) is an *explicit* relation because we have excluded x_{n+1} as an interpolating point in defining $P(x)$.

It should be noted that the range of integration (x_{n-M}, x_{n+1}) extends past the last interpolating point x_n. Hence, we must extrapolate $P(x)$ through x_{n+1} in order to integrate (9.2.4). For this reason, formula (9.2.4) is not as exact as it could be, so we call the y_{n+1} computed thereby a predicted value of $y(x_{n+1})$ and write the relation as

$$(9.2.5) \qquad y_{n+1}^p = y_{n-M} + \int_{x_{n-M}}^{x_{n+1}} P(x)\, dx.$$

Once we have computed the predicted value y_{n+1}^p by this formula (used once per step), we can evaluate $f(x_{n+1}, y_{n+1}^p)$ which we denote simply by f_{n+1}. We can then extend finite-difference table (9.2.2) by one line, obtaining

$$(9.2.6) \quad
\begin{array}{ccccccc}
x_0 & y_0 & f_0 & & & & \\
 & & & \nabla f_1 & & & \\
x_1 & y_1 & f_1 & & \nabla^2 f_2 & \cdot & \\
 & & & \nabla f_2 & & \cdot & \nabla^n f_n \\
x_2 & y_2 & f_2 & \vdots & \vdots & \cdot & \\
\vdots & \vdots & \vdots & & \nabla^2 f_n & \cdot & \nabla^n f_{n+1} \\
\vdots & \vdots & \vdots & \nabla f_n & & \cdot & \\
x_n & y_n & f_n & & \nabla^2 f_{n+1} & & \\
 & & & \nabla f_{n+1} & & & \\
x_{n+1} & y_{n+1}^p & f_{n+1} & & & &
\end{array}$$

Interpolating points $x_{n+1}, x_n, x_{n-1}, \ldots$ can now be used to define a new interpolating polynomial $P^*(x)$ whose range of definition *includes* x_{n+1} (hence resulting in an *implicit* formula for computing y_{n+1}).

A corrector formula can be obtained from the relation

$$(9.2.7) \qquad y(x_{n+1}) - y(x_{n-m}) = \int_{x_{n-m}}^{x_{n+1}} y'(x)\, dx$$

where m is some non-negative integer that may or may not be the same as the integer M used in the predictor formula. (See Milne's method [18].) It should also be noted that the number of interpolating points that define $P^*(x)$ need not be the same as used to define $P(x)$, but is chosen such that the truncation error of (9.2.8) is of the same order of magnitude as that of (9.2.5).

Substituting $P^*(x)$ for $y'(x)$ as the integrand in (9.2.7), we obtain the corrector formula (since we are interpolating thru x_{n+1})

$$(9.2.8) \qquad y_{n+1}^c = y_{n-m} + \int_{x_{n-m}}^{x_{n+1}} P^*(x)\, dx$$

which is iterated until convergence each step. Explicit formula (9.2.5) is used as the predictor, and implicit formula (9.2.8) is used as the corrector formula. We will illustrate the iterative properties of multi-step predictor-correctors in the subsequent discussion.

Assume that starting values $y_1, y_2, \ldots, y_n$ have been computed by an appropriate one-step method such as the 4th-order Runge-Kutta method. Evaluate $f(x_i, y_i)$, denoted by f_i, for $i = 0, n - 1$. Then *each* of the numerical-solution values $y_{n+1}, y_{n+2}, \ldots, y_N$ is computed by the following iterative procedure.

Step 1. Set the iteration index $i = 0$. Evaluate $f(x_n, y_n)$, which is denoted by f_n. Compute the predicted value $y_{n+1}^0 \equiv y_{n+1}^p$ using either the difference- or ordinate-form equivalent of (9.2.5).†

Step 2. Evaluate $f(x_{n+1}, y_{n+1}^i)$ which is denoted by f_{n+1}.

Step 3. Compute corrected value $y_{n+1}^{i+1} \equiv y_{n+1}^c$ using either the difference- or ordinate-form equivalent of (9.2.8).

Step 4. Test convergence of iteration of corrector.

(a) If $|y_{n+1}^{i+1} - y_{n+1}^i| > \varepsilon$, $i + 1 \to i$,   and return to Step 2.

(b) If $|y_{n+1}^{i+1} - y_{n+1}^i| \leq \varepsilon$, store y_{n+1}^{i+1} as solution value y_{n+1}; go to Step 5.

Step 5. Have all solution values been computed?

(a) If current $x_{n+1} < x_N$, replace n by $n + 1$, and return to Step 1.

(b) If current $x_{n+1} = x_N$, we have completed task of computing solution values $y_{n+1}, y_{n+2}, \ldots, y_N$.

Now that the multi-step predictor-corrector process has been outlined in somewhat general terms, we should answer the following questions to fully understand the efficiency, accuracy, and convergence properties of predictor-corrector methods.

1. What are the difference and ordinate forms of the predictor and corrector formulas?

2. Under what conditions will the iteration of the corrector converge?

3. How do we determine the step-by-step accuracy of numerical integration formulas; i.e., what is the local truncation error?

4. How is the total solution error propagated in a multi-step method? Each of these questions is discussed in detail in the next subsection.

Evaluation of the Definite Integrals in Predictor-Corrector Formulas

The definite integrals

$$\int_{x_n-M}^{x_{n+1}} P(x)\,dx \quad \text{and} \quad \int_{x_n-m}^{x_{n+1}} P^*(x)\,dx$$

---

† Difference-form equivalents of the predictor and corrector formulas are given in (9.2.10) and (9.2.12), respectively. Ordinate-form equivalents of the predictor and corrector formulas are given in (9.2.11) and (9.2.13), respectively.

which appear in the predictor and corrector expressions, can be expressed in either *difference* form or *ordinate* form. The ordinate form is more efficient because it does not require the computation of the differences $\nabla^j f_n$ for each step (predictor) or the differences $\nabla^j f_{n+1}$ for each iteration (of corrector) within each step. However, the evaluation of the integrals is simpler to explain in terms of difference formulas; the resulting predictor and corrector formulas can then in turn be expressed in terms of ordinates.

Since we are working at the end of the difference tables, Newton's backward-form (NBF) interpolating polynomials seem to be the most suitable representations for $P(x)$ and $P^*(x)$.

The *difference* forms of the predictor formula are readily obtained by using the kth-degree $(k = 1, 2, \ldots)$ NBF interpolating polynomial defined over (x_{n-k}, x_n) for $P(x)$ in formula (9.2.5) which can then be written in the form

$$(9.2.9) \quad y_{n+1}^p = y_{n-M} + \int_{x_{n-M}}^{x_{n+1}} \left[f_n + \frac{\nabla f_n}{h}(x - x_n) + \frac{\nabla^2 f_n}{2h^2}(x - x_n) \right.$$

$$\left. \times (x - x_{n-1}) + \cdots + \frac{\nabla^k f_n}{k! h^k}(x - x_n)\ldots(x - x_{n-k+1}) \right] dx$$

The integration can be simplified by introducing the variable $u = (x - x_n)/h$, where also $du = dx/h$. Formula (9.2.9) then becomes

$$y_{n+1}^p = y_{n-M} + \int_{-M}^{1} \left[f_n + \nabla f_n u + \frac{\nabla^2 f_n}{2} u(u + 1) \right.$$

$$\left. + \cdots + \frac{\nabla^k f_n}{k!} u(u + 1)\ldots(u + k - 1) \right] h \, du.$$

When this integration is performed, the predictor formula can be expressed in the form

$$(9.2.10) \quad y_{n+1}^p = y_{n-M} + h[\bar{p}_0 f_n + \bar{p}_1 \nabla f_n + \bar{p}_2 \nabla^2 f_n + \cdots + \bar{p}_k \nabla^k f_n]$$

where

$$\bar{p}_j = \frac{1}{j!} \int_{-M}^{1} u(u + 1)\ldots(u + j - 1) \, du \qquad (j = 1, 2, \ldots, k).$$

The ordinate form of the predictor can then be obtained by substituting the ordinate-difference relations (see Sec. 6.3)

$$\nabla^j f_n = f_n - \binom{j}{1} f_{n-1} + \binom{j}{2} f_{n-2} - \cdots + (-1)^j f_{n-j} \qquad (j = 1, k)$$

into the difference-form predictor expression. The resulting ordinate-form predictor can be expressed in the form

$$(9.2.11) \quad y_{n+1}^p = y_{n-M} + h[p_0 f_n + p_1 f_{n-1} + p_2 f_{n-2} + \cdots + p_k f_{n-k}].$$

EXAMPLE. For the case $M = 3$ and $k = 2$,

$$y_{n+1}^p = y_{n-3} + \int_{x_{n-3}}^{x_{n+1}} \left[f_n + \frac{\nabla f_n}{h}(x - x_n) + \frac{\nabla^2 f_n}{2h^2}(x - x_n)(x - x_{n-1}) \right] dx$$

which, after substitution of $u = (x - x_n)/h$, takes the form

$$y_{n+1}^p = y_{n-3} + \int_{-3}^{1} \left[f_n + \nabla f_n u + \frac{\nabla^2 f_n}{2} u(u + 1) \right] h \, du.$$

Integrating, we obtain the difference form of the predictor

$$y_{n+1}^p = y_{n-3} + h \left[4f_n - 4\nabla f_n + \frac{8}{3} \nabla^2 f_n \right]$$

and the corresponding ordinate form of the predictor, obtained by substituting in the ordinate-difference relations, is

$$y_{n+1}^p = y_{n-3} + \frac{4h}{3} [2f_n - f_{n-1} + 2f_{n-2}].$$

This predictor formula is known as Milne's predictor.

The difference form of the corrector can be obtained, in a manner analogous to that used to find the predictor, using a kth-degree ($k = 1, 2, \ldots$) NBF interpolating polynomial defined over (x_{n-k+1}, x_{n+1}) for $P^*(x)$ in formula (9.2.8):

$$y_{n+1}^c = y_{n-m} + \int_{x_{n-m}}^{x_{n+1}} \left[f_{n+1} + \frac{\nabla f_{n+1}}{h}(x - x_{n+1}) \right.$$
$$+ \frac{\nabla^2 f_{n+1}}{2h^2}(x - x_{n+1})(x - x_n) + \cdots$$
$$\left. + \frac{\nabla^k f_{n+1}}{k! h^k}(x - x_{n+1}) \ldots (x - x_{n-k+2}) \right] dx$$

which, after substitution of the variable $u = (x - x_{n+1})/h$, becomes

$$y_{n+1}^c = y_{n-m} + \int_{-(m+1)}^{0} \left[f_{n+1} + \nabla f_{n+1} u + \frac{\nabla^2 f_{n+1}}{2} u(u + 1) \right.$$
$$\left. + \cdots + \frac{\nabla^k f_{n+1}}{k!} u(u + 1) \ldots (u + k - 1) \right] h \, du.$$

The difference form of the corrector can then be written in the form

$$(9.2.12) \quad y_{n+1}^c = y_{n-m} + h[\bar{c}_{-1} f_{n+1} + \bar{c}_0 \nabla f_{n+1} + \bar{c}_1 \nabla^2 f_{n+1}$$
$$+ \cdots + \bar{c}_{k-1} \nabla^k f_{n+1}]$$

and the corresponding ordinate form of the corrector can be expressed as

$$(9.2.13) \quad y_{n+1}^c = y_{n-m} + h[c_{-1} f_{n+1} + c_0 f_n + c_1 f_{n-1}$$
$$+ \cdots + c_{k-1} f_{n-k+1}].$$

EXAMPLE. For the case $m = 1$ and $k = 2$,

$$y_{n+1}^c = y_{n-1} + \int_{x_{n-1}}^{x_{n+1}} \left[f_{n+1} + \frac{\nabla f_{n+1}}{h} (x - x_{n+1}) \right.$$

$$\left. + \frac{\nabla^2 f_{n+1}}{2h^2} (x - x_{n+1})(x - x_n) \right] dx$$

$$= y_{n-1} + \int_{-2}^{0} \left[f_{n+1} + \nabla f_{n+1} u + \frac{\nabla^2 f_{n+1}}{2} u(u + 1) \right] h \, du.$$

The difference form of the corrector is

$$y_{n+1}^c = y_{n-1} + h \left[2f_{n+1} - 2\nabla f_{n+1} + \frac{1}{3} \nabla^2 f_{n+1} \right]$$

and the corresponding ordinate form of the corrector is

$$y_{n+1}^c = y_{n-1} + \frac{h}{3} [f_{n+1} + 4f_n + f_{n-1}].$$

This is the corrector formula for Milne's predictor-corrector method. Note also that the corrector has the same form as Simpson's Rule (see Sec. 8.2) for numerically integrating

$$\int_{x_{n-1}}^{x_{n+1}} f(x) \, dx.$$

Convergence of the Corrector Formula

It has thus far been tacitly assumed that the corrector formula will converge. An analysis of the convergence properties of correctors reveals that this convergence depends on the step size h. A proper selection of h must therefore be made to ensure the convergence of the corrector.

Let y_{n+1}^i denote the ith approximation of the solution value corresponding to x_{n+1}. Then the corrector formula (9.2.13) can be written in the form

$$(9.2.14) \qquad y_{n+1}^{i+1} = y_{n-m} + h c_{-1} f(x_{n+1}, y_{n+1}^i) + h \sum_{j=0}^{k-1} c_j f_{n-j},$$

where i is the iteration index. Note that $y_{n+1}^0 \equiv y_{n+1}^p$, and y_{n+1}^{i+1} is the corrector of iteration i, y_{n+1}^i is the corrector of iteration $i - 1$.

If the sequence of successive approximations y_{n+1}^{i+1}, generated by iterating the corrector, *converges to a limit*, denoted y_{n+1}, then y_{n+1} satisfies the relation

$$(9.2.15) \qquad y_{n+1} = y_{n-m} + h c_{-1} f(x_{n+1}, y_{n+1}) + h \sum_{j=0}^{k-1} c_j f_{n-j}.$$

Subtracting (9.2.15) from (9.2.14), we obtain

$$y_{n+1}^{i+1} - y_{n+1} = hc_{-1}[f(x_{n+1}, y_{n+1}^i) - f(x_{n+1}, y_{n+1})]$$

which, by the mean value theorem, can be written as

$$y_{n+1}^{i+1} - y_{n+1} = hc_{-1}[y_{n+1}^i - y_{n+1}]f_y(x_{n+1}, \bar{y}_{n+1})$$

where $\bar{y}_{n+1}$ lies between y_{n+1} and y_{n+1}^i.

If $|f_y| \le C$, where C is some positive constant, in the neighborhood of (x_{n+1}, y_{n+1}), then

$$|y_{n+1}^{i+1} - y_{n+1}| \le h|c_{-1}|C|y_{n+1}^i - y_{n+1}|$$

and it follows by induction that

$$|y_{n+1}^{i+1} - y_{n+1}| \le [h|c_{-1}|C]^{i+1}|y_{n+1}^0 - y_{n+1}|.$$

Therefore, the iteration of the corrector will converge to the limit y_{n+1} provided that $h|c_{-1}|C < 1$; i.e., provided that

$$h < [|c_{-1}|C]^{-1}.$$

EXAMPLE. The corrector formula (for $m = 0, k = 1$)

$$y_{n+1}^{i+1} = y_n + \frac{h}{2}[f(x_{n+1}, y_{n+1}^i) + f(x_n, y_n)]$$

generates a sequence of successive approximations y_{n+1}^{i+1} of the solution value at x_{n+1}. If the corrector iteration converges to a limit, say y_{n+1}, the converged value satisfies the relation

$$y_{n+1} = y_n + \frac{h}{2}[f(x_{n+1}, y_{n+1}) + f(x_n, y_n)].$$

It follows then that

$$y_{n+1}^{i+1} - y_{n+1} = h\frac{1}{2}[f(x_{n+1}, y_{n+1}^i) - f(x_{n+1}, y_{n+1})]$$

which, by the mean value theorem, can be written as

$$y_{n+1}^{i+1} - y_{n+1} = \frac{h}{2}[y_{n+1}^i - y_{n+1}]f_y(x_{n+1}, \bar{y}_{n+1})$$

where $\bar{y}_{n+1}$ lies between y_{n+1} and y_{n+1}^i.

Under the assumption that $|f_y(x, y)| \le C$ in the neighborhood of (x_{n+1}, y_{n+1}), it follows that

$$|y_{n+1}^{i+1} - y_{n+1}| \le \frac{h}{2}C|y_{n+1}^i - y_{n+1}|$$

and by induction

$$|y_{n+1}^{i+1} - y_{n+1}| \le \left[\frac{h}{2}C\right]^{i+1}|y_{n+1}^0 - y_{n+1}|$$

Therefore, if $hC/2 < 1$, that is, if $h < 2/C$, then the sequence y_{n+1}^{i+1} will converge to some definite value y_{n+1}, but not necessarily to the true solution value $y(x_{n+1})$.

Local Truncation Error in Multi-Step Formulas

The error in y_{n+1}^p caused by approximating $f(x, y) \equiv y'(x)$ by an interpolating polynomial $P(x)$, referred to as the *local truncation error*, is

$$\int_{x_{n-M}}^{x_{n+1}} [y'(x) - P(x)]\, dx.$$

From Sec. 6.1, it follows that the error

$$t_k(x) = y'(x) - P(x)$$

caused by approximating $y'(x)$ by a kth-degree interpolating polynomial defined over (x_{n-k}, x_n) can be expressed in the form

$$t_k(x) = \frac{y^{(k+2)}(\bar{x})}{(k+1)!}(x - x_n)(x - x_{n-1})\ldots(x - x_{n-k}), \qquad x_{n-k} \le \bar{x} \le x_n$$

so that the local truncation error in y_{n+1}^p is

$$T_{n+1} = \int_{x_{n-M}}^{x_{n+1}} \frac{y^{(k+2)}(\bar{x})}{(k+1)!}(x - x_n)(x - x_{n-1})\ldots(x - x_{n-k})\, dx.$$

If $M = 0$, the interval of integration is (x_n, x_{n+1}). Now, since the error function $t_k(x)$ does not change sign in this interval, the second theorem of the mean for integrals† can be applied, so that

$$T_{n+1} = \frac{y^{(k+2)}(\gamma_{\bar{x}})}{(k+1)!} \int_{x_n}^{x_{n+1}} (x - x_n)(x - x_{n-1})\ldots(x - x_{n-k})\, dx.$$

Formulas for T_{n+1} for the case $M = 0$ are given in the section on Adams-Moulton methods, for $k = 1, 2, \ldots$.

If $M \ne 0$, the determination of the local truncation error is more involved. An excellent treatment of this case is given in Jennings [12, Chapter 15]. For some simpler cases, the local truncation error can be determined by application of Taylor series.

The local truncation error in the corrector y_{n+1}^c can be determined in an analogous manner where

$$t_k^*(x) = y'(x) - P^*(x).$$

Propagation of Error in a Multi-Step Method

Let Y_n denote the true solution $y(x_n)$ of the initial-value problem $y'(x) = f(x, y)$, $y(x_0) = y_0$, and let y_n denote the corresponding numerical solution calculated by a multi-step method. Then the total solution error

† The second theorem of the Mean for Integrals (Jennings [12, p. 207]): If $f(x)$ is continuous for $a \le x \le b$ and $g(x)$ does not change sign there, then there exists at least one number $\bar{x}$ between a and b such that

$$\int_a^b f(x)g(x)\, dx = f(\bar{x}) \int_a^b g(x)\, dx$$

provided only that $g(x)$ be integrable on (a, b).

at step n, due to truncation in the integration formula, roundoff in the computation, and propagated error from preceding steps, is defined as

$$(9.2.16) \qquad \varepsilon_n = Y_n - y_n.$$

In some multi-step methods, e.g., Adams-Bashforth† (see Henrici [7, p. 277]) only one recursion formula of the form

$$y_{n+1} = y_{n-m} + h \sum_{j=0}^{k} p_j f_{n-j}$$

is used, whereas the predictor-corrector methods use both this formula (for the predictor) and a corrector of the form

$$y_{n+1} = y_{n-m} + h \sum_{j=-1}^{k-1} c_j f_{n-j}.$$

To study the propagation of error in both the single-formula and the predictor-corrector multi-step methods, the two formulas above can be combined into a single general formula of the form

$$(9.2.17) \qquad y_{n+1} = y_{n-m} + h \sum_{j=-1}^{k} \beta_j f_{n-j}.$$

An even more general form of the multi-step formulas is

$$y_{n+1} = \sum_{j=0}^{k} \alpha_j y_{n-j} + h \sum_{j=-1}^{k} \beta_j f_{n-j}.$$

However, we will confine our attention to the propagation of errors in multi-step formulas of the form of (9.2.17).

The analysis of the propagation of error in a multi-step integration procedure is commonly referred to as the study of *stability* of the particular method. To analyze the stability of a numerical integration method of the form of (9.2.17), let us examine the relation between ε_{n+1} and ε_n. For us to do this, certain conditions will have to be imposed, e.g., that the partial derivative of $f(x, y)$ with respect to y, denoted f_y, is continuous and bounded in the range of values considered. Another common assumption is that the magnitude of the sum of the roundoff error and local truncation error for step n is bounded by a constant.

The solution value generated by (9.2.17) satisfies the relation

$$(9.2.18) \qquad y_{n+1} = y_{n-m} + h\beta_{-1} f(x_{n+1}, y_{n+1})$$
$$+ h \sum_{j=0}^{k} \beta_j f(x_{n-j}, y_{n-j}) - R_{n+1}$$

---

† For the Adams-Bashforth method, $m = 0$; i.e.,

$$y_{n+1} = y_n + h \sum_{j=0}^{k} p_j f_{n-j}.$$

where R_{n+1} denotes the roundoff error in evaluating (9.2.17). Note that for the predictor-corrector methods, the y_{n+1} in this formula is the *converged* value of the corrector formula.

Similarly, the true solution values Y_n satisfy the relation

$$(9.2.19) \quad Y_{n+1} = Y_{n-m} + h\beta_{-1}f(x_{n+1}, Y_{n+1})$$
$$+ h \sum_{j=0}^{k} \beta_j f(x_{n-j}, Y_{n-j}) + T_{n+1}$$

where T_{n+1} denotes the local truncation error in the integration formula, caused by approximating $y'(x)$ by a kth-degree interpolating polynomial.

Subtracting (9.2.18) from (9.2.19) and using the error definition (9.2.16), we find that

$$(9.2.20) \quad \varepsilon_{n+1} = \varepsilon_{n-m} + h\beta_{-1}[f(x_{n+1}, Y_{n+1}) - f(x_{n+1}, y_{n+1})]$$
$$+ h \sum_{j=0}^{k} \beta_j [f(x_{n-j}, Y_{n-j}) - f(x_{n-j}, y_{n-j})] + E_{n+1}$$

where $E_{n+1} = T_{n+1} + R_{n+1}$.

By application of the mean value theorem, equation (9.2.20) can be expressed in the form

$$\varepsilon_{n+1} = \varepsilon_{n-m} + h\beta_{-1}[Y_{n+1} - y_{n+1}]f_y(x_{n+1}, \bar{y}_{n+1})$$
$$+ h \sum_{j=0}^{k} \beta_j [Y_{n-j} - y_{n-j}]f_y(x_{n-j}, \bar{y}_{n-j}) + E_{n+1}.$$

And since $Y_i - y_i = \varepsilon_i$, it follows that

$$(9.2.21) \quad \varepsilon_{n+1}[1 - h\beta_{-1}f_y(x_{n+1}, \bar{y}_{n+1})]$$
$$= \varepsilon_{n-m} + h \sum_{j=0}^{k} \beta_j \varepsilon_{n-j} f_y(x_{n-j}, \bar{y}_{n-j}) + E_{n+1}.$$

In practice, $f_y(x, y)$ and E_{n+1} tend to vary slowly from step to step, so let us assume that $|f_y(x, y)| \le C$ and $|E_{n+1}| \le E$, where C and E are both positive constants.

Hildebrand [10] notes that under these assumptions, equation (9.2.21) can be replaced by the related difference equation

$$(9.2.22) \quad e_{n+1}[1 - h|\beta_{-1}|C] = e_{n-m} + hC \sum_{j=0}^{k} |\beta_j| e_{n-j} + E.$$

Further, if $|\varepsilon_{n-j}| \le e_{n-j}$ $(j = 0, k)$, where $k \ge m$, it follows from (9.2.21) that

$$|\varepsilon_{n+1}| \cdot |1 - h\beta_{-1}f_y(x_{n+1}, \bar{y}_{n+1})|$$
$$\le |\varepsilon_{n-m}| + hC \sum_{j=0}^{k} |\beta_j| |\varepsilon_{n-j}| + |E_{n+1}|$$
$$\le e_{n-m} + hC \sum_{j=0}^{k} |\beta_j| e_{n-j} + E = e_{n+1}[1 - hC|\beta_{-1}|]$$

so that, if $hC|\beta_{-1}| < 1$,

$$|\varepsilon_{n+1}| \leq e_{n+1}.$$

That is, if the magnitude of the propagated error is dominated by the solution of the related difference equation for $k + 1$ successive steps, namely $|\varepsilon_i| \leq e_i$ $(i = 0, k)$, then by induction, the same is true for all successive integral values of i. Therefore, a bound for the propagated error in the multi-step method can be determined by obtaining a solution of the related difference equation.

The solution of the homogeneous difference equation (setting $E = 0$) is of the form (Wilf [30, p. 163]).

$$(9.2.23) \qquad e_n = d_1 r_1^n + d_2 r_2^n + \cdots + d_{k+1} r_{k+1}^n$$

where the r_j $(j = 1, k + 1)$ are the roots of the characteristic equation

$$(9.2.24) \quad [1 - hC|\beta_{-1}|]r^{k+1} - r^{k-m} - hC[|\beta_0|r^k + |\beta_1|r^{k-1} + \cdots$$
$$+ |\beta_k|] = 0.$$

A particular solution of the difference equation can be obtained by assuming that $e_n = -\lambda$, from which it can be shown that [Hildebrand, 10]

$$\lambda = \frac{E}{hC \sum |\beta_j|}.$$

The general solution of the difference equation can then be written as

$$(9.2.25) \qquad e_n = d_1 r_1^n + d_2 r_2^n + \cdots + d_{k+1} r_{k+1}^n + \frac{E}{hC \sum |\beta_j|}.$$

The stability of a multi-step integration method, i.e., the behavior of the propagated error, can be defined in a number of ways, such as (simple) stability, partial stability, and relative stability. Because of the complexity of these variations, we will define only (simple) stability and refer the reader to other sources for more complicated variations.†

If $f_y(x, y) = -C$ in the region of (x, y) under consideration and $E_{n+1} = E$, where C and E are both positive constants, the multi-step integration procedure (9.2.17) is said to possess (simple) stability if the solution of the related difference equation (9.2.22) dominates the error expression (9.2.21) *and* decrease in magnitude with increasing n.

EXAMPLE. Consider the multi-step formula $(m = 0, k = 0)$

$$y_{n+1} = y_n + h\left[\frac{1}{2}f_{n+1} + \frac{1}{2}f_n\right].$$

† See Chapter 14 of Henrici [7], Chapter 6 of Hildebrand [10], and Chapter 8 of Ralston and Wilf [24] for excellent treatments of this subject.

Assuming that $f_y(x, y) = -C$ and that $E_{n+1} = E$, we can analyze the stability of this formula by examining the related characteristic equation

$$\left[1 - h\frac{1}{2}(-C)\right]r^1 - 1 - h(-C)\left[\frac{1}{2}r\right] = 0$$

which has roots

$$r_1 = \frac{(1 - hC/2)}{(1 + hC/2)}.$$

The difference equation solution is then

$$e_n = d_1\left[\frac{(1 - hC/2)}{(1 + hC/2)}\right]^n + \frac{E}{hC\left(\frac{1}{2} + \frac{1}{2}\right)}.$$

From example, on p. 328, we know that $hC/2$ must be less than 1 if convergence is to be assured. If this condition is met, then the quantity $(1 - hC/2)/(1 + hC/2) < 1$, and the solution e_n of the difference equation decreases with increasing n. Therefore, the method is stable.†

Adams-Moulton Predictor-Corrector Methods

The Adams-Moulton methods form one of the most widely used classes of multi-step predictor-corrector methods for solving the initial-value problem. Both the predictor formula and the corrector formula of a particular Adams-Moulton method are based on the relation

(9.2.26) $$y(x_{n+1}) - y(x_n) = \int_{x_n}^{x_{n+1}} y'(x)\,dx.$$

The predictor (Adams-Bashforth explicit formula) is obtained by substituting for $y'(x) \equiv f(x, y)$ the kth-degree Newton backward-form interpolating polynomial, defined over points $x_n, x_{n-1}, \ldots, x_{n-k}$. This substitution produces

$$y^p_{n+1} = y_n + \int_{x_n}^{x_{n+1}}\left[f_n + \frac{\nabla f_n}{h}(x - x_n) + \frac{\nabla^2 f_n}{2h^2}(x - x_n)(x - x_{n-1})\right.$$

(9.2.27) $$\left. + \cdots + \frac{\nabla^k f_n}{k!h^k}(x - x_n)\ldots(x - x_{n-k+1})\right]dx.$$

To simplify the integration, we introduce the variable $u = (x - x_n)/h$, where also $du = dx/h$. Formula (9.2.27) then becomes

(9.2.28) $$y^p_{n+1} = y_n + \int_0^1\left[f_n + \nabla f_n u + \frac{\nabla^2 f_n}{2}u(u + 1)\right.$$

$$\left. + \cdots + \frac{\nabla^k f_n}{k!}u(u + 1)\ldots(u + k - 1)\right]h\,du.$$

† See Hamming [5, p. 190] and McCracken and Dorn [16, p. 339] for case $f_y(x, y) = +C$, with excellent treatments of *relative* stability of this corrector formula.

When this integration is performed, we obtain the predictor formula (Adams-Bashforth)

$$(9.2.29) \quad y_{n+1}^p = y_n + h\left[f_n + \frac{\nabla f_n}{2} + \frac{5}{12}\nabla^2 f_n + \frac{3}{8}\nabla^3 f_n\right.$$

$$\left. + \frac{251}{720}\nabla^4 f_n + \frac{95}{288}\nabla^5 f_n + \cdots\right].$$

We can evaluate the differential equation at the next point, i.e., compute

$$(9.2.30) \qquad f_{n+1} = f(x_{n+1}, y_{n+1}^p).$$

Having computed f_{n+1}, we can extend the difference table by one line (obtaining a table of the form 9.2.6). A new kth-degree Newton polynomial $P^*(x)$ can then be computed using points $x_{n+1}, x_n, \ldots, x_{n+1-k}$. Substituting $P^*(x)$ for $y'(x) \equiv f(x, y)$ in (9.2.26), we obtain

$$(9.2.31) \quad y_{n+1}^c = y_n + \int_{x_n}^{x_{n+1}}\left[f_{n+1} + \frac{\nabla f_{n+1}}{h}(x - x_{n+1})\right.$$

$$+ \frac{\nabla^2 f_{n+1}}{2h^2}(x - x_{n+1})(x - x_n) + \cdots$$

$$\left. + \frac{\nabla^k f_{n+1}}{k!h^k}(x - x_{n+1})\ldots(x - x_{n-k+2})\right]dx.$$

The integral above can be simplified by introducing variable $u = (x - x_{n+1})/h$, where $du = dx/h$. The result

$$(9.2.32) \quad y_{n+1}^c = y_n + \int_{-1}^{0}\left[f_{n+1} + \nabla f_{n+1}u + \frac{\nabla^2 f_{n+1}}{2}u(u+1)\right.$$

$$\left. + \cdots + \frac{\nabla^k f_{n+1}}{k!}u(u+1)\ldots(u+k-1)\right]h\,du$$

is then integrated to obtain the Adams-Moulton corrector formula

$$(9.2.33) \quad y_{n+1}^c = y_n + h\left[f_{n+1} - \frac{1}{2}\nabla f_{n+1} - \frac{1}{12}\nabla^2 f_{n+1} - \frac{1}{24}\nabla^3 f_{n+1}\right.$$

$$\left. - \frac{19}{720}\nabla^4 f_{n+1} - \frac{3}{160}\nabla^5 f_{n+1} - \cdots\right].$$

A particular member of the Adams-Moulton class of formulas can be obtained by selecting a value of k and evaluating (9.2.29) and (9.2.33) for that value. See Figure 9-5 for a representative flow chart of the Adams-Moulton predictor-corrector method for $k = 3$. Note that starting values are computed by a Runge-Kutta subroutine.

Ordinate Form of Adams-Moulton Predictor-Corrector Formulas

The Adams-Moulton predictor-corrector methods are seldom used in the *difference-form* of formulas (9.2.29) and (9.2.33), which we restate here for reference:

$$y_{n+1}^p = y_n + h\left[1 + \frac{1}{2}\nabla + \frac{5}{12}\nabla^2 + \frac{3}{8}\nabla^3 + \frac{251}{720}\nabla^4 + \frac{95}{288}\nabla^5 + \cdots\right]f_n$$

$$y_{n+1}^c = y_n + h\left[1 - \frac{1}{2}\nabla - \frac{1}{12}\nabla^2 - \frac{1}{24}\nabla^3 - \frac{19}{720}\nabla^4 - \frac{3}{160}\nabla^5 - \cdots\right]f_{n+1}.$$

These difference-form formulas are inefficient because the finite differences must be computed for each step of the predictor and each iteration of the corrector. For this reason, the equivalent ordinate-form formulas are employed. That is, for a particular value of k, where k is the highest-order difference retained, formulas (9.2.29) and (9.2.33) are converted to equivalent ordinate-form expressions by substituting the ordinate-difference relations

$$\nabla f_{n+i} = f_{n+i} - f_{n+i-1}$$
$$\nabla^2 f_{n+i} = f_{n+i} - 2f_{n+i-1} + f_{n+i-2}$$
$$\nabla^3 f_{n+i} = f_{n+i} - 3f_{n+i-1} + 3f_{n+i-2} - f_{n+i-3}$$
$$\nabla^4 f_{n+i} = f_{n+i} - 4f_{n+i-1} + 6f_{n+i-2} - 4f_{n+i-3} + f_{n+i-4}$$
$$\nabla^5 f_{n+i} = f_{n+i} - 5f_{n+i-1} + 10f_{n+i-2} - 10f_{n+i-3} + 5f_{n+i-4} - f_{n+i-5}$$

into (9.2.29) and (9.2.33) for a particular value of k.

The ordinate-form expressions for the Adams-Moulton predictors and correctors are given below for $k = 1, 2, 3, 4, 5$.

$k = 1$ $y_{n+1}^p = y_n + \dfrac{h}{2}[3f_n - f_{n-1}]$

$\qquad$ $y_{n+1}^c = y_n + \dfrac{h}{2}[f_{n+1} + f_n]$

$k = 2$ $y_{n+1}^p = y_n + \dfrac{h}{12}[23f_n - 16f_{n-1} + 5f_{n-2}]$

$\qquad$ $y_{n+1}^c = y_n + \dfrac{h}{12}[5f_{n+1} + 8f_n - f_{n-1}]$

$k = 3$ $y_{n+1}^p = y_n + \dfrac{h}{24}[55f_n - 59f_{n-1} + 37f_{n-2} - 9f_{n-3}]$

$\qquad$ $y_{n+1}^c = y_n + \dfrac{h}{24}[9f_{n+1} + 19f_n - 5f_{n-1} + f_{n-2}]$

$k = 4$ $y_{n+1}^p = y_n + \dfrac{h}{720}[1901f_n - 2984f_{n-1} + 2616f_{n-2} - 1274f_{n-3}$
$$+ 251f_{n-4}]$$

$\qquad$ $y_{n+1}^c = y_n + \dfrac{h}{720}[251f_{n+1} + 646f_n - 264f_{n-1} + 106f_{n-2} - 19f_{n-3}]$

$k = 5$ $y_{n+1}^p = y_n + \dfrac{h}{1440}[4277f_n - 7923f_{n-1} + 9982f_{n-2} - 7298f_{n-3}$
$$+ 2877f_{n-4} - 475f_{n-5}]$$

$\qquad$ $y_{n+1}^c = y_n + \dfrac{h}{1440}[475f_{n+1} + 1427f_n - 798f_{n-1} + 482f_{n-2}$
$$- 173f_{n-3} + 27f_{n-4}]$$

Computational Summary for Adams-Moulton Method ($k = 3$)†

Step 0. Define $f(x, y)$. Input and initialization. Read $x_1, y_1 =$ initial values; $h =$ step size; $\varepsilon =$ convergence term; $n =$ number of starter values required ($n > k$); $N =$ number of solution points desired; $max =$ maximum number of corrector iterations per step.

Step 1. Call starter (Runge-Kutta) to generate starting points. Subroutine RKUTTA provides $(x_1, y_1, f_1), \ldots, (x_n, y_n, f_n)$. Initialize predictor-corrector step counter i (set $i = 0$).

Step 2. Compute $f_{n+i} = f(x_{n+i}, y_{n+i})$.

Step 3. Compute predicted solution value for step i

$$y^p_{n+1+i} = y_n + \frac{h}{24} [55 f_{n+i} - 59 f_{n+i-1} + 37 f_{n+i-2} - 9 f_{n+i-3}].$$

Step 4. (a) Reset corrector iteration counter $m = 0$.

(b) Compute $x_{n+1+i} = x_{n+i} + h$.

(c) Compute $f_{n+1+i} = f(x_{n+1+i}, y^p_{n+1+i})$.

(d) Compute corrected solution value for step i, iteration m

$$y^c_{n+1+i} = y_{n+i} + \frac{h}{24} [9 f_{n+1+i} + 19 f_{n+i} - 5 f_{n+i-1} + f_{n+i-2}].$$

(e) Compute magnitude of difference between predictor and corrector.

$$\delta y = |y^c_{n+1+i} - y^p_{n+1+i}|.$$

(f) Test convergence. If $\delta y \leq \varepsilon$, store y^c_{n+1+i} as y_{n+1+i}, go to Step 4h. If $\delta y > \varepsilon$, go to Step 4g.

(g) Test maximum number of iterations of corrector. If $m \leq max$, replace m by $m + 1$, replace y^p_{n+1+i} by y^c_{n+1+i}, and return to Step 4b. If $m > max$, go to "fails to converge" exit 5b.

(h) Test step counter. If $i < N - n$, set $i = i + 1$, return to Step 2. If $i \geq N - n$, go to Step 5a.

Step 5. (a) Write output (x_i, y_i), $i = 1, N$. Stop program.

(b) Write "fails to converge." Write solution points (x_i, y_i) computed prior to failure. Stop program.

† For other values of k, simply replace predictor and corrector formulas by those for selected k.

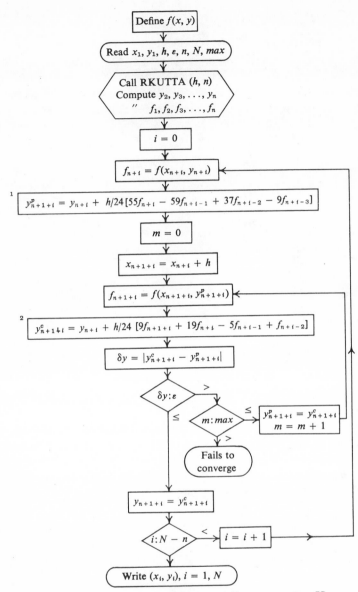

Figure 9-5. Flow chart for Adams-Moulton predictor corrector. Here $x_1, y_1 =$ initial values; $h =$ step size for Runge-Kutta *and* Adams-Moulton; $\varepsilon =$ convergence term; $n =$ number of starting values; $N =$ number of solution points; *max* = maximum number of corrector iterations. Note that this flow chart is for third-order Adams-Moulton method. To obtain flow chart for other orders, simply replace blocks 1 and 2 formulas by predictor and corrector formulas of appropriate order.

```
          PROGRAM ADAMMOUL
C         FORTRAN PROGRAM FOR ADAMS MOULTON PREDICTOR CORRECTOR
C         ORDINATE FORM EQUIVALENT OF METHOD USING 3RD DIFFERENCES
          COMMON X(100), Y(100), F(100)
C         DEFINE DERIVATIVE FUNCTION
          FCT (ARG1,ARG2) = - ARG1 * ARG2
          READ (5,100) X(1), Y(1), H, EPS, N, NC, MAX
          NCMN = NC - N
C         CALL RUNGE KUTTA TO GENERATE STARTING VALUES
          CALL RKUTTA (H, N)
C         ADAMS MOULTON PREDICTOR CORRECTOR TO CONTINUE SOLUTION
C         INITIALIZE STEP COUNTER I
          I = 0
    5     F(N+I) = FCT (X(N+I),Y(N+I))
C         COMPUTE PREDICTED VALUE YP
          NPI = N + I
          YP=Y(NPI)+(H/24.)*(55.*F(NPI) - 59.*F(NPI-1) + 37.*F(NPI-2)
          X - 9.*F(NPI-3))
          M = 0
C         COMPUTE CORRECTED VALUE YC
          X(NPI+1) = X(NPI) + H
   10     F(NPI+1) = FCT (X(NPI+1),YP)
          YC = Y(NPI) + (H/24.) * (9. * F(NPI+1) + 19. * F(NPI)
          X - 5. * F(NPI-1) + F(NPI-2))
          DELY = ABS (YC-YP)
C         TEST CONVERGENCE
          IF (DELY-EPS) 30, 30, 20
   20     IF (M-MAX) 25, 25, 97
   25     YP = YC
          M = M + 1
          GO TO 10
   30     Y(NPI+1) = YC
          IF (I - NCMN) 35, 35, 40
   35     I = I + 1
          GO TO 5
   40     WRITE (6,108)
          NP1 = N + 1
          WRITE (6,110) (X(I), Y(I), I = NP1,NC)
   99     STOP
   97     WRITE (6,113)
          ILAST = I
          WRITE (6,110) (X(I), Y(I), I = 1, ILAST)
          GO TO 99
  100     FORMAT (2F15.7, 2F10.8, 3I2)
  108     FORMAT (11X,1HX,16X,1HY,8X,24H VALUES BY ADAMS MOULTON)
  110     FORMAT (1X, 2E18.8)
  113     FORMAT (18H FAILS TO CONVERGE)
          END




          SUBROUTINE RKUTTA (H,N)
          COMMON X(100), Y(100), F(100)
          FCT (ARG1,ARG2) = - ARG1 * ARG2
          DO 10  I = 1,N
          C1 = H * FCT (X(I),Y(I))
          C2 = H * FCT (X(I)+H/2.,Y(I)+C1/2.)
          C3 = H * FCT (X(I)+H/2.,Y(I)+C2/2.)
          C4 = H * FCT (X(I)+H, Y(I)+C3)
          X(I+1) = X(I) + H
          Y(I+1) = Y(I) + (C1 + 2.*C2 + 2.*C3 + C4)/6.
   10     F(I+1) = FCT (X(I+1),Y(I+1))
          WRITE (6,108)
          WRITE (6,110) (X(I),Y(I), I=1,N)
          RETURN
  108     FORMAT (8X,1HX,16X,1HY,8X,17H VALUES BY RKUTTA)
  110     FORMAT (1X, 2E18.8)
          END
```

338

NUMERICAL EXAMPLE. Given starting values (x_i, y_i) $(i = 0, 5)$, use the Adams-Moulton predictor-corrector method to continue the solution of the initial-value problem

$$y'(x) = -xy, \qquad y(0) = 1.$$

Use (as in Table 9-2) the ordinate-form equivalent of the formulas obtained by retaining fifth-order differences to compute y_6 and y_7. Let $\varepsilon = 5(10^{-8})$.

Table 9-2

	x_i	y_i	$f(x_i, y_i) = -x_i y_i$	True Solution† $y(x_i)$
	0.0	1.000	0.0	1.0
	0.1	0.995 012 479	$-0.099\ 501\ 247\ 9$	0.995 012 479
	0.2	0.980 198 673	$-0.196\ 039\ 734\ 6$	0.980 198 673
	0.3	0.955 997 482	$-0.286\ 799\ 244\ 6$	0.955 997 482
	0.4	0.923 116 346	$-0.369\ 246\ 538\ 4$	0.923 116 346
	0.5	0.882 496 903	$-0.441\ 248\ 451\ 5$	0.882 496 903
y^p	0.6	0.835 269 370	$-0.501\ 161\ 622\ 0$	
y^c		0.835 270 273	$-0.501\ 162\ 163\ 8$	
y^c		0.835 270 255	$-0.501\ 162\ 153$	0.835 270 211
y^p	0.7	0.782 703 565	$-0.547\ 892\ 495\ 5$	
y^c		0.782 704 651	$-0.547\ 893\ 256$	
y^c		0.782 704 626		0.782 704 538

† Closed-form solution: $y(x) = e^{-x^2/2}$.

Note that the predictor y^p is used only once each step, while the corrector y^c is iterated until convergence. In this example, two iterations of the corrector were required each step.

Truncation Error in Adams-Moulton Predictor-Corrector Method

The error in a numerical-integration formula caused by replacing the integrand $y'(x)$ by an interpolating polynomial is referred to as the truncation error. An expression for the truncation error in a numerical-integration formula can be obtained by integrating the error term due to the replacement of $y'(x)$ by the interpolating polynomial. The error formulas for the predictor and the corrector in the Adams-Moulton method are derived as follows.

1. Error in predictor formula. Let $t_k(x)$ denote the error inherent in replacing $y'(x)$ by the kth-degree interpolating polynomial $P(x)$. That is,

$$(9.2.34) \qquad t_k(x) = y'(x) - P(x).$$

The following expression for this error term can be obtained in the manner described in Sec. 6.1

$$t_k(x) = \frac{y^{(k+2)}(\bar{x})}{(k+1)!} (x - x_n)(x - x_{n-1})\ldots(x - x_{n-k}), \qquad \bar{x} \in (x_{n-k}, x_n).$$
(9.2.35)

The truncation error T_{n+1} in the predictor formula is obtained by integrating $t_k(x)$ between x_n and x_{n+1}. That is,

$$T_{n+1} = \int_{x_n}^{x_{n+1}} t_k(x)\, dx = \frac{y^{(k+2)}(\bar{x})}{(k+1)!} \int_{x_n}^{x_{n+1}} (x - x_n)\ldots(x - x_{n-k})\, dx.$$
(9.2.36)

The integration of the error term is simplified by introducing the variable $u = (x - x_n)/h$. Substituting u into (9.2.36), we find

$$(9.2.37) \qquad T_{n+1} = \frac{y^{(k+2)}(\bar{x})}{(k+1)!} \int_0^1 u(u+1)\ldots(u+k)h^{k+1}h\, du.$$

The truncation error for the Adams-Moulton predictor formula of order k is obtained by evaluating (9.2.37) for k. Formulas for the truncation error for $k = 1, 2, 3, 4, 5$ are given below.

$$k = 1: T_{n+1} = \frac{h^3 y^{(3)}}{2!} \int_0^1 u(u+1)\, du = \frac{5}{12} h^3 y^{(3)}$$

$$k = 2: T_{n+1} = \frac{h^4 y^{(4)}}{3!} \int_0^1 u(u+1)(u+2)\, du = \frac{3}{8} h^4 y^{(4)}$$

$$k = 3: T_{n+1} = \frac{h^5 y^{(5)}}{4!} \int_0^1 u(u+1)(u+2)(u+3)\, du = \frac{251}{720} h^5 y^{(5)}$$

$$k = 4: T_{n+1} = \frac{h^6 y^{(6)}}{5!} \int_0^1 u(u+1)(u+2)(u+3)(u+4)\, du = \frac{95}{288} h^6 y^{(6)}$$

$$k = 5: T_{n+1} = \frac{h^7 y^{(7)}}{6!} \int_0^1 u(u+1)(u+2)(u+3)(u+4)(u+5)\, du$$

$$= \frac{19087}{60480} h^7 y^{(7)}$$

2. Truncation error in corrector formula. In a similar manner, the following expression for the truncation error in the corrector formula can be found.

$$(9.2.38) \qquad T^*_{n+1} = \frac{y^{(k+2)}(\bar{x})}{(k+1)!} \int_{-1}^{0} u(u + 1)\ldots(u + k)h^{k+1}h \, du.$$

The truncation error for the Adams-Moulton corrector formula is obtained by evaluating (9.2.38) for that value of k corresponding to the highest-order difference retained in the corrector formula. Formulas for the corrector truncation error are given below for $k = 1, 2, 3, 4, 5$.

$$k = 1: \quad T^*_{n+1} = \frac{h^3 y^{(3)}}{2!} \int_{-1}^{0} u(u + 1) \, du = -\frac{1}{12} h^3 y^{(3)}$$

$$k = 2: \quad T^*_{n+1} = \frac{h^4 y^{(4)}}{3!} \int_{-1}^{0} u(u + 1)(u + 2) \, du = -\frac{1}{24} h^4 y^{(4)}$$

$$k = 3: \quad T^*_{n+1} = \frac{h^5 y^{(5)}}{4!} \int_{-1}^{0} u(u + 1)(u + 2)(u + 3) \, du = -\frac{19}{720} h^5 y^{(5)}$$

$$k = 4: \quad T^*_{n+1} = \frac{h^6 y^{(6)}}{5!} \int_{-1}^{0} u(u + 1)(u + 2)(u + 3)(u + 4) \, du = -\frac{3}{160} h^6 y^{(6)}$$

$$k = 5: \quad T^*_{n+1} = \frac{h^7 y^{(7)}}{6!} \int_{-1}^{0} u(u + 1)(u + 2)(u + 3)(u + 4)(u + 5) \, du$$

$$= -\frac{863}{60480} h^7 y^{(7)}$$

9.3   The Euler-Romberg Method

The Euler-Romberg method for solving the initial-value problem

$$(9.3.1) \qquad y' = f(x, y), \quad y(x_0) = y_0$$

computes solution approximations $y_1, y_2, y_3, \ldots, y_N$ of the true-solution values $y(x_1), y(x_2), y(x_3), \ldots, y(x_N)$ by an iterative procedure based on (1) repeated application of Euler's method with successive halving of the interval; and (2) repeated linear interpolation (extrapolated to $h = 0$).

The procedure is self-starting and rivals the best predictor-corrector methods for efficiency and accuracy. The choice of step size is fairly arbitrary, since the method automatically applies successive halving of the interval until the required accuracy is achieved for each step of the solution.

Given (x_n, y_n) and step size h_0, the method generates approximation y_{n+1} by constructing a table of values as given in Table 9-3. The procedure is started by computing an initial y_{n+1} (denoted $Y_0{}^0$) by Euler's method, using full step h_0. That is, first calculate

$$(9.3.2) \qquad Y_0{}^0 = y_n + h_0 f(x_n, y_n).$$

Then generate the table of values.

Table 9-3

Iteration	Interval	Euler-Romberg Values $Y_m{}^k$				
	h_0	$Y_0{}^0$				
$L = 1$	$h_0/2$	$Y_0{}^1$	$Y_1{}^0$			
$L = 2$	$h_0/2^2$	$Y_0{}^2$	$Y_1{}^1$	$Y_2{}^0$		
$L = 3$	$h_0/2^3$	$Y_0{}^3$	$Y_1{}^2$	$Y_2{}^1$	$Y_3{}^0$	
$\vdots$	$\vdots$	$\vdots$	$\vdots$	$\vdots$	$\vdots$	$\vdots$ $\cdots$

An iterative procedure (iteration index L) similar to the Romberg method of numerical integration is used to generate the elements $Y_m{}^k$ of Table 9-3, called the Euler-Romberg table. Note that all the elements in a given column have the same subscript. The subscript represents the column number. In iteration L all the elements on line L are computed unless the method converges before the line is completed. Convergence of the method is determined by comparing successive values $Y_m{}^k$ and Y_m^{k-1} in column m. The converged value $Y_m{}^k$ is the desired numerical solution value y_{n+1}.

Repeated Application of Euler's Method with Successive Interval Halving

The first element $Y_0{}^0$ in column 0 is an approximation to true solution value $y(x_{n+1})$. This approximation is computed by Euler's method, using full step h_0; i.e.,

$$Y_0{}^0 = y_n + h_0 f(x_n, y_n).$$

The remaining elements in column 0 are approximations to $y(x_{n+1})$ computed by Euler's method with successive halving of the interval in each succeeding iteration. These approximations are illustrated below in Figure 9-6.

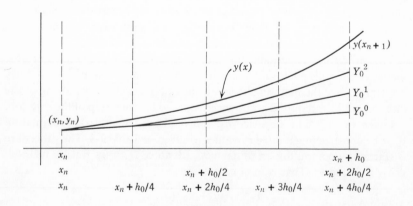

Figure 9-6

The successive Euler values depicted in Figure 9-6 are calculated as follows:

For $L = 1$, use interval $h_1 = h_0/2$ and apply Euler formula *twice*.

$$
\begin{aligned}
X_0 &= x_n & Y_0 &= y_n & f_0 &= f(X_0, Y_0) \\
X_1 &= x_n + h_1 & Y_1 &= Y_0 + h_1 f_0 & f_1 &= f(X_1, Y_1) \\
X_2 &= x_n + 2h_1 & Y_2 &= Y_1 + h_1 f_1 & & \text{Euler } Y_0^1 = Y_2.
\end{aligned}
$$

For $L = 2$, use interval $h_2 = h_0/2^2$, and apply Euler formula *four* times.

$$
\begin{aligned}
X_0 &= x_n & Y_0 &= y_n & f_0 &= f(X_0, Y_0) \\
X_1 &= x_n + h_2 & Y_1 &= Y_0 + h_2 f_0 & f_1 &= f(X_1, Y_1) \\
X_2 &= x_n + 2h_2 & Y_2 &= Y_1 + h_2 f_1 & f_2 &= f(X_2, Y_2) \\
X_3 &= x_n + 3h_2 & Y_3 &= Y_2 + h_2 f_2 & f_3 &= f(X_3, Y_3) \\
X_4 &= x_n + 4h_2 & Y_4 &= Y_3 + h_2 f_3 & & \text{Euler } Y_0^2 = Y_4.
\end{aligned}
$$

This procedure of applying Euler's formula with successive halving of the interval can be continued to obtain $Y_0^3, Y_0^4, \ldots, Y_0^L$.

Repeated (Iterated) Linear Interpolation in Variable h

Elements Y_m^k in column m ($m = 1, 2, 3, \ldots$) are obtained by applying the basic linear interpolation formula (6.4.10*) in variable h to the (h, Y) pairs† (h_k, Y_{m-1}^k) and (h_{k+m}, Y_{m-1}^{k+1}) by simply replacing x by h and y by Y. These substitutions in (6.4.10*) give us

$$(9.3.3) \qquad Y_m^k(h) = \frac{Y_{m-1}^k[h - h_{k+m}] - Y_{m-1}^{k+1}[h - h_k]}{h_k - h_{k+m}}.$$

Now, evaluate this formula at $h = 0$, denoting $Y_m^k(0)$ by Y_m^k. That is, we *extrapolate* to $h = 0$, obtaining

$$(9.3.4) \qquad Y_m^k = \frac{Y_{m-1}^k[0 - h_0/2^{k+m}] - Y_{m-1}^{k+1}[0 - h_0/2^k]}{(h_0/2^k) - (h_0/2^{k+m})}.$$

If $h_0/2^{k+m}$ is factored from both the numerator and denominator, this formula can be reduced to the form

$$(9.3.5) \qquad Y_m^k = \frac{2^m Y_{m-1}^{k+1} - Y_{m-1}^k}{2^m - 1}.$$

Formula (9.3.5) is the basic recursion formula for generating the elements Y_m^k in the mth column of the Euler-Romberg table, for $m = 1, 2, 3, \ldots$.

Figure 9-7 illustrates the application of the linear interpolation formula to number pairs (h_k, Y_{m-1}^k) and (h_{k+m}, Y_{m-1}^{k+1}).

† Here, $h_k = h_0/2^k$ and $h_{k+m} = h_0/2^{k+m}$.

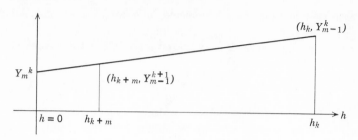

Figure 9-7. Here Y_m^k is the intercept of the line determined by the two points $(h_k,\ Y_{m-1}^k)$ and $(h_{k+m},\ Y_{m-1}^{k+1})$, with the line extrapolated to $h = 0$.

The Euler-Romberg Iterative Procedure

The repeated application of Euler's method can be combined with the iterated interpolation principle to produce the Euler-Romberg iterative method for generating elements Y_m^k. Bauer et al.† state that the elements Y_m^k of any column m converge to $y(x_{n+1})$ provided that $f(x, y)$ is continuous in both variables and satisfies the condition of Lipschitz [7, p. 63].

Using Y_0^0 as a starter, the iterative Euler-Romberg procedure generates for iteration L, $(L = 1, 2, \ldots)$ all elements Y_m^k $(m = 0, 1, \ldots, L)$ $(k = L, L - 1, \ldots, 0)$ on line L of the triangular array, unless the procedure converges before line L is completed. When the procedure has converged, i.e., when $|Y_m^k - Y_m^{k-1}| \le \varepsilon$, the converged value Y_m^k is used as the numerical solution value y_{n+1}, since each column converges to $y(x_{n+1})$, assuming that $y_n = y(x_n)$. However, since y_n is only an approximation of $y(x_n)$, then y_{n+1} is only an approximation of $y(x_{n+1})$.

To compute the next numerical solution point (x_{n+2}, y_{n+2}), step index n is replaced by $n + 1$, and the entire process is repeated. This procedure is continued until we have $y_N \doteq y(x_N)$, where x_N is the abscissa of the last required solution value.

† *Proceedings of Symposia of Applied Mathematics*, Vol. XV, Amer. Math. Soc., 1963.

Computational Summary for Euler-Romberg Method

Step 0. Define $f(x, y)$. Input and initialization. Read initial values x_1, y_1; step size h; convergence term ε; maximum subdivisions per step *Lmax*; number of solution steps N.

Set step counter $n = 1$.

Note. In Steps 1 and 2 the elements $Y_m{}^k$ of the Euler-Romberg table (Table 9-4) will be generated. In iteration L all elements on line L will be generated. Table on left is in Euler-Romberg notation; to conserve storage elements generated in each iteration, L will be stored in one-dimensional array T_K, as shown in table on right.

Table 9-4

Step size	Euler-Romberg elements $Y_m{}^k$ notation						Line	Step size	Euler-Romberg elements in array T_K					
h	$Y_0{}^0$							h	T_1					
$h/2$	$Y_0{}^1$	$Y_1{}^0$					$L = 1$	$h/2$	T_2	T_1				
$h/2^2$	$Y_0{}^2$	$Y_1{}^1$	$Y_2{}^0$				$L = 2$	$h/2^2$	T_3	T_2	T_1			
$h/2^3$	$Y_0{}^3$	$Y_1{}^2$	$Y_2{}^1$	$Y_3{}^0$			$L = 3$	$h/2^3$	T_4	T_3	T_2	T_1		
$\vdots$	$\vdots$	$\vdots$	$\vdots$	$\vdots$	$\vdots$	$\vdots$	$\vdots$	$\vdots$	$\vdots$	$\vdots$	$\vdots$	$\vdots$	$\vdots$	$\vdots$

Step 1. Approximate y_{n+1} by the Euler method, using full step h

$$Y_0{}^0 = y_n + hf(x_n, y_n) \qquad \text{Store } Y_0{}^0 \text{ as } T_1.$$

Step 2. Generate elements of the Euler-Romberg table. Initialization: Set the iteration counter $L = 1$.

(a) Set $X = x_n$, set $Y = y_n$.
Approximate y_{n+1} by the Euler method, using successive halving of the interval. Denote by T_{L+1} the Euler value.

(b) Set column counter $m = 1$. Set T array index $K = L$.

(c) Compute the Euler-Romberg element $Y_m{}^k$ (denoted T_K in the flow chart, Figure 9-8)

$$Y_m{}^k = \frac{2^m Y_{m-1}^{k+1} - Y_{m-1}^k}{2^m - 1} \quad \text{i.e., } T_K = \frac{2^m T_{K+1} - T_K}{2^m - 1}$$

(d) If $K > 1$, go to Step 2e. If $K \le 1$, go to Step 2f.

(e) Test convergence (compare successive terms in column m). If $|T_K - T_{K-1}| \le \varepsilon$, go to Step 2g. If $|T_K - T_{K-1}| > \varepsilon$, set $m = m + 1$, set $K = K - 1$, return to Step 2c.

(f) If $L < Lmax$, set $L = L + 1$, return to Step 2a. If $L \ge Lmax$, go to "fails to converge" exit.

(g) Set $x_{n+1} = x_n + h$, set $y_{n+1} = T_K$, go to Step 2h.

(h) If $n < N$, set $n = n + 1$, return to Step 1. If $n \ge N$, go to Step 3.

Step 3. Output solution points (x_n, y_n) $n = 1, N$.

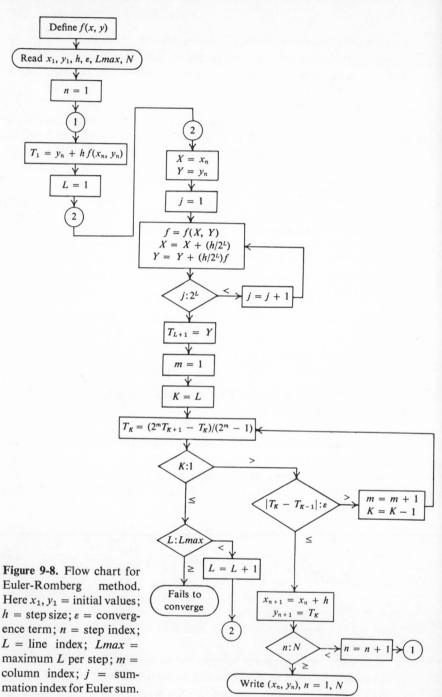

Figure 9-8. Flow chart for Euler-Romberg method. Here x_1, y_1 = initial values; h = step size; ε = convergence term; n = step index; L = line index; $Lmax$ = maximum L per step; m = column index; j = summation index for Euler sum.

346

```
      PROGRAM EULRBG
C     FORTRAN PROGRAM FOR EULER-ROMBERG METHOD
      DIMENSION X(100), Y(100), T(20)
C     DEFINE DERIVATIVE FUNCTION
      FCT(ARG1,ARG2) = ARG1**2 * ARG2
C     INPUT INITIAL VALUES, STEP SIZE, CONVERGENCE TERM,
C     MAXIMUM SUBDIVISIONS PER STEP, TOTAL NUMBER OF STEPS
      READ (5,100) X(1), Y(1), H, EPS, LMAX, NC
C     COMPUTE NUMERICAL SOLUTION VALUE FOR EACH STEP N
      DO 55 N = 1, NC
C     COMPUTE VALUES BY EULER METHOD
    5 T(1) = Y(N) + H * FCT(X(N),Y(N))
      L = 1
   10 XC = X(N)
      YC = Y(N)
      LIM = 2**L
      DO 15 J = 1, LIM
      F = FCT (XC, YC)
      XC = XC + H/FLOAT(LIM)
   15 YC = YC + (H/FLOAT(LIM)) * F
      T(L+1) = YC
C     COMPUTE EULER ROMBERG TABLE VALUES T(K)
      M = 1
      K = L
   20 T(K) = (2.**M * T(K+1) - T(K))/(2, **M - 1.)
      IF(K-1) 35, 35, 25
   25 IF (ABS(T(K) - T(K-1)) -EPS) 50, 50, 30
   30 M = M +1
      K = K - 1
      GO TO 20
   35 IF (L-LMAX) 40, 75, 75
   40 L = L + 1
      GO TO 10
   50 X(N+1) = X(N) + H
   55 Y(N+1) = T(K)
C     OUTPUT SOLUTION POINTS COMPUTED BY EULER ROMBERG
   60 WRITE (6, 110) (X(N), Y(N), N = 1, NC)
      GO TO 99
C     OUTPUT SOLUTION POINTS PRIOR TO CONVERGENCE FAILURE
   75 WRITE (6, 113)
      N2 = N
      WRITE (6, 110) (X(N), Y(N), N = 1, N2)
   99 STOP
  100 FORMAT (2F15.7, 2F10.8, 2I2)
  110 FORMAT (1X, 2E17.7)
  113 FORMAT (20H FAILURE TO CONVERGE)
      END
```

Summary

The one-step methods referred to as Euler, improved Euler, and improved Euler predictor corrector were presented to acquaint the reader with simple numerical methods for solving the initial-value problem of ordinary differential equations. Further, the latter method provides a simple example of the predictor-corrector concept.

The Taylor algorithm was, until recently, considered impractical for computational work because of the difficulty of evaluating high-order derivatives. However, Ramon E. Moore [38, p. 61] has shown that recursion formulas can be developed for these derivatives, at least for rational functions, so that interest in direct application of the Taylor algorithm may revive.

A quadratic Runge-Kutta formula was derived to illustrate the development of these formulas by indirect use of the Taylor algorithm. The quartic Runge-Kutta methods have long been among the most widely used one-step methods, especially the Gill modification to minimize storage requirements. The Runge-Kutta methods are still used as the starting technique for most multi-step methods. An excellent derivation of quartic Runge-Kutta formulas is given in Ralston and Wilf [24, pp. 110–120]. See also SHARE routine, SDA 3010.

The Milne method illustrates how a particular method is derived from the general predictor-corrector formulas. This method is *not* recommended for automatic computation because of its instability.

The Adams-Moulton ordinate-form predictor-corrector methods offer an excellent combination of accuracy, efficiency, and stability. T. E. Hull and A. L. Creemer ["Efficiency of Predictor-Corrector Procedures," *J. ACM* 10, 1963, p. 291] recommend these formulas highly, especially those formulas with truncation error of order h^8 or h^9 for accuracy, with two repetitions per step of the corrector for stability.

The Euler-Romberg method can also be highly recommended for automatic computation because of its simplicity of recursion formulas, its stability, and its automatic control of accuracy; i.e., its ability to automatically determine the step size required to attain a specified accuracy. Since this method is essentially a one-step method, it has the advantages of self-starting and variable size of steps.

DERIVATION EXERCISES

1. Show that the predictor formula

$$y_{n+1}^p = y_{n-1} + 2hf(x_n, y_n)$$

has a truncation error of order $O(h^3)$.

Hint: Use two Taylor series in powers of h and $-h$.

Using this predictor and the improved Euler corrector (9.1.18), together with a suitable one-step starter, flow chart the resulting algorithm for starting and continuing the solution of the initial-value problem. Note that this algorithm is superior to the improved Euler predictor-corrector method because the predictor and corrector have truncation errors of the same order. (See McCracken and Dorn [16, p. 257] for a discussion of why you should choose a predictor and a corrector with truncation errors of the same order.)

2. Derive a cubic Runge-Kutta algorithm by assuming that y_{n+1} can be computed by an algorithm of the form

$$y_{n+1} = y_n + a_1 k_1 + a_2 k_2 + a_3 k_3$$

with

$$k_1 = hf(x_n, y_n)$$
$$k_2 = hf(x_n + \alpha h, y_n + \alpha k_1)$$
$$k_3 = hf(x_n + \beta h, y_n + \beta k_2)$$

which is equivalent to a cubic algorithm (Taylor)

$$y_{n+1} = y_n + h\left[f(x_n, y_n) + \frac{h}{2} f'(x_n, y_n) + \frac{h^2}{2} f''(x_n, y_n) \right]$$

where

$$f' = f_x + f f_y \quad \text{and} \quad f'' = f_{xx} + 2 f f_{xy} + f^2 f_{yy} + f_y f'.$$

(See Stanton [27, p. 151] for an analogous derivation of a quartic Runge-Kutta algorithm.)

3. For $k = 1$, the Adams-Moulton corrector formula is the corrector for the improved Euler predictor-corrector method

$$y_{n+1}^c = y_n + \frac{h}{2} [f(x_{n+1}, y_{n+1}^p) + f(x_n, y_n)]$$

and the converged value y_{n+1} of this corrector satisfies the relation

$$y_{n+1} = y_n + \frac{h}{2} [f(x_{n+1}, y_{n+1}) + f(x_n, y_n)] - R_{n+1}$$

while the true solution values Y_n satisfy the relation

$$Y_{n+1} = Y_n + \frac{h}{2} [f(x_{n+1}, Y_{n+1}) + f(x_n, Y_n)] + T_{n+1}.$$

Show that if $|f_y(x, y)| \leq C$ and $|E_{n+1}| \leq C$, where C and E are both positive constants, then the expression for the propagated error in the improved Euler predictor corrector will have the related difference equation

$$e_{n+1}\left(1 - \frac{hC}{2}\right) = e_n\left(1 + \frac{hC}{2}\right) + E.$$

4. Let $\varepsilon_n' = Y_n' - y_n'$, where $Y_n' = f(x_n, Y_n)$ and $y_n' = f(x_n, y_n)$. Use the relation (see Hamming [5, p. 197])

$$\varepsilon_n' = f_y(x_n, \bar{y}_n)\varepsilon_n$$

to determine an expression for the propagated error in the Adams-Moulton correctors:

(a) $$y_{n+1}^c = y_n + \frac{h}{2}[f_{n+1} + f_n]$$

(b) $$y_{n+1}^c = y_n + \frac{h}{12}[5f_{n+1} + 8f_n - f_{n-1}].$$

(See Chapter 15 of Hamming [5] for an excellent treatment of the stability of multi-step methods employing formulas in ε_n and ε_n'.)

COMPUTATIONAL EXERCISES

1. Calculate three steps of the solution of the initial-value problem

$$y' = y, \quad y(0) = 1$$

with $h = 0.1$, using (a) Euler's method; (b) improved Euler method; (c) improved Euler predictor-corrector method with $\varepsilon = 0.001$; (d) quartic Runge-Kutta method. Compare the results in each case with the exact-solution values $y(0.1)$, $y(0.2)$, $y(0.3)$.

2. Use the numerical solution values computed in exercise 1d as starting values, and calculate two more steps of the solution of exercise 1, using (a) Milne's method; (b) Adams-Moulton third-order method, with $\varepsilon = 0.00001$. Compare the results in each case with the exact-solution values $y(0.4)$, $y(0.5)$.

3. The same as exercise 1 for the initial-value problem

$$y' = x + y, \quad y(0) = 1.$$

4. The same as exercise 2 for the initial-value problem

$$y' = x + y, \quad y(0) = 1.$$

5. The same as exercise 1 for the initial-value problem

$$y' = -xy, \quad y(0) = 1.$$

6. The same as exercise 2 for the initial-value problem

$$y' = -xy, \quad y(0) = 1.$$

7. Calculate the approximations of $y(0.1)$ for the initial-value problem

$$y' = xy, \quad y(0) = 1$$

by Euler's method, using (a) one solution step with $h = 0.1$; (b) two solution steps with $h = 0.05$; (c) four solution steps with $h = 0.025$; (d) eight solution steps with $h = 0.0125$.

8. Use the Euler values approximating $y(0.1)$ computed in exercise 7 for first column of the Euler-Romberg table, and compute the Euler-Romberg approximation of $y(0.1)$ with $\varepsilon = 5(10^{-5})$.

PROGRAMMING EXERCISES

1. Write a FORTRAN program to solve the initial-value problem for a single first-order differential equation, using Euler's method.

2. Write a FORTRAN program to solve the initial-value problem

$$y' = x^3 + y^2, \quad y(0) = 1$$

using the Taylor algorithm of order 4.

3. Write a FORTRAN program to solve the initial-value problem for a single first-order differential equation, using the improved Euler method (see flow chart, p. 311).

4. Write a FORTRAN program to solve the initial-value problem for a single first-order differential equation, using the quartic Runge-Kutta method as a starter and the fourth-order Adams-Moulton method (in ordinate form) to continue the solution.

Bibliography

1. Arden, B. W., *An Introduction to Digital Computing*, Addison Wesley, Reading, Mass., 1963.
2. Bodewig, E., *Matrix Calculus*, North-Holland Publishing Co., Amsterdam, 1959.
3. Booth, K. H. V., *Programming for an Automatic Digital Calculator*, Academic Press, New York, 1958.
4. Fadeeva, V. N., *Computational Methods in Linear Algebra*, Dover, New York, 1959.
5. Hamming, R. W., *Numerical Methods for Scientists and Engineers*, McGraw-Hill, New York, 1962.
6. Hartree, D. P., *Numerical Analysis*, Clarendon Press (Oxford), 1952.
7. Henrici, P., *Elements of Numerical Analysis*, John Wiley and Sons, New York, 1964.
8. ———, *Discrete Variable Methods in Ordinary Differential Equations*, John Wiley and Sons, New York, 1962.
9. Herriot, J. G., *Methods of Mathematical Analysis and Computation*, John Wiley and Sons, New York, 1963.
10. Hildebrand, F. B., *Introduction to Numerical Analysis*, McGraw-Hill, New York, 1956.
11. Householder, A. S., *Principles of Numerical Analysis*, McGraw-Hill, New York, 1953.
12. Jennings, W., *First Course in Numerical Methods*, Macmillan, New York, 1964.
13. Kunz, K. S., *Numerical Analysis*, McGraw-Hill, New York, 1957.
14. Macon, N., *Numerical Analysis*, John Wiley and Sons, New York, 1963.
15. McCracken, D. D., *A Guide to FORTRAN Programming*, John Wiley and Sons, New York, 1961.
16. McCracken, D. D., and W. S. Dorn, *Numerical Methods and FORTRAN Programming*, John Wiley and Sons, New York, 1964

17. McCormick, J. M., and M. G. Salvadori, *Numerical Methods in FORTRAN*, Prentice-Hall, Englewood Cliffs, N.J., 1964.
18. Milne, W. E., *Numerical Calculus*, Princeton University Press, Princeton, N.J., 1949.
19. Nielsen, K. L., *Methods in Numerical Analysis* (2nd Edition), Macmillan, New York, 1964.
20. Organick, E. I., *A FORTRAN Primer*, Addison Wesley, Reading, Mass., 1962.
21. Pennington, R. H., *Introductory Computer Methods and Numerical Analysis*, Macmillan, New York, 1965.
22. Perlis, S., *Theory of Matrices*, Addison Wesley, Reading, Mass., 1952.
23. Powell, J. E., and C. P. Wells, *Differential Equations*, Ginn and Company, Boston, Mass., 1950.
24. Ralston, A., and H. S. Wilf, *Mathematical Methods for Digital Computers*, John Wiley and Sons, New York, 1960.
25. Scarborough, J. B., *Numerical Mathematical Analysis* (5th Edition), Johns Hopkins Press, Baltimore, 1962.
26. Singer, J., *Elements of Numerical Analysis*, Academic Press, New York, 1964.
27. Stanton, R. G., *Numerical Methods for Science and Engineering*, Prentice-Hall, Englewood Cliffs, N.J., 1961.
28. Todd, J., *Survey of Numerical Analysis*, McGraw-Hill, New York, 1962.
29. Vilenkin, N., *Successive Approximation*, Macmillan, New York, 1964.
30. Wilf, H. S., *Mathematics for the Physical Sciences*, John Wiley and Sons, New York, 1962.
31. Wilkinson, J. H., *Rounding Errors in Algebraic Processes*, Prentice-Hall, Englewood Cliffs, N.J., 1963.
32. Stiefel, E. L., *An Introduction to Numerical Mathematics*, Academic Press, New York, 1963.
33. Kopal, Z., *Numerical Analysis* (2nd Edition), John Wiley and Sons, New York, 1961.
34. Uspensky, J. V., *Theory of Equations*, McGraw-Hill, New York, 1948.
35. Birkhoff, G., and S. MacLane, *A Survey of Modern Algebra*, Macmillan, New York, 1953.
36. Lanczos, C., *Applied Analysis*, Prentice-Hall, Englewood Cliffs, N.J., 1956.
37. Froberg, C. E., *Introduction to Numerical Analysis*, Addison Wesley, Reading, Mass., 1965.
38. Rall, L. B., (Editor) *Errors in Digital Computation*, Vol. 1, Proceedings of An Advanced Seminar conducted by Mathematics Research Center, U. S. Army, at Univ. of Wisconsin, 1964; John Wiley and Sons, New York, 1965.

Index

355